THE
HANDY
ANATOMY
ANSWER
BOOK

About the Authors

Patricia Barnes-Svarney is a science and science fiction writer. Over the past few decades, she has written or coauthored more than 35 books, including *When the Earth Moves: Rogue Earthquakes, Tremors, and Aftershocks* and the award-winning *New York Public Library Science Desk Reference,* along with authoring several hundred magazine articles.

Thomas E. Svarney is a scientist who has written extensively about the natural world. His books, with Patricia Barnes-Svarney, include Visible Ink Press' *The Handy Dinosaur Answer Book, The Handy Math Answer Book,* and *The Handy Nutrition Answer Book*; in addition, they have written *Skies of Fury: Weather Weirdness around the World* and *The Oryx Guide to Natural History.* You can read more about their work and writing at www.pattybarnes.net.

Please visit the "Handy" series website at www.handyanswers.com

Also from Visible Ink Press

The Handy African American History Answer Book
by Jessie Carnie Smith
ISBN: 978-1-57859-452-8

The Handy Answer Book for Kids (and Parents), 2nd edition
by Gina Misiroglu
ISBN: 978-1-57859-219-7

The Handy Art History Answer Book
by Madelynn Dickerson
ISBN: 978-1-57859-417-7

The Handy Astronomy Answer Book, 3rd edition
by Charles Liu
ISBN: 978-1-57859-190-9

The Handy Bible Answer Book
by Jennifer Rebecca Prince
ISBN: 978-1-57859-478-8

The Handy Biology Answer Book, 2nd edition
by Patricia Barnes Svarney and Thomas E. Svarney
ISBN: 978-1-57859-490-0

The Handy Chemistry Answer Book
by Ian C. Stewart and Justin P. Lamont
ISBN: 978-1-57859-374-3

The Handy Civil War Answer Book
by Samuel Willard Crompton
ISBN: 978-1-57859-476-4

The Handy Dinosaur Answer Book, 2nd edition
by Patricia Barnes-Svarney and Thomas E. Svarney
ISBN: 978-1-57859-218-0

The Handy English Grammar Answer Book
by Christine A. Hult, Ph.D.
ISBN: 978-1-57859-520-4

The Handy Geography Answer Book, 2nd edition
by Paul A. Tucci
ISBN: 978-1-57859-215-9

The Handy Geology Answer Book
by Patricia Barnes-Svarney and Thomas E. Svarney
ISBN: 978-1-57859-156-5

The Handy History Answer Book, 3rd edition
by David L. Hudson, Jr.
ISBN: 978-1-57859-372-9

The Handy Hockey Answer Book
by Stan Fischler
ISBN: 978-1-57859-569-3

The Handy Investing Answer Book
by Paul A. Tucci
ISBN: 978-1-57859-486-3

The Handy Islam Answer Book
by John Renard, Ph.D.
ISBN: 978-1-57859-510-5

The Handy Law Answer Book
by David L. Hudson Jr.
ISBN: 978-1-57859-217-3

The Handy Math Answer Book, 2nd edition
by Patricia Barnes-Svarney and Thomas E. Svarney
ISBN: 978-1-57859-373-6

The Handy Military History Answer Book
by Samuel Willard Crompton
ISBN: 978-1-57859-509-9

The Handy Mythology Answer Book,
by David A. Leeming, Ph.D.
ISBN: 978-1-57859-475-7

The Handy Nutrition Answer Book
by Patricia Barnes-Svarney and Thomas E. Svarney
ISBN: 978-1-57859-484-9

The Handy Ocean Answer Book
by Patricia Barnes-Svarney and Thomas E. Svarney
ISBN: 978-1-57859-063-6

The Handy Personal Finance Answer Book
by Paul A. Tucci
ISBN: 978-1-57859-322-4

The Handy Philosophy Answer Book
by Naomi Zack
ISBN: 978-1-57859-226-5

The Handy Physics Answer Book, 2nd edition
By Paul W. Zitzewitz, Ph.D.
ISBN: 978-1-57859-305-7

The Handy Politics Answer Book
by Gina Misiroglu
ISBN: 978-1-57859-139-8

The Handy Presidents Answer Book, 2nd edition
by David L. Hudson
ISB N: 978-1-57859-317-0

The Handy Psychology Answer Book
by Lisa J. Cohen
ISBN: 978-1-57859-223-4

The Handy Religion Answer Book, 2nd edition
by John Renard
ISBN: 978-1-57859-379-8

The Handy Science Answer Book, 4th edition
by The Carnegie Library of Pittsburgh
ISBN: 978-1-57859-321-7

The Handy Supreme Court Answer Book
by David L Hudson, Jr.
ISBN: 978-1-57859-196-1

The Handy Technology Answer Book
by by Naomi Balaban and James Bobick
ISBN: 978-1-57859-563-1

The Handy Weather Answer Book, 2nd edition
by Kevin S. Hile
ISBN: 978-1-57859-221-0

Please visit the "Handy" series website at www.handyanswers.com

THE HANDY ANATOMY ANSWER BOOK

INCLUDES PHYSIOLOGY

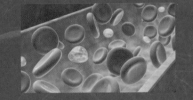

Second Edition

Patricia Barnes-Svarney and Thomas E. Svarney

VISIBLE
INK
PRESS

Detroit

THE HANDY ANATOMY ANSWER BOOK

Visible Ink Press®
43311 Joy Rd., #414
Canton, MI 48187–2075

Visible Ink Press is a registered trademark of Visible Ink Press LLC.

Most Visible Ink Press books are available at special quantity discounts when purchased in bulk by corporations, organizations, or groups. Customized printings, special imprints, messages, and excerpts can be produced to meet your needs. For more information, contact Special Markets Director, Visible Ink Press, www.visibleink.com, or 734-667-3211.

Managing Editor: Kevin S. Hile
Art Director: Mary Claire Krzewinski
Typesetting: Marco DiVita
Proofreaders: Larry Baker and Sharon R. Gunton
Indexer: Shoshana Hurwitz
Cover images: Shutterstock.

Library of Congress Cataloging-in-Publication Data

Barnes-Svarney, Patricia L.
  The handy anatomy answer book / Patricia Barnes-Svarney, Thomas E. Svarney. — Second edition.
        pages cm. — (The handy answer book series)
  Includes bibliographical references and index.
  ISBN 978-1-57859-542-6 (paperback)
1. Human anatomy—Miscellanea. 2. Physiology—Miscellanea. I. Svarney, Thomas E. II. Title.
  QM23.2.B62 2016   612—dc23
                                                           2015029143

Contents

ACKNOWLEDGMENTS ... *ix*
PHOTO CREDITS ... *x*
INTRODUCTION ... *xi*

HISTORY OF ANATOMY ... 1

Defining Anatomy (1) ... Studies in Anatomy (2) ... Comparing Other Organisms (7)

ANATOMY AND BIOLOGY BASICS ... 9

Human Anatomical Terminology (9) ... Chemistry in Biology and Anatomy (12) ... Biological Compounds and the Human Body (15) ... Comparing Other Organisms (18)

LEVELS OF ORGANIZATION ... 21

Defining Levels of Organization in Anatomy (21) ... Cells (22) ... Tissues (31) ... Organs and Organ Systems (38) ... Homeostasis (40) ... Comparing Other Organisms (41)

SENSORY SYSTEM ... 43

Smell (46) ... Taste (48) ... Hearing (50) ... Vision (55) ... Comparing Other Organisms (63)

INTEGUMENTARY SYSTEM ... 67

Introduction (67) ... Skin Structure (68) ... Skin Function (75) ... Nails (78) ... Hair (79) ... Accessory Glands (82) ... Comparing Other Organisms (84)

SKELETAL SYSTEM ... 87

Introduction (87) ... Bone Basics (89) ... Axial Skeleton (93) ... Appendicular Skeleton (98) ... Joints (102) ... Comparing Other Organisms (107)

MUSCULAR SYSTEM ... 109

Introduction (109) ... Organization of Muscles (112) ... Muscle Structure (116) ... Muscle Function (120) ... Comparing Other Organisms (126)

NERVOUS SYSTEM ... 129

Introduction (129) ... Neuron Function (131) ... Central Nervous System (136) ... The Brain (139) ... Spinal Cord (145) ... Peripheral Nervous System: Somatic

Nervous System (146) ... Peripheral Nervous System: Autonomic Nervous System (150) ... Learning and Memory (151) ... Sleep and Dreams (154) ... Comparing Other Organisms (159)

ENDOCRINE SYSTEM ... 163

Introduction (163) ... Hormones (164) ... Pituitary Gland (167) ... Thyroid and Parathyroid Glands (171) ... (Adrenal Glands (176) ... Pancreas (178) ... Pineal Gland (182) ... Reproductive Organs (182) ... Other Sources of Hormones (184) ... Comparing Other Organisms (186)

CARDIOVASCULAR SYSTEM ... 189

Introduction (189) ... Blood (190) ... The Heart (198) ... Blood Vessels (202) ... Circulation (205) ... Comparing Other Organisms (208)

LYMPHATIC SYSTEM ... 211

Introduction (211) ... Lymphatic Vessels and Organs (214) ... Nonspecific Defenses (217) ... Specific Defenses (218) ... Allergies (226) ... Comparing Other Organisms (229)

RESPIRATORY SYSTEM ... 231

Introduction (231) ... Structure and Function (232) ... Respiration and Breathing (239) ... Sound Production (243) ... Comparing Other Organisms (244)

DIGESTIVE SYSTEM ... 247

Introduction (247) ... Upper Gastrointestinal Tract (248) ... Lower Gastrointestinal Tract (255) ... Accessory Glands (260) ... Metabolism and Nutrition (265) ... Comparing Other Organisms (272)

URINARY SYSTEM ... 275

Introduction (275) ... Kidneys (276) ... Accessory Organs (281) ... Urine and Its Formation (282) ... Comparing Other Organisms (283)

REPRODUCTIVE SYSTEM ... 285

Introduction (285) ... Male Reproductive System (286) ... Female Reproductive System (292) ... Conception (298) ... Sexually Transmitted Diseases (301) ... Comparing Other Organisms (302)

HUMAN GROWTH AND DEVELOPMENT ... 303

Introduction (303) ... Prenatal Development—Embryonic Period (304) ... Prenatal Development—Fetal Stage (309) ... Birth and Lactation (313) ... Postnatal Development (318) ... Comparing Other Organisms (322)

HELPING HUMAN ANATOMY ... 325

Anatomy and Imaging Techniques (325) ... Diagnostic Techniques for Various Systems (329) ... Operations, Procedures, and Transplants (335) ... Comparing Other Organisms (340)

FURTHER READING ... 341
GLOSSARY ... 345
INDEX ... 351

Acknowledgements

We are indebted to the authors of *The Handy Anatomy Answer Book*'s first edition, Naomi Balaban and James Bobick. Their knowledge and research for the book made the task of revising that much easier. We would also, as always, like to thank Roger Jänecke for all his help, patience, and consideration, and especially for asking us to revise another Handy Answer book. Also, thanks to typesetter Marco Di-Vita, cover and page designer Mary Claire Krzewinski, indexer Shoshana Hurwitz, and proofreaders Larry Baker and Sharon R. Gunton. Our editor, Kevin Hile, always gets special kudos from us—he's the best. And an extra special thanks to Agnes Birnbaum, who not only helps us with the intricacies of publishing but has been a wonderful friend for many, many years (and we hope many more).

We would also like to express our thanks to all the scientists, researchers, and health care professionals who deal with anatomy of the human body. We are not all alike, and trying to figure out what is best for "everyone's internal [some say infernal] configuration" is an immense task. We know so much more about human anatomy than we did even five years ago. And we have no doubt we will know more in the near and far future thanks to their expertise and efforts.

Photo Sources

Backrach (Wikicommons): p. 40.

Bruce Blaus (Wikicommons): pp. 35, 255.

OpenStax College: pp. 33, 102.

H. Pollock: p. 6.

Shutterstock: pp. 1, 9, 14, 17, 20, 21, 24, 27, 29, 37, 43, 44, 46, 48, 51, 54, 56, 58, 61, 63, 65, 67, 68, 70, 73, 74, 75, 78, 79, 81, 83, 84, 87, 88, 91, 93, 95, 96, 97, 99, 100, 106, 109, 111, 113, 115, 117, 119, 121, 123, 125, 127, 129, 130, 132, 133, 137, 140, 142, 143, 145, 146, 149, 152, 153, 156, 158, 159, 163, 164, 167, 168, 172, 174, 177, 179, 181, 184, 185, 189, 191, 192, 194, 198, 201, 203, 205, 207, 211, 212, 214, 217, 219, 222, 224, 227, 228, 231, 232, 234, 236, 237, 239, 243, 247, 249, 250, 252, 253, 257, 261, 264, 268, 270, 271, 273, 275, 277, 278, 281, 285, 286, 287, 289, 291, 293, 295, 297, 299, 300, 303, 305, 309, 311, 314, 315, 319, 321, 325, 326, 328, 330, 331, 333, 335, 337, 338, 339,

Welcome Trust: p. 2.

Public domain: pp. 4, 23, 317.

Introduction

The human body is a wondrous, incredible, natural, compact unit of muscle, bone, fat, cartilage, tissues, cells, genes, and a multitude of other major and minor elements. Bodies vary from person to person, thanks to genetics and how your cells grew from a single egg to the person you are today—but overall, they still have certain distinct characteristics.

At this time, understanding all the characteristics of the human body is not possible. But the continuing goal of scientists and health care professionals is to unravel more of the intricacies of the body to truly see what makes us tick. Two major byproducts of this understanding are to eradicate certain devastating diseases and to prolong our lives, while remaining healthy, of course.

This book presents some of the latest in these anatomical studies. It includes all the systems of the human body (and some interesting comparisons between humans and other animal species). It offers some of the history behind anatomic and physiologic studies, and the basics of anatomy. It gives the most recent techniques and research about how technology is helping to understand our insides, and even the future of those studies. And, finally, it offers a glossary of terms, further reading, and websites you can examine to expand your knowledge of your own anatomy.

You'll find the answers to over 1,200 questions, such as how many bones are there in the human body? What were "anatomy theatres"? Why do we die without oxygen? What are DNA and RNA? Is all cartilage the same? How does the brain detect different smells? What causes color blindness? What's the purpose of "goose bumps"? How big are capillaries? What are the most common heart surgeries? Do identical twins share the same fingerprints?

We hope you will enjoy perusing this book and find the answers to many of your anatomy questions. You can also use this book as a platform to discover other books or Internet sites concerning the human body. And overall, we hope this book will help you understand how everything in your body works together to make you you.

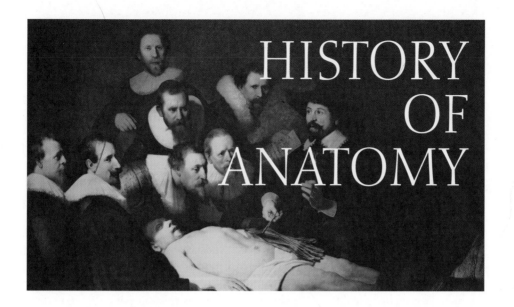

HISTORY OF ANATOMY

DEFINING ANATOMY

What is anatomy?

Anatomy is the study of an organism's internal structure or any of its parts—including the major organs, tissues, and cells—along with how the parts are organized within the organism. An "organism" includes plants, animals (for example, humans), bacteria, or other living entities.

How is the field of anatomy divided?

The field of anatomy has many different divisions depending on various criteria. For example, anatomy can be generally divided into macroscopic, or gross anatomy (not requiring a microscope), and microscopic anatomy. Gross anatomy includes the subdivisions of regional anatomy, systemic anatomy, developmental anatomy, clinical anatomy, and comparative anatomy. Regional anatomy studies specific regions of the body, such as the head and neck or lower and upper limbs. Systemic anatomy studies different body systems, such as the digestive system and reproductive system.

Developmental anatomy describes the changes that occur from conception through physical maturity. Clinical anatomy includes medical anatomy (anatomical features that change during illness) and radiographic anatomy (anatomical structures seen using various imaging techniques). Comparative anatomy examines the similarities and differences between structures and organizations of organisms.

The two major subdivisions of microscopic anatomy are cytology and histology. Cytology (from the Greek *cyto*, meaning "cell") is the study and analysis of the internal structure of individual cells. Histology (from the Greek *histos*, meaning "web") is the study and examination of tissues.

1

What is the difference between anatomy and physiology?

The scientific disciplines of anatomy and physiology study the human body. Anatomy (from the Greek *ana* and *temnein*, meaning "to cut up") is the study of the structure of the body parts, including their form and organization. Physiology (from the Latin *physiologia*, meaning "the study of nature") is the study of the function of the various body parts and organs. Anatomy and physiology are often studied together to achieve a complete understanding of the human body.

What are some specialties within the field of physiology?

Specialties and subdivisions of physiology include cell physiology, special physiology, systemic physiology, and pathological physiology, often called simply pathology. Cell physiology is the study of the functions of cells, including both chemical processes within cells and chemical interactions between cells. Special physiology is the physiological study of specific organs, such as cardiac physiology, which is the study of heart function. Systemic physiology is comparable to systemic anatomy since it is the study of the functions of different body systems, such as renal physiology and neurophysiology. Pathology (from the Greek *pathos*, meaning "suffering" or "disease") is the study of the effects of diseases on organs or systems and diseased cells and tissues.

What are the four levels of structural organization in animals, including humans?

Animals, including humans, have four levels of hierarchical organization: cell, tissue, organ, and organ system. Each level in the hierarchy is of increasing complexity, and all organ systems work together to form an organism. (For more about cells, tissues, organs, and organ systems, see the chapter "Levels of Organization.")

STUDIES IN ANATOMY

What is biology and who first coined the term?

Anatomy is a subdivision in the field of biology, which is often called "the science of life." Biology includes everything from an organism's conception to its death. In 1802 French biologist Jean-Baptiste Pierre Antoine de Monet de Lamarck (1744–1829) coined the term "biology" (from the Greek terms *bios*, meaning "life," and *logy*, meaning "study of") to describe the science of life.

Jean-Baptiste Pierre Antoine de Monet de Lamarck (1744–1829) was the French biologist who came up with the term for his own occupation.

What were some of the earliest studies in human anatomy?

Greek physician Diocles (c. 4th century–?), a student of Greek philosopher and scientist Aristotle (384–322 B.C.E.), is thought to have written the first anatomy books and the first book of herbal remedies. Greek physicians Herophilus (c. 335 B.C.E.–c. 280 B.C.E.; thought by some as the "father of anatomy") and Erasistratus (fl. c. 250 B.C.E.), both of whom lived in Alexandria, performed dissections in public. Both physicians described human organs such as the liver, brain, eye, ovaries, prostate gland, and spleen. They also determined that the place of reasoning was in the brain, not the heart.

When did the study of anatomy and physiology first become accepted as sciences?

Anatomy and physiology were first accepted as sciences during ancient Greek times. In particular, Aegean physician Hippocrates (c. 460–c. 377 B.C.E.) established medicine as a science, separating it from religion and philosophy. His application of logic and reason to medicine was the beginning of observational medicine.

Who were Hippocrates and Galen?

Aegean physician Hippocrates first defined the profession of physicians, making up the Hippocratic Oath, which included urging physicians to separate medicine from religion. Thus, he is considered by some as the "father of medicine." Roman physician and anatomist Galen (c. 129–c. 199 C.E.; for more about Galen, see below) was one of the first to present detailed studies of human anatomy. Not only was he the first to use the human pulse as a diagnostic aid, but he also wrote all medical knowledge that existed at the time into one systematic treatment—one that would be used by physicians until the end of the Middle Ages.

What were Aristotle's contributions to anatomy?

Greek philosopher and scientist Aristotle wrote several works laying the foundations for comparative anatomy, taxonomy, and embryology. He investigated carefully all kinds of animals, including humans. His works on life sciences, *On Sense and Sensible Objects, On Memory and Recollection, On Sleep and Waking, On Dreams, On Divination by Dreams, On Length and Shortness of Life, On Youth and Age,* and *On Respiration*, are collectively called *Parva Naturalia*.

Who is considered the father of physiology?

Greek physician and anatomist Erasistratus is considered the "father of physiology." Based on his numerous dissections of human cadavers, he accurately described the brain (including its cavities and membranes), stomach muscles, and the differences between motor and sensory nerves. He understood correctly that the heart served as a pump to circulate blood. Anatomical research ended with Erasistratus until the thirteenth century, in a large part because of public opinion against the dissection of human cadavers.

3

Whose work during the Roman era became the authoritative source for anatomy?

Galen, a Greek physician, anatomist, and physiologist living during the time of the Roman Empire, was one of the most influential and authoritative authors on medical subjects. His writings include *On Anatomical Procedures, On the Usefulness of the Parts of the Body, On the Natural Faculties,* and hundreds of other treatises. Since human dissection was forbidden, Galen made the most of his observations on different animals. He correctly described bones and muscles and observed muscles working in contracting pairs. He was also able to describe heart valves and structural differences between arteries and veins. While his work contained many errors, he provided many accurate anatomical details that are still regarded as classics. Galen's writings were the accepted standard text for anatomical studies for close to 1,400 years.

What was one of the first books in human anatomy?

In 1543, Flemish anatomist Andreas Vesalius (1514–1564) wrote the first accurate work on human anatomy, titled *De humani corporis fabrica* (*On the Structure of the Human Body*). Many scientists call it one of the books that changed the world. The book contained exceptional illustrations (usually attributed to painter Jan Stephan van Calcar and techniques drawn from Leonardo da Vinci [see sidebar on the next page]) and showed the true anatomy of the human body, all reproduced with extreme accuracy (previous hand-copied books had many inaccuracies). The book also caused a scandal. After it was published, accusations of body snatching and heresy were directed against Vesalius, mainly by physicians who were not as progressive or had older ideas of the human body, much of it based on Roman physician and anatomist Galen. As a result of the accusations, Vesalius never did any research again and remained a court physician for the rest of his life.

What were iatromathematics and iatrophysics?

Iatromathematics, or medical astrology, was an ancient belief that various parts of the body, along with human diseases—both mental and physical conditions—were all controlled and influenced by the sun, moon, planets, and the zodiac (or astrological signs). Iatrophysics (or iatromechanics) was practiced in the sixteenth through eighteenth centuries. Physicians who practiced iatrophysics believed that the human body, no matter if it was ill or

A portrait of Flemish anatomist Andreas Vesalius by Pierre Poncet.

> ## Who was Leonardo da Vinci?
>
> Leonardo da Vinci (1452–1519) was not only known for his artistry, scientific expertise, and inventions, but also for his hundreds of drawings made from human dissections. His methods of drawing anatomy were borrowed from another field of art: architecture. In 1489, he drew a series of studies of the human skull based on the architectural techniques of plane, elevation, and perspective. And by 1510, he drew a detailed and accurate series of dissections based on the entire human body. Leonardo's studies were not published in his lifetime, but his methods of illustration were widely known. Thus Leonardo's works are thought to have influenced the first accurate illustrated book on human anatomy by Andreas Vesalius.

well, obeyed the laws of physics. For example, Italian physician and scientist Giorgio Baglivi (also Gjuro Baglivi; 1668–1707) believed that parts of a human behaved like a machine. He used the example of the human arm, which he said acted as a lever, and compared the human chest to a blacksmith's bellows and the heart to a pump.

What were "anatomy theatres"?

Anatomical theatres were established in the late 1500s by educational institutions in order to teach anatomy. The idea was loosely based on the Roman amphitheatres, such as the Roman Coliseum. In the anatomical theatres, students and other observers could easily observe a dissection taking place, as the seats were arranged to allow the audience an unrestricted view.

The first anatomical theatre was built in 1594 at the University of Padua in Italy. In the United States, several such theatres were built, including one at the University of Virginia by Thomas Jefferson. He designed the theatre in 1825, a year before his death, and it was completed in 1827. The building went through several different uses and was finally demolished in 1939 when the Alderman Library was constructed.

Who improved the microscope in a way that greatly impacted anatomy and physiology studies?

Antoni van Leeuwenhoek (1632–1723) was a Dutch microscopist and scientist. Although he did not invent the microscope, he greatly improved the capability of the instrument. His expert skill in grinding lenses achieved a magnification of 270 times, which was far greater than any other microscope of the era. He was able to observe bacteria, striations in muscles, blood cells, and spermatozoa.

Who was Nehemiah Grew?

Nehemiah Grew (1641–1712) was an English plant anatomist and physiologist. He was known as the "father of plant anatomy" and was also the first to coin the term "com-

parative anatomy" in 1676. This term was not only used in plant anatomy, but, eventually, referred to the comparative anatomy of other organisms, including humans.

What discovery in the seventeenth century helped establish the science of physiology?

In 1628 English physician William Harvey (1578–1657) published *On the Movement of the Heart and Blood in Animals*. This important medical treatise proved that blood continuously circulated within a body's vessels. In particular, his hypothesis stated that the heart is a pump for the circulatory system, with blood flowing in a closed circuit. He also correctly demonstrated that when an artery is slit, all the blood in the system empties, and that the valves in the veins serve to make sure the blood returns to the heart. Harvey's discoveries were often ignored by his contemporaries, as they contradicted many beliefs about blood circulation dating back to the time of Galen. He is also considered "the father of modern physiology" for introducing the experimental method of scientific research.

Who was William Hunter?

Scottish anatomist and physician William Hunter (1718–1783) was one of the most important anatomy teachers of his time. His *Anatomy of the Human Gravid Uterus* in 1774 was one of his greatest works in anatomy and contained twenty-four masterpieces of anatomical illustrations. He was eventually appointed professor of anatomy at the Royal Academy of London.

Who is considered the founder of experimental medicine and physiology?

French physiologist Claude Bernard (1813–1878) is credited with originating the experimental approach to medicine and establishing general physiology as a distinct discipline. His classic work, *Introduction to the Study of Experimental Medicine*, was published in 1865. He was elected to the Académie Française in 1869 for this work.

Who was William E. Horner?

William E. Horner (1793–1853) was an American anatomist who wrote the first text on anatomy published in the United States. His *A Treatise on Pathological Anatomy* was published in 1829.

COMPARING OTHER ORGANISMS

What were some of the earliest studies in animal anatomy (other than humans)?

No one can truly pinpoint when the study of animal anatomy began. Some researchers believe religious sacrifices of animals may have led to the first knowledge of animal anatomy, especially how animals' structures compared to humans. Early Egyptians, and several other cultures, also apparently had some crude internal knowledge of certain animals (for example, organs and veins), especially those creatures that were embalmed and buried with the human dead. Greek philosopher and scientist Aristotle is thought to be one of the first to understand animal anatomy, as he dissected animals and described many of their physical attributes. Such dissections eventually led to more knowledge about human anatomy.

Who first developed a method for classifying animals based on structure?

French naturalist and anatomist Georges Léopold Chrétien Frédéric Dagobert Cuvier (1769–1832) was the first to use a specific method to classify mammals. This method also established a system of zoological classification, dividing the animals into four categories based on their structures. His four categories were Vertebrata, Mollusca, Articulata, and Radiata. In 1805, Cuvier's *Leçons d'anatomie comparée* (*Lessons on Comparative Anatomy*) was published, establishing the field of comparative anatomy. Thus, he is often considered by many to be the "founder of comparative anatomy."

What are the remains of ancient organisms?

The remains of ancient organisms that have been preserved in the Earth close to their original shape are called fossils (from the Latin *fossilis*, meaning "something dug up"). The different types of fossils depend on the remains and conditions present at the time the organism died. Fossils may be formed from the hard parts of an organism, such as teeth, shells, bones, or wood. They may also be unchanged from their original structures and features. In some cases, the entire organism is replaced by minerals such as calcite or pyrite. Organisms can also be preserved as fossils in other materials besides stone, including ice, tar, peat, and the resin of ancient trees.

What are some of the largest fossil bones ever discovered?

Some of the largest fossil bones ever discovered belong to the ancient dinosaurs. They are considered an extremely diverse group of animals of the clade (a type of classification) Dinosauria. They first appeared about 230 million years ago and died out "suddenly" (in terms of geologic time) about 65 million years ago. The dinosaurs

What were the first dinosaur bones uncovered in the United States and Europe?

The first recorded description of a dinosaur bone was done in 1676 by English naturalist Robert Plot (1640–1696), a professor at the University of Oxford, England, in his book *The Natural History of Oxfordshire*. Although he correctly determined that the object was a broken piece of a giant bone, Plot did not know the bone came from a dinosaur. Instead, he felt it belonged to a giant man or women, citing mythical, historical, and biblical sources. In 1763, the same bone fragment was named *Scrotum humanum* (in a book caption) by Richard Brookes to describe its appearance, but the name never gained wide acceptance. Based on Plot's illustration, modern scientists believe the bone fragment was actually the lower end of a thigh bone from a Megalosaurus, a meat-eating dinosaur from the middle Jurassic period that roamed the area now known as Oxfordshire.

In 1787, American physician Caspar Wistar (1761–1818) and Timothy Matlack (1730–1829), a brewer, soldier, and scribe of the Declaration of Independence, presented a paper about a large fossil bone—they called it a "large thigh bone"—found in the upper Cretaceous rock layers in New Jersey. Their finding was largely ignored and unverified, but many historians believe it may have been the first dinosaur bone ever collected in North America.

dominated the land and are generally divided into carnivorous (meat-eating) and herbaceous (plant-eating) animals. It is thought that all true dinosaurs lived on land, not in the air or the oceans. But in 2014, scientists believe they found the first dinosaur thought to have lived in water. Called the *Spinosaurus aegyptiacus*, the 50-foot (15.24 meters) carnivorous dinosaur allegedly lived mostly in a river or lake, but such a claim is still being debated.

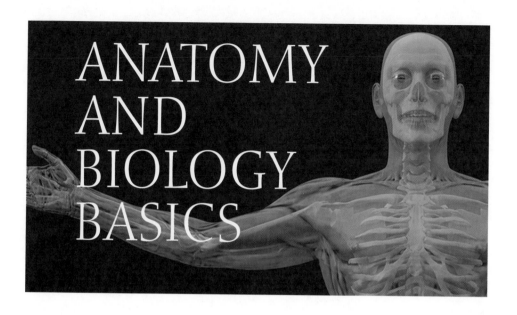

ANATOMY AND BIOLOGY BASICS

HUMAN ANATOMICAL TERMINOLOGY

What is the anatomical position?

Anatomists universally defined the anatomical position as the body standing erect, facing forward, feet together and parallel to each other, and arms are at the side of the body with the palms facing forward. All directional terms that describe the relationship of one body part to another assume the body is in the anatomical position.

What directional terms describe the location of body parts in relation to other body parts?

There are standard directional terms used to describe the location of one body part in relation to another body part. Most directional terms occur as pairs with one term of the pair having the opposite meaning of the other term. The following lists the common directional terms:

Directional Terms of the Body

Term	Definition	Example
Superior (cranial or cephalic)	Toward the head	The head is superior to the neck
Inferior (caudal)	Away from the head; toward the feet	The neck is inferior to the head
Anterior (ventral)	Toward the front	The toes are anterior to the heel
Posterior (dorsal)	Toward the back	The heel is posterior to the toes
Medial	Toward the midline	The nose is medial to the eyes
Lateral	Away from the midline of the body; towards the sides	The eyes are lateral to the nose

Directional Terms of the Body

Term	Definition	Example
Proximal	Toward the trunk of the body; nearer the attachment of an extremity to the trunk	The shoulder is proximal to the elbow
Distal	Away from the trunk of the body; further from the attachment of an extremity to the trunk	The wrist is distal to the shoulder
Superficial	Near the surface of the body	The skin is superficial (external) to the muscles
Deep (internal)	Farther from the surface of the body	The heart is deeper than the ribs

What are the two basic regions of the body?

The body is divided into two basic regions: the axial and the appendicular. The axial part of the body consists of the head, neck, and trunk, including the thorax (chest), abdomen, and pelvis. The appendicular region consists of the upper and lower extremities.

Why are body planes important for identifying anatomical structure?

In order to observe and study the structural arrangement of the internal organs, the body may be divided and sectioned (or cut) along three fundamental planes. The midsagittal (median) plane divides the body lengthwise into right and left sides. A sagittal section placed off-center divides the body into asymmetrical right and left sides. The coronal (frontal) plane divides the body into front (anterior) and back (posterior) sections. The transverse (horizontal) plane divides the body into upper (superior) and lower (inferior) sections. It is at right angles to the sagittal and frontal planes.

What are the divisions of the head and neck regions of the body?

The head is divided into the facial region and cranium. The facial region includes the eyes, nose, and mouth. The cranium is the part of the head that covers the brain. The neck is also referred to as the cervix or cervical region.

How is the abdomen divided into nine regions?

The abdomen is divided into nine regions with two vertical lines and two horizontal lines. The two vertical lines are drawn downward from the center of the collarbones. One horizontal line is placed at the lower edge of the rib cage and the other is placed at the upper edge of the hip bones. The umbilical region, containing the navel, is the center of the abdomen.

What are the major regions of the trunk?

The trunk includes the thorax (often called the chest), abdomen, and pelvis. The following lists the major regions of the human trunk:

Region	Location
Anterior trunk	
Pectoral	The chest
Abdominal	Area between the lowest ribs and the pelvis
Pelvic	The area surrounded by the pelvic bones
Inguinal	The groin; the junction of the thighs to the anterior trunk
Posterior trunk	
Dorsum	Posterior surface of the thorax
Vertebral	Region over the vertebral column
Lumbar	Lower back region between the lowest ribs and the pelvis
Sacral	Region over the sacrum and between the buttocks
Gluteal	The buttocks
Lateral trunk	
Axillary	The armpits
Coxal	The hips
Inferior trunk	
Genital	External reproductive organs
Perineal	Small region between the anus and external reproductive organs

What are the areas of the upper and lower extremities?

The upper extremities and lower extremities form the appendicular region of the body. The upper extremities include the shoulders, upper arms, forearms, wrists, and hands. The lower extremities include the thighs, legs, ankles, and feet. The following describes the major regions of the upper and lower extremities:

Anatomical Term	Common Term
Upper Extremity	
Antebrachial	Forearm
Brachial	Upper arm
Antecubital	Anterior portion of the elbow joint
Cubital	Posterior portion of the elbow joint
Digital	Fingers
Palmar	Palm of the hand
Lower Extremity	
Femoral	Thigh
Patellar	Anterior portion of the knee joint
Popliteal	Posterior portion of the knee joint
Tarsal	Ankle
Pedal	Foot
Digital	Toes
Plantar	Sole of the foot

What is the function of cavities in the human body?

The human body cavities house and protect the internal organs. There are two main body cavities: the dorsal cavity and the ventral cavity. The dorsal or posterior cavity contains the cranial cavity and the spinal cavity. The cranial cavity houses and protects the brain, while the spinal cavity houses and protects the spinal cord.

The ventral or anterior cavity is separated into the thoracic cavity and abdominopelvic cavity. The thoracic cavity contains the heart and lungs. It is protected by the rib cage. The abdominopelvic cavity is further divided into the abdominal cavity and the pelvic cavity. The stomach, intestines, liver, gall bladder, pancreas, spleen, and kidneys are in the abdominal cavity. The urinary bladder, internal reproductive organs, sigmoid colon, and rectum are in the pelvic cavity.

What structure separates the thoracic cavity from the abdominopelvic cavity?

The diaphragm separates the thoracic cavity from the abdominopelvic cavity in the ventral cavity. It is a thin, dome-shaped sheet of muscle. (For more about the diaphragm, see the chapter "Respiratory System.")

CHEMISTRY IN BIOLOGY AND ANATOMY

Why is chemistry important for understanding the human body?

The universe and everything in it is composed of matter, or anything that occupies space and has mass. Overall, the ninety-two naturally occurring chemical elements are the fundamental forms of matter.

In the human body there are twenty-six different chemical elements. The continually ongoing reactions in the body that involve these chemicals—in different numbers and proportions—underlie all physiological processes of the body. This includes human movement, digestion, the pumping of the heart, respiration, and sensory and neural processes.

Why do we die without oxygen?

Most living organisms are aerobic; that is, they require oxygen to complete the total breakdown of glucose for the production of adenosine triphosphate (ATP), the energy for life. Many people think that humans need oxygen to breathe, but actually people need oxygen to recycle the body's spent electrons and hydrogen ions (H). (For more about ATP, see the chapter "Muscular System.")

What are some important elements in living systems?

The most important elements in living systems—from humans to fungi and certain bacteria—include oxygen, carbon, hydrogen, nitrogen, calcium, phosphorus, potassium, sulfur, sodium, chlorine, magnesium, and iron. These elements are essential to life because of how they function within an organism's cells. For example, the following lists the common and important chemical elements in the human body:

Element	% of Humans by Weight	Functions in Life
Oxygen	65	Part of water and most organic molecules; essential to physiological processes
Carbon	18	Basic component of organic molecules
Hydrogen	10	Part of most organic molecules and water
Nitrogen	3	Component of proteins and nucleic acids
Calcium	2	Component of bones; essential for nerves and muscles; blood clotting
Phosphorus	1	Part of cell membranes and energy storage molecules; component of bones, teeth, and nerve tissues
Potassium	0.3	Important for nerve function, muscle contraction, and water-ion balance in body fluids
Sulfur	0.2	Structural component of some proteins
Sodium	0.1	Primary ion in body fluids; essential for nerve function
Chlorine	0.1	Major ion in body fluids
Magnesium	Trace	Cofactor for enzymes; important to muscle contraction and nerve transmission
Iron	Trace	Basic component of hemoglobin

What is the water content of various tissues of the human body?

Water accounts for approximately 62 percent of the total body weight of a human. It is found is every tissue. The following lists the tissue, percent of the total body weight, percent water, and water in terms of quarts and liters:

Tissue	% Body Weight	% Water	Quarts of Water (Liters)
Muscle	41.7	75.6	23.35 (22.1)
Skin	18	72	9.58 (9.07)
Blood	8	83	4.91 (4.65)
Skeletal	15.9	22	2.59 (2.45)
Brain	2	74.8	1.12 (1.05)
Liver	2.3	68.3	1.16 (1.1)
Intestines	1.8	74.5	0.99 (0.94)
Fat tissue	8.5	10	0.74 (0.7)
Lungs	0.7	79	0.41 (0.39)
Heart	0.5	79.2	0.3 (0.28)
Kidneys	0.4	82.7	0.24 (0.23)
Spleen	0.2	75.8	0.12 (0.11)

What is meant by pH?

The term "pH" is taken from the French phrase *la puissance d'hydrogen*, meaning "the power of hydrogen." Scientifically, pH refers to the −log of the H^+ (positive hydrogen). The mathematical equation to determine pH is usually written as follows: $pH = -\log [H^+]$. For example, if the hydrogen ion concentration in, say, a solution is 1/10,000,000 or 10^{-7}, then the pH value is 7.

The composition of water can also be used to understand the concept of pH: Water is composed of two hydrogen atoms bonded covalently to an oxygen atom. In a solution of water, some water molecules (H_2O) will break apart into the component ions—H^+ and OH^- ions; and it is the balance of these two ions that determines pH. When there are more H^+ ions than OH^- ions, the solution is an acid, and when there are more OH^- ions than H^+ ions, the solution is a base.

What is the pH scale?

The pH scale is the measurement of the H^+ concentration (hydrogen ions) in an aqueous solution and is used to measure the acidity or alkalinity of that solution. The pH scale ranges from zero to fourteen. A neutral solution has a pH of seven; a solution with a pH greater than seven is basic (or alkaline); and a solution with a pH less than seven is acidic. In other words, the lower the pH number the more acidic the solution; the higher the pH number the more basic the solution. As the pH scale is logarithmic, each whole number drop on the scale represents a tenfold increase in acidity (meaning the concentration of H^+ increases tenfold); and of course, each whole number rise on the scale represents a tenfold increase in alkalinity.

What are some examples in terms of the pH scale?

The following lists some examples of certain solutions in terms of their pH (note: the pH of some solutions ranges in value):

pH Value	Examples of Solutions
1	hydrochloric acid (HCl), battery acid
1 to 3	stomach acid
2.3	lemon juice

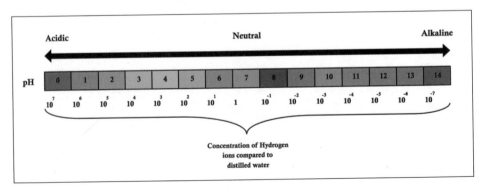

pH Value	Examples of Solutions
3	vinegar, wine, some acid rain
3.5	orange juice
4	tomatoes, grapes
4.6	banana
5	black coffee, most shaving lotions, bread, normal rainwater
5 to 6	urine
6.2 to 7.4	saliva
6.6 to 6.8	milk (cow's)
7	pure water (distilled)
7.1 to 8.2	digestive juice of the pancreas
7.3 to 7.5	blood
7.6 to 8.6	bile (liver secretion that aids in fat digestion)
8	egg white
7.8 to 8.3	seawater
9	baking soda, phosphate detergents, Clorox
10	soap solutions, milk of magnesia
10.5 to 11.9	household ammonia
11	non-phosphate detergents
12	washing soda (sodium carbonate)
13	hair remover, oven cleaner
14	lye

BIOLOGICAL COMPOUNDS AND THE HUMAN BODY

What are the major bioorganic molecules in humans?

The major bioorganic molecules are carbohydrates, lipids, proteins, and nucleic acids. These molecules are characteristic of life and have basic roles such as storing and producing energy, providing structural materials, or storing hereditary information.

What are carbohydrates?

Carbohydrates are organic compounds composed of carbon, hydrogen, and oxygen. The general chemical formula for carbohydrates is CH_2O, indicating there is twice as much hydrogen as oxygen. Carbohydrates are the major source of energy for cells and cellular activities.

How are carbohydrates classified?

Carbohydrates are classified in several ways. Monosaccharides (single unit sugars) are grouped by the number of carbon molecules they contain: triose has three, pentose has five, and hexose has six. Carbohydrates are also classified by their overall length (monosaccharide, disaccharide, and polysaccharide) or function. Examples of functional definitions are storage polysaccharides (glycogen and starch), which store energy, and structural polysaccharides (cellulose and chitin).

What are some of the uses of carbohydrates by the body?

Carbohydrates are mainly used as an energy source by the body, with the various carbohydrates having different functions. The following chart identifies some common carbohydrates and their uses in the human body:

Carbohydrate Name	Type	Use by the Body
Deoxyribose	Monosaccharide	DNA; constituent of hereditary material
Fructose	Monosaccharide	Important in cellular metabolism of carbohydrates
Galactose	Monosaccharide	Found in brain and nerve tissue
Glucose	Monosaccharide	Main energy source for the body
Ribose	Monosaccharide	Constituent of RNA
Lactose	Disaccharide	Milk sugar; aids the absorption of calcium
Sucrose	Disaccharide	Produces glucose and fructose upon hydrolysis
Cellulose	Polysaccharide	Not digestible by the body, but is an important fiber that provides bulk for the proper movement of food through the intestines
Glycogen	Polysaccharide	Stored in the liver and muscles until needed as energy source and is then converted to glucose
Heparin	Polysaccharide	Prevents excessive blood clotting
Starch	Polysaccharide	Chief food carbohydrate in human nutrition

What are lipids?

Lipids are organic compounds composed mainly of carbon, hydrogen, and oxygen, but they also may contain other elements, such as phosphorus and nitrogen. Lipids usually have more than twice as many hydrogen atoms as oxygen atoms. They are insoluble in water but can be dissolved in certain organic solvents such as ether, alcohol, and chloroform. Lipids include fats, oils, phospholipids, steroids, and prostaglandins.

What is the difference between fats and lipids?

Fats are one category of lipids. Each fat molecule is comprised of a glycerol (alcohol) molecule and at least one fatty acid (a hydrocarbon chain with an acid group attached). Fats are energy-rich molecules important as a source of reserve food for the body. They are stored in the body in the form of triacylglycerols, also known as triglycerides. Fats also provide the body with insulation, protection, and cushioning (especially of the body's organs).

What is cholesterol?

Cholesterol belongs to a category of lipids known as steroids. Steroids have a unique chemical structure. They are built from four carbon-laden ring structures that are fused together. The human body uses cholesterol to maintain the strength and flexibility of cell membranes. Cholesterol is also the molecule from which steroid hormones and bile acids are built.

What is an enzyme?

An enzyme is a protein that acts as a biological catalyst. It decreases the amount of energy needed (activation energy) to start a metabolic reaction. Different enzymes work in different environments due to changes in temperature and acidity. For example, the amylase that is active in the mouth cannot function in the acidic environment of the stomach; pepsin, which breaks down proteins in the stomach, cannot function in the mouth. In fact, without enzymes, the stomach would not be able to obtain energy and nutrients from food.

What are some enzyme deficiencies in humans?

Lactose intolerance, a condition that results from the inability to digest lactose—the sugar present in milk—is one of the most common enzyme deficiencies. According to the National Institutes of Health, it is estimated that around 90 percent of adults of East Asian descent have lactose intolerance, and overall, about 65 percent of the human population has a reduced ability to digest lactose after infancy.

Another less common but more serious enzyme deficiency is glucose–6-phosphate dehydrogenase deficiency, which is linked to the bursting of red blood cells (hemolysis). This deficiency is found in more than two hundred million people, mainly in Mediterranean, West African, Middle Eastern, and Southeast Asian populations.

What are proteins and what is their purpose?

Proteins are large, complex molecules composed of smaller structural subunits called amino acids. All proteins contain carbon, hydrogen, oxygen, and nitrogen, and sometimes sulfur, phosphorus, and iron. Human life could not exist without proteins.

The enzymes that are required for all metabolic reactions are proteins. These proteins also are important to structures like muscles, and they act as both trans-

Lactose intolerance occurs when a person cannot digest the sugar lactose. Symptoms include cramping, gas, bloating, diarrhea, and sometimes vomiting.

porters and signal receptors. The following lists the type of proteins and examples of their functions:

Type of Protein	Examples of Functions
Defensive	Antibodies that respond to invasion
Enzymatic	Increase the rate of reactions; build and break down molecules
Hormonal	Insulin and glucagon, which control blood sugar
Receptor	Cell surface molecules that cause cells to respond to signals
Storage	Store amino acids for use in metabolic processes
Structural	Major components of muscles, skin, hair
Transport	Hemoglobin carries oxygen from lungs to cells

COMPARING OTHER ORGANISMS

In general, how do scientists determine the anatomy of ancient animals?

Determining the anatomy of ancient animals is not an easy task. Scientists look for similarities between modern and ancient species' bone structures and compare ancient animal skeletons in a "death pose" that are close to a more recent, similar organism. One of the best examples of determining anatomy and bone positions of ancient animals comes from the study of dinosaur fossils (for more about fossils and dinosaurs, see the chapter "History of Anatomy.")

The positions of dinosaur bones are determined using an "anatomical direction system." This system uses pairs of names to determine certain directions based on the average (or standard) posture of tetrapods, with the back up, belly down, head pointing forward, and all four legs on the ground. Each pair of names denotes opposite directions, similar to when we refer to north and south. The following are four examples of such paired names:

Anterior and posterior: The direction of anterior is toward the tip of the snout, while the posterior direction is toward the tip of the tail. This is analogous to front and back, respectively.

Dorsal and ventral: Dorsal means toward and beyond the spine, while ventral means toward and beyond the belly. These are analogous to up and down, respectively.

Medial and lateral: These are directions referenced to an imaginary plane located in the center of the body, running from tail to snout. Medial means closer to this central reference; lateral means farther out.

Proximal and distal: These are normally used to indicate directions in the limbs and sometimes the tail. Proximal means closer to the trunk or base of a limb, while distal means farther out from the trunk or from the base of the limb.

Like humans, can other animals be grouped according to body symmetry?

Yes, animals are often divided into two groups according to their symmetry, or the arrangement of body structures in relation to the axis of the body. For example, the bodies of most primitive animals such as jellyfish, sea anemones, and starfish have radial symmetry, or a body in the form of a wheel or cylinder, with similar structures arranged as spokes from a central axis. Animals with bilateral symmetry have right and left halves that are mirror images of each other; they also have top (dorsal) and bottom (ventral) portions and a front (anterior) end and back (posterior) end. More sophisticated animals fall into this category, such as flatworms. Some organisms even exhibit both radial and bilateral symmetry, such as the echinoderms that have bilateral symmetry as larvae and revert to radial symmetry as adults.

Do animals need certain elements to stay alive?

Yes, most animals need certain elements to stay alive. In particular, close to 99 percent of all animals need the elements oxygen, carbon, hydrogen, and nitrogen for survival (exclusions include some bacteria and fungi). For example, most animals need a supply of oxygen in order to stay alive. And because animals are made up of mostly water (or H_2O, hydrogen and oxygen), the element oxygen is essential to most life. Most organisms absorb oxygen through the lungs. It is then inhaled, picked up by the red blood cells, and carried to the various parts of the body, thus aiding the cell and organ processes that keep the organism alive.

Do all animals need oxygen?

It was once thought that only some single-celled organisms such as prokaryotes and protozoa could live without oxygen. In 2010 researchers found the first multicellular animals that can survive and reproduce without oxygen in a deep hypersaline basin in the Mediterranean Sea. They found three distinct species of the multicellular animals—from a phylum called Loricifera—about two miles below the sea's surface. One reason for not requiring oxygen is the organisms do not have mitochondria, but have organelles in their cells that are similar that use anaerobic processes. (For more about cells, see the chapter "Levels of Organization.")

Why is water important to living organisms?

Water serves many purposes in the functioning of the human body. For example, in digestion it serves as a solvent to break down large compounds into smaller ones. Water is also a transporter of nutrients, waste products, blood, and materials within cells. It is also very important in temperature regulation through perspiration and evaporation. Finally, water is a main component of synovial fluid, the lubricating fluid that helps joints move smoothly and easily.

Water is important to all animals. It is essential in digestion and to aid in the transport of nutrients throughout the body.

Do animals need enzymes?

Yes, all known living organisms—from animals and plants to fungi and bacteria—cannot survive without enzymes. These enzymes carry out a plethora of different chemical reactions. Every one of the chemical reactions is made possible by some specific enzyme or enzymes working together. And each one of the enzymes reacts to a certain stimulus, such as body temperature, chemical, or digestion.

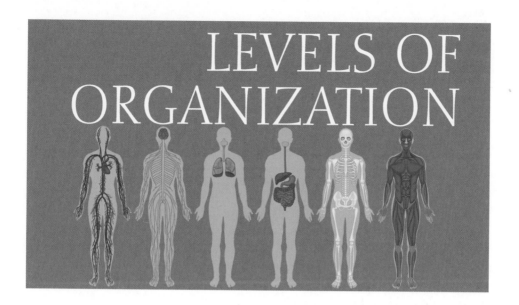

LEVELS OF ORGANIZATION

DEFINING LEVELS OF ORGANIZATION IN ANATOMY

What are the levels of structural organization in vertebrate animals, including humans?

Every vertebrate animal has four major levels of hierarchical organization. Each level in the hierarchy is of increasing complexity, and all organ systems work together to maintain life. The following lists and briefly explains the major levels

- *Cell*: A cell is a membrane-bound unit that contains hereditary material (DNA, or deoxyribonucleic acid) and cytoplasm; it is the basic structural and functional unit of all forms of life.

- *Tissue*: A tissue (from the Latin *texere*, meaning "to weave") is a group of similar cells that perform a specific function. The four major types of tissue are epithelial, connective, muscle, and nerve, with each type of tissue performing different functions.

- *Organ*: An organ is a group of several different tissues working together as a unit to perform a specific function or functions.

- *Organ system*: An organ system is a group of organs working together to perform a vital body function.

What are two additional levels of organization often mentioned by some scientists?

Some scientific literature mention the four main levels of structural organization in the human body but also add two more levels: the chemical and organismal levels. The chemical level is the simplest level and is listed before the cellular level. It includes atoms, the smallest building blocks of matter that combine to form molecules. These in turn form the organelles within the cells Cellulae. The organismal level is the highest (or last) level and is considered to be the sum total of all the structural levels as they work together—or the human being as a whole.

CELLS

What is a cell?

In general, a cell is the basic unit of all forms of life. Cells are considered to be specialized depending on their function, such as tissue or organ cells. They range in size from microscopic to the size of a chicken egg. In addition, organisms can be called single-celled, such as bacteria, and multicelled, such as humans.

What is the origin of the term *cell*?

In 1665, British physicist Robert Hooke (1635–1703) first used the term *cell* to describe the divisions he observed in a slice of cork. Using a microscope that magnified thirty times, Hooke identified little chambers or compartments in the cork that he called *cellulae* (a Latin term meaning "little rooms") because they reminded him of the small monastery cells inhabited by monks. (He further calculated that one square inch [6.45 square centimeters] of cork would contain 1,259,712,000 of these tiny chambers or "cells.") "Cellulae" eventually evolved into the modern term *cell*.

What discoveries led to the cell theory?

Many discoveries happened along the way in the modern study of cells. In 1831, British botanist Robert Brown (1773–1858) discovered and named the cell nucleus in plant cells; and German botanist Matthias Schleiden (1804–1881) named the nucleolus (the structure within the nucleus now known to be involved in the production of ribo-

What is the cell theory?

The cell theory states that the cell is the fundamental component of all life and all organisms are made up of cells. There are three basic principles to the cell theory. First, the cell is the simplest collection of matter that can live. There are diverse forms of life existing as single-celled organisms. More complex organisms, including plants and animals, are multicellular cooperatives composed of diverse, specialized cells that could not survive for long on their own. Secondly, all cells come from preexisting cells and are related by division to earlier cells that have been modified in various ways during the long evolutionary history of life on earth. Finally, all of the life processes of an organism occur fundamentally at the cellular level.

somes) around that same time. Working independently, Schleiden and German physiologist Theodor Schwann (1810–1882) described preliminary forms of the general cell theory in 1839, the former stating that cells were the basic unit of plants and Schwann extending the idea to animals. In 1855, Polish-German embryologist, physiologist, and neurologist Robert Remak (1815–1865) became the first to describe cell division. Shortly after Remak's discovery, German doctor, anthropologist, pathologist, biologist, and politician Rudolph Virchow (1821–1902) stated that all cells come from preexisting cells. Thus, the work of Schleiden, Schwann, and Virchow firmly established the cell theory.

British botanist Robert Brown discovered the nucleus in plant cells.

What are the largest, smallest, and longest cells in the human body?

Not counting our own cells, some of the smallest cells belong to the many bacteria that live inside and outside the human body, ranging in size from 0.0079 to 0.012 inches (0.2 to 0.3 millimeters) in diameter. The largest human cell is the female ovum, or egg, which measures between 120 and 150 micrometers across; the smallest human cell is the sperm cell (spermatozoon), which measures 60 micrometers across. Neurons, or nerve cells, are the longest cells in the body; some measure around 39 inches (99 centimeters) long. (For more about neurons, see the chapter "Nervous System.")

What is the chemical composition of a typical mammalian cell?

The following lists the molecular components and percent of total cell weight for an average mammalian cell (including humans):

Molecular Component	% of Total Cell Weight
Water	70
Proteins	18
Phospholipids and other lipids	5
Miscellaneous small metabolites	3
Polysaccharides	2
Inorganic ions (sodium, potassium, magnesium, calcium, chlorine, etc.)	1
RNA	1.1
DNA	0.25

What are the two major types of cells in nature?

In 1937, French marine biologist Édouard Chatton (1883–1947) first proposed the terms *procariotique* and *eucariotique* (French for prokaryotic and eukaryotic, respectively) to differentiate between cells in certain organisms. Prokaryotic, meaning "before nucleus," was used to describe bacteria, while eukaryotic, meaning "true nucleus," was used to describe all other cells. Today, the terms are more well-defined: eukaryotic cells are much more complex than prokaryotic cells, having compartmentalized interiors and membrane-contained organelles (small structures within cells that perform dedicated functions; for more about organelles, see below) within their cytoplasm. The major feature of a eukaryotic cell is its membrane-bound nucleus, the active part of the cell that contains genetic information; prokaryotic cells do not have a nuclear membrane (the membrane that surrounds the nucleus of the cell). The cells also differ in size: eukaryotic cells are generally much larger and more complex than prokaryotic cells. In fact, most eukaryotic cells are one hundred to a thousand times the volume of typical prokaryotic cells.

What are organelles?

Organelles, frequently called "little organs," are found in all eukaryotic cells; they are specialized, membrane-bound, cellular structures that perform a specific func-

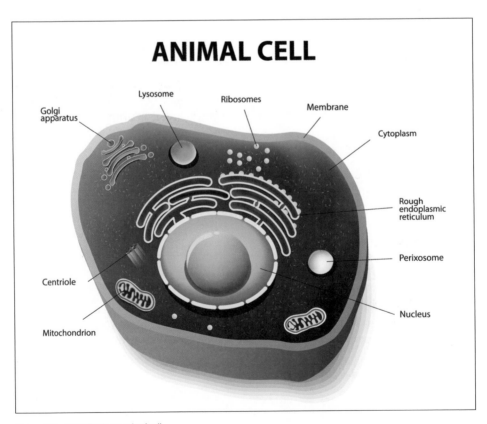

Some of the organelles in an animal cell.

tion. Eukaryotic cells contain several kinds of organelles, including the nucleus, mitochondria, chloroplasts, endoplasmic reticulum, and Golgi apparatus.

What are the major components of the eukaryotic cell?

The major components of the eukaryotic cell are as follows:

Structure	Description
Cell Nucleus	
Nucleus	Large structure surrounded by double membrane
Nucleolus	Special body within nucleus; consists of RNA and protein
Chromosomes	Composed of a complex of DNA and protein known as chromatin; resemble rodlike structures after cell division
Cytoplasmic Organelles	
Plasma membrane	Membrane boundary of living cell
Endoplasmic reticulum (ER)	Network of internal membranes extending through cytoplasm
Smooth endoplasmic reticulum	Lacks ribosomes on the outer surface
Rough endoplasmic reticulum	Ribosomes stud outer surface
Ribosomes	Granules composed of RNA and protein; some are attached to ER and some are free in cytosol
Golgi complex	Stacks of flattened membrane sacs
Lysosomes	Membranous sacs (in animals)
Vacuoles	Membranous sacs (mostly in plants, fungi, and algae)
Microbodies (for example, peroxisomes)	Membranous sacs containing a variety of enzymes
Mitochondria	Sacs consisting of two membranes; inner membrane is folded to form cristae and encloses matrix
Plastids (for example, chloroplasts)	Double membrane structure enclosing internal thylakoid membranes; chloroplasts contain chlorophyll in thylakoid membranes
The Cytoskeleton	
Microtubules	Hollow tubes made of subunits of tubulin protein
Microfilaments	Solid, rod-like structures consisting of actin protein
Centrioles	Pair of hollow cylinders located near center of cell; each centriole consists of nine microtubule triplets (called 9 X 3 structure)
Cilia	Relatively short projections extending from surface of cell; covered by plasma membrane; made of two central and nine peripheral microtubules (called 9 + 2 structure)
Flagella	Long projections made of two central and nine peripheral microtubules (9 + 2 structure); extend from surface of cell; covered by plasma membrane

How much DNA is in a typical human cell?

If the DNA (deoxyribonucleic acid) molecules in a single human cell were stretched out and laid end-to-end, they would measure approximately 6.5 feet (2 meters). The average human body contains 10 to 20 billion miles (16 to 32 billion kilometers) of DNA distributed among trillions of cells. If the total DNA in all the cells from one human were unraveled, it would stretch to the sun and back more than 500 times.

Do all cells in the human body have a nucleus?

Most eukaryotic cells in the human body have a single organized nucleus. The red blood cell is the only mammalian (and thus human) cell that does not have a nucleus.

What are the main components of the nucleus?

The nucleus, the largest organelle in a eukaryotic cell, is the repository for the cell's genetic information and the control center for the expression of that information. The boundary around the nucleus consists of two membranes (an inner one and an outer one) that form the nuclear envelope. Nuclear pores are small openings in the nuclear envelope that permit molecules to move between the nucleus and the cytoplasm. The nucleolus is a prominent structure within the nucleus. The nucleoplasm is the viscous liquid contained within the nucleus. In addition, the DNA-bearing chromosomes of the cell are found in the nucleus.

What are DNA and RNA?

DNA (deoxyribonucleic acid) and RNA (ribonucleic acid) are both nucleic acids formed from the repetition of simple building blocks of life called nucleotides. A nucleotide consists of a phosphate (PO_4), sugar, and a nitrogen base, of which there are five types: adenine (A), thymine (T), guanine (G), cytosine (C), and uracil (U). In a DNA molecule, this basic unit is repeated in a double helix structure made from two chains of nucleotides linked between the bases; these links are either between A and T or between G and C. (The structure of the bases does not allow other kinds of links.)

RNA is also a nucleic acid, but it consists of a single chain instead of a double and the sugar is ribose rather than deoxyribose. The bases are the same as in DNA, except that the thymine (T) is replaced by another base called uracil (U), which, like the thymine in DNA, links to adenine (A). All RNA exists in three different forms and is formed in the nucleus (in eukaryotic cells) or in the nucleoid region (in prokaryotic cells).

How is DNA organized in the nucleus?

Within the nucleus, DNA (deoxyribonucleic acid) is organized with proteins into a fibrous material called chromatin. As a cell prepares to divide or reproduce, the thin chromatin fibers condense, becoming thick enough to be seen as separate structures, which are called chromosomes.

What are chromosomes and genes?

A chromosome is the threadlike part of a cell that contains DNA and carries the genetic material of a cell. In prokaryotic cells, chromosomes consist entirely of DNA and are not enclosed in a nuclear membrane. In eukaryotic cells, the chromosomes are found within the nucleus and contain both DNA and RNA (ribonucleic acid). The human genome contains twenty-four distinct, physically separate units called chromosomes. Arranged linearly along the chromosomes are tens of thousands of genes.

A gene is one of the complex protein molecules that are associated with chromosomes. As a unit or in certain biochemical combinations, they are responsible for the transmission of certain inherited characteristics from the parent to the offspring. The term *gene*, from the Greek term *genos*, meaning "to give birth to," was first used in 1909 by Danish botanist Wilhelm Johannsen (1857–1927), who is considered to be one of the founders of modern genetics.

Can a person see a gene or a chromosome?

A gene cannot be seen because it is submicroscopic, whereas a chromosome (containing genes) can be seen. In fact, scientists can pinpoint the location of a gene on a chromosome, but the actual gene cannot be seen.

What did scientists recently discover about the DNA helix?

The DNA double helix is not the only genetic code in our cells. Researchers have often discussed a triple helix, but have now uncovered a "quadruple helix" in human cells, called the G-quadruplex ("G" refers to guanine, one of the four bases in DNA). The G-quadruplex seems to form in human DNA where the base guanine exists in large quantities, and it occurs more frequently during the so-called s-phase, or when a cell copies its DNA just before the cell divides. Some scientists believe this quadruple structure may also be responsible for the development of some cancers, but more studies need to be conducted to determine a true connection.

What is a mutation?

A mutation is an alteration in the DNA sequence of a gene. Mutations are a source of variation to a population, but they can have harmful effects in that they may cause diseases and disorders. One example of a disease caused by a mutation is sickle cell disease, in which

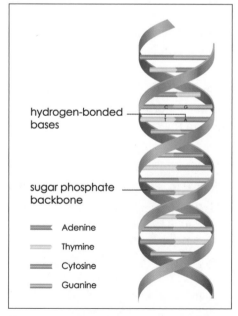

hydrogen-bonded bases

sugar phosphate backbone

Adenine
Thymine
Cytosine
Guanine

The basic structure of the DNA molecule is a double helix joined together like a ladder by pairs of molecules.

there is a change in the amino acid sequence that makes up the oxygen-carrying protein known as hemoglobin found in blood. Mutations are not all bad. Although people may use the term *mutant* in a disparaging manner, mutations are important because of the variation they contribute to a population's gene pool. Without mutations, there would be no variations and no natural selection within the population.

What are telomerases and telomeres?

The human body consists of fifty to one hundred trillion cells. Each cell has forty-six chromosomes, which are the structures in the nucleus containing hereditary material, also called DNA or deoxyribonucleic acid (for more about DNA, see above). The ends of all chromosomes are protected by so-called telomeres, which function like the plastic covers that protect the end of a shoelace. Each time the cell divides, the telomeres get shorter until the cell will not divide anymore.

The enzyme that conserves the ends of the telomeres and affects the number of times a cell can divide is telomerase, one of the basic enzymes in cell biology. Not all cells stop dividing. Some special cells in the body can activate telomerase, which can once again elongate the telomeres. One of the reasons for the interest in the telomerase and telomeres is their connection as catalysts that may play an active role in at least 85 percent of all cancers.

What are lysosomes?

Lysosomes, first observed by Belgian biochemist Christian de Duve (1917–2013) in the early 1950s, are single, membrane-bound sacs that contain digestive enzymes. The digestive enzymes break down all the major classes of macromolecules including proteins, carbohydrates, fats, and nucleic acids. Throughout a cell's lifetime, the lysosomal enzymes digest old organelles to make room for newly formed organelles. The lysosomes allow cells to continually renew themselves and prevent the accumulation of cellular toxins.

What are mitochondria?

A mitochondrion (singular form) is a self-replicating, double-membraned body found in the cytoplasm of all eukaryotic cells. The outer membrane of a mitochon-

---

Have scientists decoded and mapped the structure of telomerases?

Yes, researchers have decoded and mapped the structure of telomerases. In 2008, scientists at the University of Pennsylvania were the first to decode the structure of telomerase, the enzyme that conserves the ends of chromosomes (telomeres). The scientists looked at a part of a beetle telomerase to determine its structure. And in 2013, an international team of researchers mapped telomerase for the first time, thought to be a major step toward the fight against some cancers.

drion is smooth, while the inner membrane is folded into numerous layers that are called cristae. Mitochondria are the location for much of the metabolism necessary for protein synthesis and for the production of both ATP and lipids (for more about ATP, see below).

How many mitochondria are there in a cell?

The number of mitochondria varies according to the type of cell. The number ranges between 1 and 10,000, but averages about 200. Each cell in the human liver has over 1,000 mitochondria. Cells with high energy requirements, such as muscle cells, may have many more mitochondria.

What is ATP?

ATP (adenosine triphosphate) is the universal energy currency of a cell. Its secret lies in its structure. ATP contains three negatively charged phosphate groups. When the bond between the outermost two phosphate groups is broken, ATP becomes ADP (adenosine diphosphate). This reaction releases 7.3 kilocalories/mole of ATP, which is a great deal of energy by cell standards.

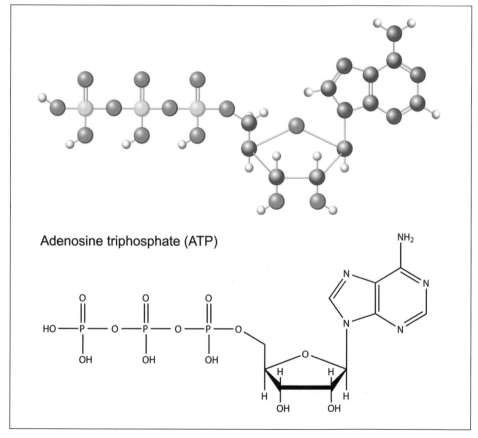

A diagram of the molecular structure of adenosine triphosphate (ATP).

29

The three main types of blood cells include the red blood cells (carry oxygen throughout the body), white blood cells, including lymphocytes (protect the body from infection), and platelets (help blood to clot normally). All blood cells, no matter what type, are produced in the bone marrow. (For more about blood, see the chapter "Circulatory System.")

Overall, each cell in the human body is estimated to use between one and two billion ATPs per minute, which comes to about 1×10^{23} for a typical human body. Thus, in a span of 24 hours, the body's cells produce about 441 pounds (200 kilograms) of ATP. (For more about how ATP is used in the body, see the chapter "Muscular System.")

What is the average lifespan of various cells in the human body?

The human body is self-repairing and self-replenishing. According to one estimate, almost 200 billion cells die each hour. In a healthy body, dying cells are simultaneously replaced by new cells, but the replenishment depends on the type of cell.

Overall, the lifespan of certain types of cells vary greatly. In general, most living cells do not live longer than a month. Even cells that live longer, such as liver, brain, and some blood cells, constantly renew their components so that no part of the cell is really more than a month old. There are also certain short-lived cells, such as plasma cells (cells that immediately fight off antigens a body encounters) that die off in about ten to seventeen days. The following lists some cell types in the human body and their average lifespan:

Cell Type	Average Lifespan
Blood cells: Red blood cells	120 days
Blood cells: Lymphocytes	Over 1 year
Blood cells: Other white cells	10 hours
Blood cells: Platelets	10 days
Bone cells	25–30 years
Brain cells*	Lifetime
Colon cells	3–4 days
Liver cells	500 days
Skin cells	19–34 days
Spermatozoa	2–3 days
Stomach cells	2 days

*Brain cells are the only cells that do not divide further during a person's lifetime. They either last the entire lifetime, or if they die during a person's lifetime they are not replaced.

How many types of cells are found in the human body?

It is estimated that the human body has fifty to one hundred trillion cells, all of them differing in size and shape, from round and flat to star-like and cubed. Scientists have organized these many cells into around 200 major types. Each cell type performs a specific task. For example, cells in the human body include heart cells, nerve cells, kidney cells, and white and red blood cells.

TISSUES

What are the general characteristics of the different types of tissue?

Each of the four major types of tissue has different functions. They are located in different parts of the body and have certain distinguishing features. The table below explains these differences between the body's tissues:

Characteristics of Tissues

Tissue	Function	Location	Distinguishing Features
Epithelial	Protection, secretion, absorption, excretion	Covers body surfaces, covers and lines internal organs, composes glands	Lacks blood vessels
Connective	Bind, support, protect, fill spaces, store fat, produce blood cells	Widely distributed throughout the body	Matrix between cells, good blood supply
Muscle	Movement	Attached to bones, in the walls of hollow internal organs, heart	Contractile
Nervous	Transmits impulses for coordination, regulation, integration, and sensory reception	Brain, spinal cord, nerves	Cells connect to each other and other body parts

Where is epithelial tissue found?

Epithelial tissue, also called epithelium (from the Greek *epi*, meaning "on," and *thele*, meaning "nipple"), covers every surface, both external and internal, of the body. The outer layer of the skin, the epidermis, is one example of epithelial tissue. Other examples of epithelial tissue are the lining of the lungs, kidney tubules, and the inner surfaces of the digestive system, including the esophagus, stomach, and intestines. Epithelial tissue also includes the lining of parts of the respiratory system.

What are the different shapes and functions of the epithelium?

Epithelial tissue consists of densely packed cells. It is either simple or stratified, based on the number of cell layers. Simple epithelium has one layer of cells, while stratified epithelium has multiple layers. Epithelial tissue may have squamous- (flat and square), cuboidal- (box or cube), or columnar-shaped cells. The epithelium forms a barrier, allowing the passage of certain substances, while impeding the passage of other substances.

Where are different types of epithelial tissues found in the body?

The different types of epithelial tissue are located in different parts of the body according to their specialization. The following lists the major epithelial tissue, location, and functions:

Type of Epithelial Tissue	Major Locations	Major Functions
Simple squamous	Lining of lymph vessels, blood vessels, heart, glomerular capsule in kidneys, alveoli (air sacs in lungs), serous membranes lining peritoneal, pleural, pericardial, and scrotal cavities	Permits diffusion or filtration through selectively permeable surfaces
Simple cuboidal	Lining of many glands and their ducts, surface of ovaries, inner surface of eye lens, pigmented epithelium of eye retina	Secretion and absorption
Simple columnar	Stomach, intestines, digestive gland, and gall bladder	Secretion, absorption, protection, lubrication; cilia and mucus combine to sweep away foreign substances
Stratified squamous	Epidermis, vagina, mouth and esophagus, anal canal, distal end of urethra	Protection
Stratified cuboidal	Ducts of sweat glands, sebaceous glands, and developing epithelium in ovaries and testes	Secretion
Stratified columnar	Moist surfaces such as larynx, nasal surface of soft palate, parts of pharynx, urethra, and excretory ducts of salivary and mammary glands	Secretion and movement

Do epithelial tissues contain blood vessels?

Epithelial tissues are avascular, meaning they do not contain blood vessels. Oxygen and other nutrients diffuse through the permeable basement membranes from capillaries in the underlying connective tissue, while wastes diffuse into connective tissue capillaries.

How often is the epithelium replaced?

Epithelial cells are constantly being replaced and regenerated throughout a person's lifetime. Depending on the type of epithelial cell, replacement or regeneration varies greatly. For example, the epidermis (outer layer of the skin) is renewed every two

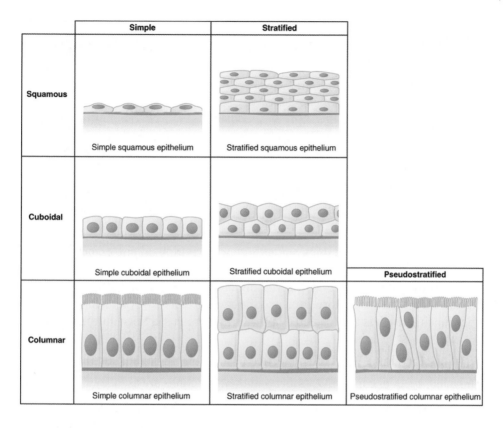

	Simple	Stratified	
Squamous	Simple squamous epithelium	Stratified squamous epithelium	
Cuboidal	Simple cuboidal epithelium	Stratified cuboidal epithelium	**Pseudostratified**
Columnar	Simple columnar epithelium	Stratified columnar epithelium	Pseudostratified columnar epithelium

The different types of epithelial tissue include squamous, cuboidal, and columnar.

weeks, while the epithelial lining of the stomach is replaced every two to three days. The lining of the respiratory tract is only replaced every five to six weeks. Various epithelial cells regenerate at different rates, too. For example, the liver—a gland consisting of epithelial tissue—easily regenerates after portions are removed surgically.

What is a gland?

Glands are secretory cells or multicellular structures that are derived from epithelium and often stay connected to it. They are specialized for the synthesis, storage, and secretion of chemical substances. Glands are classified as either endocrine or exocrine glands. Endocrine glands do not have ducts, but they release their secretions directly into the extracellular fluid. The secretions pass into capillaries and are then transported by the bloodstream to target cells elsewhere in the body. Exocrine glands have ducts that carry the secretions to some body surface. Mucus, saliva, perspiration, earwax, oil, milk, and digestive enzymes are examples of exocrine secretions. (For more about glands, see the chapter "Endocrine System.")

How are exocrine glands classified?

Exocrine glands may be unicellular (single-celled) or multicellular. Multicellular exocrine glands may be either simple or compound glands. Simple glands are glands with only one unbranched duct, while those with more than one branch are compound glands. There is only one unicellular exocrine gland in the human body: the goblet, or mucus cell, is found in the lining of the intestines and other parts of the digestive tract, the respiratory tract, and the conjunctiva of the eye. Goblet cells produce mucin, which is then secreted in the form of mucus, a thick, lubricating fluid.

What are the major types of connective tissues and their functions?

The major types of connective tissues and their functions are as follows:

- Loose connective tissue, also called areolar tissue (from the Latin *areola*, meaning "open place"), is a mass of widely scattered cells whose matrix is a loose weave of fibers. Many of the fibers are strong protein fibers called collagen. Loose connective tissue is found beneath the skin and between organs. It is a binding and packing material whose main purpose is to provide support to hold other tissues and organs in place.

- Adipose tissue consists of adipose cells in loose connective tissue. Each adipose cell stores a large droplet of fat that swells when fat is stored and shrinks when fat is used to provide energy. Adipose tissue provides padding, absorbs shocks, and insulates the body to slow heat loss.

- Blood is a loose connective tissue whose matrix is a liquid called plasma. Blood consists of red blood cells (erythrocytes), white blood cells (leukocytes), and platelets (thrombocytes), which are tiny pieces of bone marrow cell. Plasma also contains water, salts, sugars, lipids, and amino acids. Blood is approximately 55 percent plasma and 45 percent formed elements. Blood transports substances from one part of the body to another and plays an important role in the immune system. (For more about blood, see the chapter "Cardiovascular System.")

- Collagen (from the Greek *kola*, meaning "glue," and *genos*, meaning "descent") is a dense connective tissue, also known as fibrous connective tissue. It has a matrix of densely packed collagen fibers. There are two types of collagen: regular and irregular. The collagen fibers of regular dense connective tissue are lined up in parallel. Tendons, which bind muscle to bone, and ligaments, which join bones together, are examples of dense regular connective tissue. The strong covering of various organs, such as kidneys and muscle, is dense irregular connective tissue.

- Cartilage (from the Latin, meaning "gristle") is a connective tissue with an abundant number of collagen fibers in a rubbery matrix. It is both strong and flexible. Cartilage provides support and cushioning. It is found between the discs of the vertebrae in the spine, surrounding the ends of joints such as knees, and in the nose and ears.

- Bone is a rigid connective tissue that has a matrix of collagen fibers embedded in calcium salts. It is the hardest tissue in the body, although it is not brittle. Most of the skeletal system is comprised of bone, which provides support for muscle attachment and protects the internal organs.

Where is adipose tissue found?

Adipose tissue is abundant in the body and constitutes 18 percent of an average person's body weight. Adipose tissue is found under the skin of the groin, sides, buttocks, and breasts. It is found behind the eyeballs, surrounding the kidneys, and in the abdomen and hips.

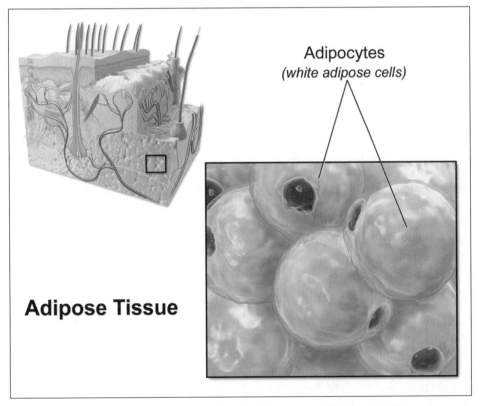

Adipose tissue—what we usually call fat—is found just about everywhere in the body. About one fifth of the average person's body is adipose tissue.

How does brown fat differ from white fat?

White fat (or adipose tissue) stores nutrients. Brown fat, also called brown adipose tissue, consumes its nutrient stores to generate heat to warm the body. It is called brown fat because it has a deep, rich, dark color that is derived from the numerous mitochondria in each individual cell. Brown adipose tissue is found in infants and very young children between the shoulder blades, around the neck, and in the anterior abdominal wall. Older children and adults rely on shivering to warm the body.

What types of cancers develop and grow in which types of tissues?

Different types of cancers develop and grow in the different types of tissue. Carcinomas, perhaps the most common type of cancer, are cancers of the epithelial tissue. Sarcomas are cancers in the muscle and connective tissue. Leukemias are cancers of the blood. Lymphomas are cancers of the reticular connective tissue.

Is all the cartilage in the body the same?

There are three types of cartilage in the human body: Hyaline cartilage (from the Greek *hyalos*, meaning "glass") is the most common type of cartilage in the body. It has a translucent, pearly, blue-white appearance resembling glass. Hyaline cartilage provides stiff but flexible support and reduces friction between bony surfaces. It is found between the tips of the ribs and the bones of the sternum, at the end of the long bones, at the tip of the nose, and throughout the respiratory passages. Elastic cartilage is similar to hyaline cartilage except it is very flexible and resilient. It is ideal for areas that need repeated bending and stretching. Elastic cartilage forms the external flap of the outer ear and is found in the auditory canal and epiglottis. Fibrocartilage is often found where hyaline cartilage meets a ligament or tendon. It is found in the pads of the knees, between the pubic bones of the pelvis, and between the spinal vertebrae. It prevents bone-to-bone contact.

What conditions result from the buildup of fluids in the body?

Edema—once known as dropsy or hydrosy—is characterized by swelling of the affected area usually caused by fluid retention, or excessive fluid trapped in the body's tissues. Most edemas commonly occur in the hands, arms, legs, ankles, and feet, and are often associated with the lymphatic or venous (circulatory) system. It can also include an abnormal buildup of fluids in the ventral body cavity caused by infection or chronic irritation. For example, pleurisy is an inflammation of the pleural cavity; pericarditis is an inflammation of the pericardium; and peritonitis is an inflammation of the peritoneum.

What are the three types of muscle tissue?

There are three types of muscle tissue in the body: 1) smooth muscle; 2) skeletal muscle; and 3) cardiac muscle. Muscle tissue, consisting of bundles of long cells called muscle fibers, is specialized for contraction. It enables body movements, as well as the movement of substances within the body. (For more about muscles, see the chapter "Muscular System.")

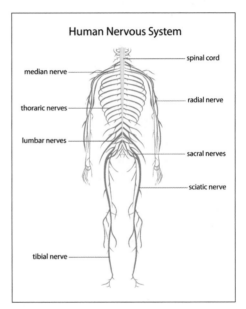

The major human nerves of the body.

Does exercise increase the number of muscle cells?

Adults have a fixed number of skeletal muscle cells, so exercise does not increase their number. Exercise, however, does enlarge the existing skeletal muscle cells. There are two factors involved. Hypertrophy is an increase in the muscle's size because of an increase in the size of the muscle fibers, and hyperplasia is an increase in the number of muscle fibers (also called filaments). Both are reasons why a person can "increase" their muscle size.

What type of cell is found in nerve tissue?

Neurons are specialized cells that produce and conduct "impulses," or nerve signals. Neurons consist of a cell body, which contains a nucleus and two types of cytoplasmic extensions, dendrites, and axons. Dendrites are thin, highly branched extensions that receive signals. Axons are tubular extensions that transmit nerve impulses away from the cell body, often to another neuron. Nerve tissue also has supporting cells, called neuroglia or glial cells, which nourish the neurons, insulate the dendrites and axons, and promote quicker transmission of signals. (For more about neurons, see the chapter "Nervous System.")

How many different types of neurons are found in nerve tissue?

There are three main types of neurons: 1) sensory neurons; 2) motor neurons; and 3) interneurons (also called association neurons). Sensory neurons conduct impulses from sensory organs (eyes, ears, and the surface of the skin) into the central nervous system. Motor neurons conduct impulses from the central nervous system to muscles or glands. Interneurons are neither sensory neurons nor motor neurons. They permit elaborate processing of information to generate complex behaviors. Interneurons comprise the majority of neurons in the central nervous system.

Which type of tissue accounts for the greatest amount of body weight?

Muscle tissue accounts for approximately 50 percent of body weight and connective tissue accounts for 45 percent of total body weight. The remaining 5 percent is divided between epithelium and glands (3 percent) and neural tissue (2 percent). Tissues combine to form all the organs and systems of the human body.

Is it possible to repair damaged tissue?

Tissue responds to injury or other damage with a two-step process: 1) inflammation; and 2) regeneration to restore homeostasis. Inflammation, or the inflammatory response, produces swelling, redness, warmth, and pain in the area of injury. The injured area is isolated while damaged cells and dangerous microorganisms are destroyed. During the second process, regeneration, the damaged tissues are replaced or repaired to restore normal function. Regeneration begins while the cleanup processes of inflammation are still in process.

What are pus and an abscess?

Lysosomes are responsible for releasing enzymes that destroy the injured cells and attack surrounding tissues. Pus is an accumulation of debris, fluid, dead and dying cells, and necrotic tissue. An abscess is an accumulation of pus in an enclosed tissue space.

How does a scar form?

A scar forms when the dense mass of fibrous connective tissue that fills in the gap after an injury is deep or large. A scar also may form when the cell damage was extensive and the dense fibrous mass remains and is not replaced by normal tissue.

ORGANS AND ORGAN SYSTEMS

What is an organ?

An organ is a group of several different tissues working together as a unit to perform a specific function or functions. Each organ performs functions that none of the component tissues can perform alone. This cooperative interaction of different tissues is a basic feature of animals, including humans. The heart is an example of an organ. It consists of cardiac muscle wrapped in connective tissue. The heart chambers are lined with epithelium. Nerve tissue controls the rhythmic contractions of the cardiac muscles.

How many organs are in the human body?

About seventy-eight organs make up the human body, each one with a different function (or sometimes the same, as with two kidneys), size, and sometimes shape. The largest organ in the human body, with respect to size and weight, is the skin. Not everyone has organs in the same place, either; for example, sometimes organs such as the kidney may be located closer to the pelvis, or only one kidney may be present. These differences can be due to genetics, differences in the organ's cell growth, or even disease.

What is an organ system?

An organ system is a group of organs working together to perform a vital body function. There are twelve major organ systems in the human body. The following lists the organ systems, their components, and functions:

Organ System	Components	Functions
Cardiovascular and circulatory	Heart, blood, and blood vessels	Transports blood throughout the body, supplying nutrients and carrying oxygen to the lungs and wastes to kidneys
Digestive	Mouth, esophagus, stomach, intestines, liver, and pancreas	Ingests food and breaks it down into smaller chemical units
Endocrine	Pituitary, adrenal, thyroid, and other ductless glands	Coordinates and regulates the activities of the body
Excretory	Kidneys, bladder, and urethra	Removes wastes from the bloodstream
Immune	Lymphocytes, macrophages, and antibodies	Removes foreign substances
Integumentary	Skin, hair, nails, and sweat glands	Protects the body
Lymphatic	Lymph nodes, lymphatic capillaries, lymphatic vessels	Captures fluid and returns it to the cardiovascular system
Muscular	Skeletal muscle, cardiac muscle, and smooth muscle	Allows body movements
Nervous	Nerves, sense organs, brain, and spinal cord	Receives external stimuli, processes information, and directs activities
Reproductive	Testes, ovaries, and related organs	Carries out reproduction
Respiratory	Lungs, trachea, and other air passageways	Exchanges gases—(captures oxygen [O_2] and disposes of carbon dioxide [CO_2])
Skeletal	Bones, cartilage, and ligaments	Protects the body and provides support for locomotion and movement

What is not an "organ system" but often indirectly incorporates an organ system?

The sensory system is part of the human body but is not considered an organ system. The senses often indirectly incorporate parts of the organ systems. For example, the skeletal system's skull contains open sinuses where the olifactory (smell) receptors are located; the skin also contains sensory receptors. (For more about the senses, see the chapter "Sensory System.")

What are vestigial organs and structures?

Vestigial organs and structures are considered those organs that have no real function in the human body. Thus, removing such organs would have no detrimental effect on the body. For example, the tailbone (or coccyx) is thought to be a remnant of a lost ancestral tail—assisting in balance and mobility—lost when humans began to walk upright.

Recently, scientists have discovered that many organs thought of as vestigial actually may have major functions in the body. For example, the appendix is thought to be useful to a human fetus by producing important compounds within the fetus. For adult humans, the appendix may help with immune functions and may contain useful bacteria.

HOMEOSTASIS

What is homeostasis?

Homeostasis (from the Greek *homois*, meaning "same," and *stasis*, meaning "standing still") is the state of inner balance and stability maintained by the human body despite constant changes in the external environment. Nearly everything that occurs in the body helps to maintain homeostasis: kidneys filtering the blood and removing a carefully regulated amount of water and wastes; lungs working together with the heart, blood vessels, and blood to distribute oxygen throughout the body and remove wastes.

Who coined the term "homeostasis"?

American physiologist Walter Bradford Cannon (1871–1945), who elaborated on French physiologist Claude Bernard's

American physiologist Walter Bradford Cannon described the body's ability to regulate itself as "homeostasis."

(1813–1878) concept of the *milieu intérieur* (interior environment), used the term "homeostasis" to describe the body's ability to maintain a relative constancy in its internal environment.

What are the three components necessary to maintain homeostasis?

The three components of homeostasis are sensory receptors, integrators, and effectors. These three components interact to maintain the state of homeostasis. Sensory receptors are cells that can detect a stimulus that signals a change in the environment. The brain is the integrator that processes the information and selects a response. Muscles and glands are effectors that carry out the response.

How does the term "negative feedback" apply to homeostasis?

Negative feedback is a cellular process that is similar to the manner in which an air conditioner operates: an air conditioner is set to a specific temperature, and it shuts off when the surrounding air reaches the set temperature. In the human body, negative feedback is part of the homeostatic process through which cells conserve energy by synthesizing products only for their immediate needs. For example, maintaining the normal blood sugar level of the body is an example of negative feedback. When the blood sugar level decreases, the body responds to raise the level. If the blood sugar level increases, the body acts to lower the level. Each of these responses is a negative action, since the response does the opposite of the initial stimulus.

What is a "positive feedback" system?

Positive feedback systems are stimulatory, since the initial stimulus is reinforced rather than reversed. The stimulus continues to increase rapidly until the process is stopped. Positive feedback is relatively uncommon in the human body, since it disrupts homeostasis. For example, if there were a positive feedback response to blood sugar level decreasing, the blood sugar level would continue to decrease, without ceasing, until the person died.

One example of positive feedback in humans is during childbirth when the woman experiences an increase in the number of uterine contractions. The positive feedback response is to increase the frequency of uterine contractions. The birth of a baby stops the positive feedback response.

COMPARING OTHER ORGANISMS

How big are animal cells?

Most animal cells are very small, and most cannot be seen without a microscope. In fact, most cells are smaller than the period at the end of this sentence. Two exceptions are bird and frog egg cells. Both are larger cells readily observable with the unaided eye. For example, the chicken egg is actually one single cell. It contains the nucleus, cytoplasm, and cellular membrane. Within the membrane is an abundance of yolk and albumin, the nutrients needed for the developing chick (embryo)—making this cell much larger than the normally functioning cells within the chicken.

What is a knockout mouse?

A knockout mouse is a mouse that has specific genes mutated ("knocked out") so that the lack of production of the gene product can be studied in an animal model. Knockout mice are frequently used in pharmaceutical studies to test the potential for a particular human enzyme as a therapy target. Mice have identical or nearly identical copies of human proteins, making them useful in studies that model effects in humans.

What tissues do almost all animals have?

In general—and similar to humans—there are four types of tissues found in animals: epithelial, connective, nerve, and muscle tissue. But some animals do not have the same types of tissues. For example, sponges do not have true tissues (or organs), as their cells are organized in a different configuration than most animals.

What is the largest organ in most animals?

In most animals, the liver is commonly the largest internal organ (in other words, not counting the skin). Although its shape and form vary between species, the liver of most animals usually has two sides (left and right lobes); the shape and location are determined by the surrounding organs. There are exceptions. For example, in a frog, the liver has three lobes; a fetal pig liver has five lobes, called the right lateral, right central, left central, left lateral, and caudate lobes.

Do any animals have vestigial organs or structures?

Yes, the list of such vestigial organs and structures is long. For example, whales have small leg bones that are found in the back of the body. These bones are thought to be the remnants of their land-dwelling ancestors' legs (several water-based animals once lived on land and eventually returned to the oceans). Penguins also carry vestigial structures in their flippers, which contain hollow bones usually reserved for flying birds (less weight allows the bird to stay in the air). This is thought to be vestigial remnants of their flying ancestors.

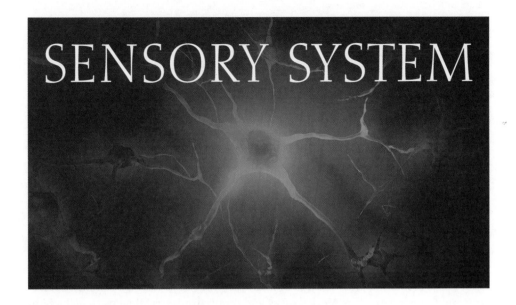

SENSORY SYSTEM

What are the major senses?

As early as the days of Greek philosopher and scientist Aristotle (384–322 B.C.E.), the five senses were recognized to include smell, taste, sight, hearing, and touch. More recently, scientists categorize the senses into two major groups. One group is the special senses, which are produced by highly localized sensory organs and include the senses of smell, taste, sight, hearing, and balance. The other group is the general senses, which are more widely distributed throughout the body and include such senses as touch, pressure, pain, temperature, and vibration.

What are sensory receptors?

Sensory receptors are structures in the skin and other tissues that detect changes in the internal or external environment. These receptors consist of specialized neuron endings or specialized cells in close contact with neurons that convert the energy of the stimulus (sound, color, odor, etc.) to electrical signals within the nervous system. Sensory receptors, together with other cells, compose the major sense organs, including eyes, ears, nose, and taste buds.

How many types of sensory receptors have been identified?

Five types of sensory receptors, each responding to a different type of stimulus, have been identified. They include the following receptor types:

- *Chemoreceptors*—Respond to chemical compounds such as odor molecules.
- *Photoreceptors*—Respond to light.
- *Thermoreceptors*—Respond to changes in temperature.
- *Mechanoreceptors*—Respond to changes in pressure or movement.
- *Pain receptors*—Respond to stimuli that result in the sensation of pain.

What is synesthesia?

Synesthesia, or cross perception (from the Greek *syn*, or "together" and *aisthesis*, or "perception"), is a condition in which a person perceives stimuli not only with the sense for which it is intended, but with others as well. It can also be when a person puts together objects such as shapes and numbers, with a certain sensory perception, such as color or smell. For example, a synesthete may see musical notes as color hues or feel flavors as different textures on the skin. Most commonly, letters, numbers, or periods of time evoke specific colors. One in every 500,000 people has this condition, which appears to be inherited and is more common in women than in men.

What structures in the body are associated with the receptors for the general senses?

Receptors for the general senses, including touch, pressure, pain, temperature, and vibration, are usually associated with the skin. The skin has many receptors, especially those associated with touch. For example, free nerve endings primarily detect pain and temperature. The Merkel disks and root hair plexus allow us to feel a light touch, as do the Meissner corpuscles (which have the highest sensitivity to light touch). Pacinian corpuscles detect deep pressure and vibrations, while the Ruffini endings and Krause end bulbs help us sense continuous pressure. (For more about the human body's other receptors, see the chapter "Sensory System." For more about the skin, see the chapter "Integumentary System.")

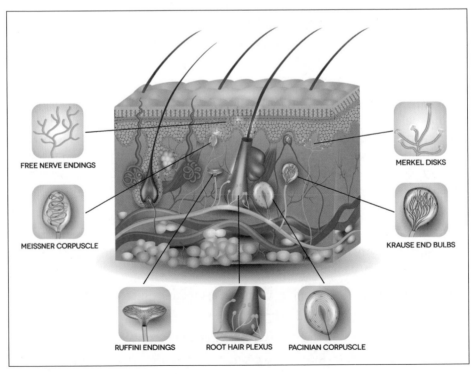

FREE NERVE ENDINGS

MERKEL DISKS

MEISSNER CORPUSCLE

KRAUSE END BULBS

RUFFINI ENDINGS   ROOT HAIR PLEXUS   PACINIAN CORPUSCLE

Sensory receptors in the skin.

Other receptors are associated with deeper structures, such as tendons, ligaments, joints, muscles, and viscera.

Is the tongue considered to be a sensitive organ?

Yes. In humans, the tongue—a muscle that manipulates food so it can be chewed and allows the food to be tasted—is one of the most sensitive organs in the human body. (Some scientists believe it is the most sensitive, but others mention, for example, the skin.) The tongue is especially sensitive to touch, temperature, and pain.

What sense is most closely associated with emotions?

Although most people associate a sound, such as a song or voice, with emotions, it is actually the sense of smell that is most closely tied to our emotions. It is physiological: some of the nerves that travel to the brain from olfactory (smell) receptors must pass through the limbic system (brain structures involved in emotions and feelings), thus stimulating it and its centers of emotion (and sexuality) each time a smell is received.

What is phantom limb pain?

Phantom pain is perceived in tissue that is no longer present. Phantom limb pain is a complex phenomenon that occurs when nerves that would normally supply the missing limb cause pain. This name was attached to the phenomenon by a physician during the American Civil War, when a veteran with amputated legs asked for someone to massage his cramped leg muscle. The explanation for phantom limb pain is debated, but one explanation is that the nerves remaining in the stump may generate nerve impulses that are transmitted to the brain and interpreted as arising from the missing limb. Other theories propose that the phantom sensation might be produced by brain reorganization caused by the absence of the sensations that would normally arise from the missing limb.

Still others believe the reason for phantom limb pain does not lie in the brain, but it originates in the nervous system. For example, in 2013 researchers studied several people with leg amputations. The researchers anesthetized the area where the nerves from the patients' amputated legs entered the spinal cord in the lower back and the pain was temporarily reduced or eliminated within minutes.

45

How are variations in temperature detected by the body?

Temperature sensations are detected by specialized free nerve endings called cold receptors and warm receptors. Cold receptors respond to decreasing temperature and warm receptors respond to increasing temperatures. Cold receptors are most sensitive to temperatures between 50°F (10°C) and 68°F (20°C). Temperatures below 50°F (10°C) stimulate pain receptors, producing a freezing sensation. Warm receptors are most sensitive to temperatures above 77°F (25°C) and become unresponsive at temperatures above 113°F (45°C). Temperatures near and above 113°F (45°C) stimulate pain receptors, producing a burning sensation. Both warm and cold receptors rapidly adapt. Within about a minute of continuous stimulation, the sensation of warmth or cold begins to fade.

SMELL

How does the sense of smell work?

The sense of smell is associated with sensory receptor cells in the upper nasal cavity. The smell, or olfactory, receptors are chemoreceptors. Chemicals that stimulate olfactory receptors enter the nasal cavity as airborne molecules called gases. They must dissolve in the watery fluids that surround the cilia of the olfactory receptor cells before they can be detected. These specialized cells, the olfactory receptor neurons, are

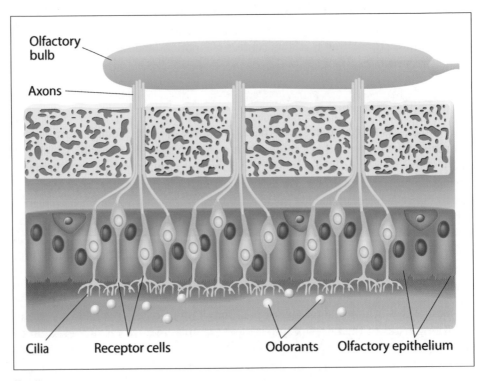

Olfactory bulb

Axons

Cilia   Receptor cells   Odorants   Olfactory epithelium

What is the common chemical sense?

The common chemical sense includes thousands of nerve endings and influences our sense of smell. This sense is found on the moist surfaces of several parts of the body, especially the eyes, nose, mouth, and throat. They help sense the irritating substances the body encounters, such as the coolness of peppermint or the power of an onion that makes a person's eyes tear.

the only parts of the nervous system that are in direct contact with the outside environment. The odorous gases then waft up to the olfactory cells, where the chemicals bind to the cilia that line the nasal cavity. That action initiates a nerve impulse being sent through the olfactory cell, into the olfactory nerve fiber, to the olfactory bulb, and to the brain. The brain then knows what the chemical odors are.

How does the brain detect different smells?

The exact mechanism of olfaction is unclear, but some studies have been completed that attempt to classify smells into groups such as floral, mushy, pungent, and so on, which are related to molecules of a particular shape and size. One hypothesis suggests that the shapes of gaseous molecules fit complementary shapes of membrane receptor sites on those olfactory receptor cells.

How does age affect the sense of smell?

Since the olfactory receptor neurons are exposed to the external environment, they are subject to damage over time. This is because some environmental pollutants and other toxic elements—including if a person smokes or is exposed to second-hand smoke—can damage the olfactory receptors in a person's nose.

It is thought that the average human smelling ability peaks around the age of eight. From that time, people typically experience a progressive diminishing of olfactory sense with age. Although it is estimated that an individual loses about one percent of the olfactory receptors every year, not everyone agrees. Some researchers suggest that human smell sensitivity starts to deteriorate from the early twenties. Still other researchers believe the deterioration begins around fifteen years of age. And of course, the ability to smell is also tied to one's mental and physical health. In fact, many healthy elderly people have just as much ability to smell as a young person in his or her twenties.

Do males or females have a better sense of smell?

In general, females have a better smelling ability than males. In 2014, researchers in Brazil measured the number of cells in certain brain structures called the olfactory bulb, the first part of the brain to receive olfactory information from the nose. They found that women had an average of 43 percent more cells in the olfactory bulb than men.

Without being able to smell, is it possible to tell the difference between an onion and an apple?

The special senses of smell and taste are very closely related, both structurally and functionally. Experimental evidence shows that taste is partially dependent on the sense of smell. Most subjects are unable to distinguish between an onion and an apple on a blind taste test when their sense of smell is blocked. This also explains why food is "tasteless" when a person has a cold, because the olfactory receptor cells are covered with a thick mucus (in other words, a runny nose from the cold) blocking the sense of smell. It is also why when people hold their breath, they cannot taste the food they are eating.

What is anosmia?

Anosmia is a partial or complete loss of smell, either on a temporary basis or permanently. It may result from a variety of factors, including inflammation of the nasal cavity lining due to a respiratory infection, excessive tobacco smoking, head trauma, exposure to chemicals (such as working in certain factories or industries), or from the use of certain drugs such as cocaine. In young people, the loss of smell is most often linked to a viral infection, while in the elderly it more commonly follows a head injury. Because they cannot smell, many people who have anosmia often lose their appetites or eat "strange" combinations of food because they cannot smell or taste what they are eating.

TASTE

What are the special organs of taste?

The special organs of taste are the taste buds located primarily on the surface of the tongue, where they are associated with tiny elevations called papillae surrounded by deep folds. A taste bud is a cluster of approximately 100 taste cells representing all taste sensations and 100 supporting cells that separate the taste cells. Taste buds can also be found on the roof of the mouth and in the throat. An adult has approximately 9,000 to 10,000 taste buds; the average human taste bud lives for seven to ten days.

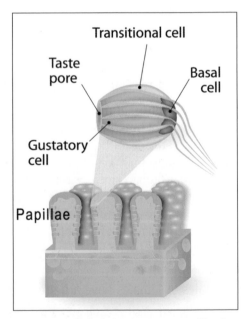

The taste buds (papillae) of the tongue are covered in cells that have pores to perceive salty, sweet, bitter, and sour flavors. Taste buds are also found on the inside of the cheeks, upper esophagus, soft palate, and epiglottis.

How do taste buds function?

The taste cells that comprise each taste bud act as receptors. Taste cells and adja-

What basic "taste" was only recently confirmed?

Many people remember illustrations of the cross-section of the human tongue from grade school days—the image of the tongue was split into salty, sweet, bitter, and sour tastes. Now researchers know that humans have taste buds all over their tongue, mouth, and even the pharynx, all representing these tastes in various proportions.

But a "newer" taste was uncovered in the early 1900s, when Japanese physical chemist Kikunae Ikeda (1864–1936) extracted a white compound from giant sea kelp. He further used this to give broths a savory and meaty taste, even without meat in the pot. It was chemically a glutamate (a type of amino acid); he called it umami, after the taste it produced, or "savory." It took until around 2000 before molecular biologists discovered a receptor for the substance in the mouth (the umami receptors are actually located in the pharynx), and umami is often considered the fifth basic taste. In America, if a stew or soup is described as "robust" or "full," it usually means "umami"; it has also been described as the "beef taste" of steak, the tang of aging cheese, and the flavor of MSG (the food additive monosodium glutamate).

cent epithelial cells comprise a spherical structure with small projections called taste hairs that protrude from the taste cells. The taste hairs are the sensitive part of each receptor cell. A network of nerve fibers surrounds and connects all of the taste cells. Stimulation of a receptor cell triggers an impulse on a nearby nerve fiber, and the impulse then travels to the brain via a cranial nerve for interpretation.

How many basic taste sensations are recognized?

It has been believed generally that there are only four basic taste sensations: sweet, sour, salty, and bitter. Some other taste sensations that are frequently mentioned are alkaline, metallic, and umami (see above). Different tastes are experienced by combining the four basic taste sensations. Some individuals claim that with the senses of smell and taste working together, an individual can experience 10,000 different combinations.

Are certain areas of the tongue associated with a particular taste sensation?

All taste buds are able to detect each of the four (or more) basic taste sensations; however, each taste bud is usually most sensitive to one type of taste stimuli. Scientists know that taste buds representing the various taste sensations are all over the mouth and tongue, but in general, the stimulus type to which each taste bud responds most strongly is related to its position on the tongue. Sweet receptors are concentrated at the tip of the tongue, while sour receptors are more common at the sides of the tongue. Salt receptors occur most frequently at the tip and front edges of the tongue. Bitter receptors are most numerous at the back of the tongue.

Does the sense of taste diminish with age like the sense of smell?

The sense of taste diminishes with age, but not as significantly as the sense of smell. The number of taste buds begins to decrease around age sixty and, therefore, taste and flavor perception declines. On the average, as a person ages, salty and sweet tastes are lost first, then bitter and sour tastes. This decline may be accelerated in an older person if he or she smokes, is exposed to harmful chemicals, has a disease, or takes medication that can affect taste.

HEARING

What two functions are performed by the ear?

The ear has three major parts: the external, middle, and inner ear. It also has two functions: hearing and maintaining equilibrium or balance (the eyes also often help assist the ears with balance). These two functions rely on certain special nerve receptors that respond to sound waves (hearing) or changes in movement of the body (balance).

What are the parts of the external ear?

The external ear is the visible part of the ear. It consists of an outer, funnel-shaped structure called the auricle (pinna) and a tube called the auditory canal that leads inward for about 1 inch (2.5 centimeters). It ends at the eardrum (tympanic membrane).

What structures comprise the middle ear?

The middle ear consists of the tympanic membrane, or the eardrum (it is about 0.00435 inches [about 0.11 millimeters] thick), tympanic cavity (an air-filled space in the temporal bone), and three small bones called auditory ossicles. The tympanic cavity is connected to the nasopharynx (the region linking the back of the nasal cavity and the back of the oral cavity) by the auditory (Eustachian) tube.

What are the three bones in the middle ear?

The three bones, or auditory ossicles, in the middle ear are the malleus (hammer), the incus (anvil), and stapes (stirrup). Tiny ligaments attach them to the wall of the tympanic cavity, and they are covered by mucous membranes. A special muscle, the stapedius, is attached to the stapes and can dampen its vibrations. These bones bridge the eardrum and the inner ear, transmitting vibrations.

What is the auditory tube and its function?

The auditory tube (Eustachian tube) connects each middle ear to the throat. This tube conducts air between the tympanic cavity and the outside of the body by way of the throat and mouth. It also helps maintain equal air pressure on both sides of the eardrum, which is necessary for normal hearing. The function of the auditory tube can be experienced during rapid change in altitude. As a person moves from a high altitude to a lower one, the air pressure on the outside of the membrane becomes

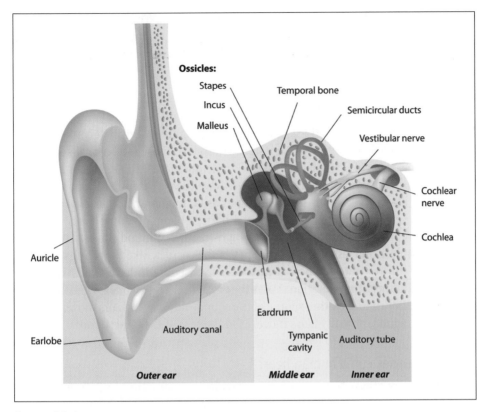

Anatomy of the human ear.

greater and greater. As a result, the eardrum may be pushed inward, out of its normal position, and hearing may be impaired. When the air pressure difference is great enough, some air may force its way up through the auditory tube into the middle ear. This allows the pressure on both sides of the eardrum to equalize, and the drum moves back to its regular position. An individual usually hears a popping sound at this time, and normal hearing is restored. A reverse movement of air occurs when a person moves from a low altitude to a higher one.

What is the labyrinth?

The labyrinth is a complex system of chambers and tubes in the inner ear. There are actually two labyrinths in each ear: the osseous—or bony—labyrinth and the membranous labyrinth. The three regions of the bony labyrinth are the vestibule, the cochlea, and the semicircular canals. There are two membranous sacs within the vestibule, the saccule and utricle, which contain receptors that respond to linear acceleration (for example, the pull of gravity, acceleration in a vehicle, and changes in head position).

What are the basic stages of sensing sound?

A sound wave is a vibration in the air that enters the ear canal. The sound strikes the eardrum, causing it to vibrate. Behind the vibrating eardrum, in the middle ear, are

51

three small bones that move in response to the eardrum. These bones transfer the vibrations to the cochlea, traveling through the cochlear duct toward the auditory nerve. Nerve impulses travel to the brain, which translates them into a sound you can understand.

What is the organ of Corti?

The organ of Corti, located in the cochlear duct, is the auditory organ. It contains about 20,000 hearing receptor cells and many supporting cells. These receptor cells are called hair cells. The organ of Corti sits on the basilar membrane, a flexible, fibrous structure on the floor of the cochlear duct. As a pressure wave travels through the cochlear duct it causes the basilar membrane to vibrate. The basilar membrane is narrow and stiff at the base of the cochlea (like the strings of a harp or piano used in playing the high notes), where it resonates in response to high-frequency sound waves. The basilar membrane is wide and less stiff near the apex of the cochlea (like the strings of a harp or piano for the low notes), where it resonates in response to lower-frequency pressure waves. This resulting vibration causes the organ of Corti to vibrate, which is sensed by hair cells. Depending on the volume of the sound, either a few hairs move, as in the case of a soft sound, or many hairs move, as in the case of the loud sound.

What frequencies can a human hear?

Sound waves are alternating zones of high and low pressure traveling through air or water and characterized by their frequency or intensity. Frequency is measured in hertz (Hz), which represents cycles per second (cps). The frequency range of human hearing is from 20 to 20,000 Hz. Sounds with a frequency lower than 20 Hz cannot be heard by humans and are referred to as infrasonic. Such signals start below 20 Hz, but can be detected at frequencies as low as a hundredth or even a thousandth of a hertz. Human ears are most sensitive to frequencies between 1,500 and 4,000 Hz. Within that range, an individual can distinguish frequencies that differ by only 2 or 3 Hz.

What are some common levels of sound and how do they affect hearing?

Some common levels of sound and their effects on hearing are listed in the following table:

Sound	Decibel Level (dB)	Effects on Hearing
Lowest audible sound	0	None
Rustling leaves	20	None
Quiet library or office	30–40	None
Normal conversation; refrigerator running; light road traffic at a distance	50–60	None
Busy car traffic; vacuum cleaner; noisy restaurant	70	None
Heavy city traffic; subway; shop tools; power lawn mower	80–90	Some damage if continuous for 8 hours or more

What is sound and what unit measures it?

Sound is the vibration of air or other matter. Sound intensity refers to the energy of the sound waves, and loudness refers to the interpretation of sound as it reaches the ears. Both intensity and loudness are measured in logarithmic units called decibels (dB). A sound that is barely audible has an intensity of zero decibels. Every 10 decibels indicates a tenfold increase in sound intensity. A sound is ten times louder than threshold at 10 dB and 100 times louder at 20 dB.

Sound	Decibel Level (dB)	Effects on Hearing
Chain saw	100	Some damage if continuous for 2 hours
Rock concert	110–120	Definite risk of permanent hearing loss
Gunshot	140	Immediate danger of hearing loss
Jet engine	150	Immediate danger of hearing loss
Rocket launching pad	160	Hearing loss inevitable

What are the two types of deafness?

The two types of deafness are conduction deafness and sensorineural, or perceptive, deafness. In conduction deafness, the transmission of sound waves through the middle ear is impaired. In sensorineural deafness, the transmission of nerve impulses anywhere from the cochlea to the auditory cortex of the brain is impaired.

What are some causes of hearing loss and deafness?

Deafness may be caused by dysfunction of either the sound-transmitting mechanism of the outer, middle, or inner ear, or the sound-receiving mechanism of the inner ear. Causes of dysfunction include disease, toxic exposure, injury (including exposure to loud noise such as heavily amplified music through headphones), or genetic disorders.

What is presbycusis?

Presbycusis is the scientific name for age-related sensorineural (or perceptive) hearing loss. The first symptom is an inability to hear sounds at the highest frequencies and can occur as early as age twenty. Around age sixty, there is considerable variation in how well people hear. Some have had significant loss of hearing since age fifty, while others have no hearing problems into their 90s. In general, men seem to experience hearing loss more often and more severely than women. One explanation for this difference may be that men's occupations are usually associated with prolonged exposure to louder noises.

What is tinnitus?

Tinnitus is the perception of sound in the ears or head where no external source is present. In almost all cases, tinnitus is a subjective noise, meaning that only the person who has tinnitus can hear it. It is often referred to as "ringing in the ears." Persistent tinnitus usually indicates the presence of hearing loss. The exact cause of tinnitus is not known, but there are

A hearing test conducted by an audiologist can determine the extent of hearing loss in a patient and what can be done to correct it.

several likely sources, all of which are known to trigger or worsen the condition. They include noise-induced hearing loss, wax build-up in the ear canal, medicines that are toxic to the ear, ear or sinus infections, head or neck trauma, or jaw misalignment (usually TMD, or temporomandibular disorders; for more about TMD, see the chapter "Skeletal System").

Where are the organs of equilibrium located?

The organs of equilibrium are located in the inner ear. The otolith organs are located in the vestibule of the membranous labyrinth. They consist of sheets of hair cells covered by a membrane that contains otoliths (Greek for "ear stones"), which are calcium carbonate crystals. The otolith organs sense linear acceleration of the head in any direction, such as acceleration due to changing the position of your head relative to gravity or acceleration in a car or amusement ride. The inner ear also contains horizontal, posterior, and anterior semicircular canals, which sense angular motions (acceleration) of the head. Each semicircular canal has a specialized sensory region that contains hair cells, and each canal is important for sensing rotation of the head in a different primary direction. For example, the horizontal semicircular canal receptors are sensitive to rotating the head leftward and rightward.

What is Ménière's disease?

Ménière's disease, named after French physician Prosper Ménière (1799–1862), who first described it in 1861, is a disorder characterized by recurring attacks of disabling vertigo (a whirling sensation), hearing loss, and tinnitus. It is thought to be caused by an imbalance in the fluid that is normally present in the inner ear. Either an increase in the production of inner ear fluid or a decrease in its reabsorption results in an imbalance of fluid, but why this happens is not known. It most often occurs in middle age and is more common in men than women.

What is motion sickness?

Motion sickness (also known as car, sea, train, or air sickness) occurs when the body is subjected to accelerations of movement in different directions or under conditions where visual contact with the actual outside horizon is lost. The brain receives contradictory information from its motion sensors such as the eyes or semicircular canals in the middle ears that provide information about body position. Symptoms include dizziness, fatigue, and nausea, which may progress to vomiting. Prevention is best accomplished by seeking areas of lesser movement in an interior location of a ship or by facing forward and looking outside an airplane. Various prescription and over-the-counter medications—and even natural herbs such as ginger—may prevent or limit the symptoms of motion sickness.

What is cerumen, also known as ear wax?

Cerumen is an oily, fatty substance produced by the ceruminous glands in the outer portion of the ear canal. This compound is commonly referred to as ear wax and, together with hairs in the auditory canal, helps prevent foreign objects from reaching the delicate eardrum. Ear wax also keeps the ear lubricated and helps to regulate bacterial growth (it contains a special enzyme called lysozyme that breaks down the cell walls of bacteria). Dust, dirt, bacteria, fungi, and other foreign dangers to the body all stick to the wax and do not enter the ear.

Should ear wax be removed?

In most individuals, the ear canal is self-cleansing and there is no need to remove ear wax. However, ear wax may be impacted due to such things as poor attempts at cleaning the ear, which is why physicians warn their patients not to stick any foreign objects in the ears, such as bobby pins or cotton swabs. If a person's ear has impacted ear wax—which can cause temporary hearing loss, ringing in the ears, infection, or ear pain—the wax should be removed by a healthcare professional.

VISION

What are the parts of the eye and their functions?

The major parts of the eye and their functions are summarized in the following chart:

Structure	Function
Sclera	Maintains shape of eye; protects eyeball; site of eye muscle attachment; it is also referred to as "the white of the eye"
Cornea	Refracts incoming light; focuses light on the retina
Pupil	Admits light
Iris	Regulates amount of incoming light
Lens	Refracts and focuses light rays
Aqueous humor	Helps maintain shape of eye; maintains introcular pressure; nourishes and cushions cornea and lens

Structure	Function
Ciliary body	Holds lens in place; changes shape of lens
Vitreous humor	Maintains intraocular pressure; transmits light to retina; keeps retina firmly pressed against choroids
Retina	Absorbs light; stores vitamin A; forms impulses that are transmitted to brain
Optic nerve	Transmits impulses to the brain
Choroid	Absorbs stray light; nourishes retina

The accessory structures of the eye include the eyebrows, eyelids, eyelashes, conjunctiva, and lacrimal apparatus. These structures have several functions, including protecting the anterior portion of the eye, preventing the entry of foreign particles, and keeping the eyeball moist.

Do the eyes grow like other organs?

Unlike most other organs, the eyes do not grow very much from infancy to adulthood. The average diameter of the eyeball is about 0.68 inches (17 millimeters) at birth and 0.84 inches (21 millimeters) in adulthood. However, since new lens fibers are produced throughout life, the thickness of the lens varies with age. At birth, the thickness measures from 0.14 inches (3.5 millimeters) to 0.16 inches (4 millimeters) and at age 95 it may be 0.19 inches (4.75 millimeters) to 0.20 inches (5 millimeters) thick.

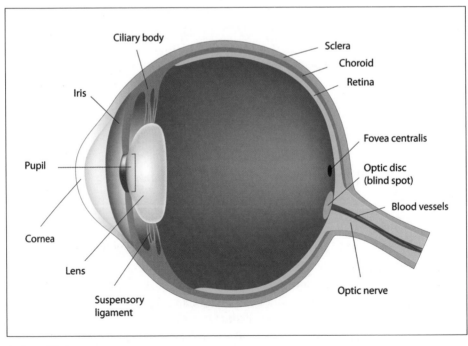

Anatomy of the human eye

Do blue-eyed humans have a single, common ancestor?

Yes, it is thought that blue-eyed humans have a single, common ancestor. In 2008, researchers in Denmark found that this single human ancestor had a genetic mutation. Originally, all humans had brown eyes. A person who lived between 6,000 to 10,000 years ago is thought to have carried a genetic mutation that changed the production of melanin, the pigment that gives color to our eyes. And this one mutation is thought to be the cause of all blue-eyed humans living on the planet today.

What determines eye color?

Variations in eye color range from light blue to dark brown and are inherited. Eye color is chiefly determined by the amount and distribution of melanin within the irises. If melanin is present only in the epithelial cells that cover the posterior surface of the iris, the iris appears blue. When this condition exists together with denser-than-usual tissue within the body of the iris, the eye color looks gray. When melanin is present within the body of the iris, as well as the epithelial covering, the iris appears brown.

What are the floaters that move around on the eye?

Floaters are semi-transparent specks perceived to be floating in the field of vision. Some originate with red blood cells that have leaked out of the retina. The blood cells swell into spheres, some forming strings, and float around the areas of the retina. Others are shadows caused by the microscopic structures in the vitreous humor, a jellylike substructure located behind the retina.

What are vitreous and retinal detachments?

Vitreous detachment occurs mainly in older persons. The gel-like substance called the vitreous humor fills the space between the lens and the retina of the eyeball. Also within the vitreous are millions of fine fibers that are attached to the surface of the retina. When a person ages, the vitreous shrinks, causing the fibers to pull on and away from the retinal surface. Symptoms include floaters that are specks or long and cobweb-like, and occasionally semicircle flashes of light when the eye moves quickly (usually best seen in a darkened room). It is, in most cases, not sight-threatening. Retinal detachment is more serious in terms of vision, and if not treated could cause permanent vision loss. The detachment occurs when the retina is pulled or lifted from its normal position. It can occur after a head trauma, or if there is already a small retinal tear or break. Symptoms include a sudden or gradual increase in floaters or light flashes in the eye, or in some cases, the sudden appearance of shadows or blind spots, or blurred vision.

What are the two layers of the retina?

The two layers that comprise the retina are an outer pigmented layer called the pigment epithelium, which adheres to the choroid, and an inner layer of nerve tissue called the sensory (or neural) retina. The inner layer of nerve tissue consists of three separate layers of neurons. The first and closest to the choroid is a layer of sensory receptors, the photoreceptors cells called rods and cones, and various other neurons. Next is a layer of bipolar neurons, the nerve cells that receive impulses generated by the rod and cone cells. The third or inner layer consists of ganglionic neurons attached directly to the optic nerve.

Why does diabetes cause blindness?

Diabetic retinopathy is the major cause of blindness in the United States among adults ages 20 to 65. The high blood sugar level in diabetes weakens blood vessel walls in the retina and choroids. This increases susceptibility to hemorrhaging, scarring, and retinal detachment—and thus can cause blindness.

What is the difference in the functions of the rods and cones found in the eyes?

Rods and cones are photoreceptor cells that convert light first into chemical energy and then into electrical energy for transmission to the vision centers of the brain via the optic nerve. Rods are specialized for vision in dim light; they cannot detect color, but they are the first receptors to detect movement and register shapes. There are about 125 million rods in a human eye. They contain a pigment called rhodopsin. Cones provide acute vision, functioning best in bright daylight. They allow us to see colors and fine detail. Cones are divided into three different types that contain cyanolabe, chlorolabe, or erythrolabe. These photopigments absorb wavelengths in

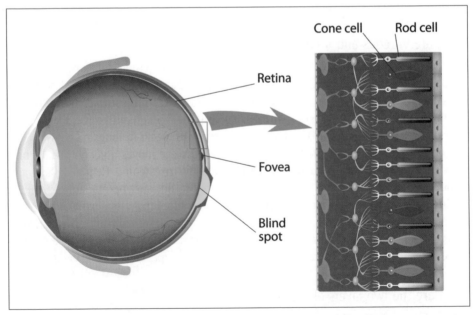

There are two types of photoreceptors in the eye: cones can detect colors, while rods are best for detecting objects in dim light but cannot perceive colors. This is why you do not see colors well when there is little light.

Does eating carrots and/or blueberries improve a person's ability to see at night?

Healthy rods in a person's retina mean they usually have adequate night vision. When the rods detect movement and dim light, rhodopsin is released, which splits into the protein opsin and the pigment retinal. Because retinal is derived from vitamin A, and carrots do contain significant quantities of vitamin A, it has often been thought that carrots help a person with their night vision. But in reality, although eating carrots does keep a person healthy, it does not allow them to see better in the dark (although if a person is deficient in vitamin A, it might make it harder for them to see in the dark). The idea was started as a ruse during World War II by the British Air Ministry. They didn't want the Germans to know about using radar to intercept bombers on night raids, so they sent out press releases that their pilots were eating carrots to give them exceptional night vision. Thus, the myth was born.

And although blueberries have more recently been touted to help a person see better at night, a 2014 study in the *Journal of Agricultural and Food Chemistry* did not agree. People who consumed blueberries did not see better at night, but the berries' anthocyonins (a healthy flavonoid) still promoted heart health, better memory, and digestion.

the short (blue), middle (green), and long (red) ranges, respectively. There are about seven million cones in each eye.

How long does it take for a person to adapt to dim light?

Rods give us vision in dim light but not in color and not with sharp detail. They are hundreds of times more sensitive to light than cones, letting us detect shape and movement in dim light. This type of photoreceptor takes about fifteen minutes to fully adapt to very dim light.

Is night blindness dangerous or serious?

Night blindness is a condition in which the rods in the retina are seriously damaged due to vitamin A deficiency. This results in the inability to drive safely at night. Vitamin A supplements can reverse this condition if administered before degeneration of the rods.

What are the different types of cone cells in humans?

Color perception depends on cones. Humans have three types of cones: blue, green, and red. Each contains a slightly different photopigment. Although the retinal portion of the pigment molecule is the same as in rhodopsin, the opsin protein differs slightly in each type of photoreceptor. Each type of cone responds to light within a range of wavelengths but is named for the ability of its pigment to absorb a wavelength more strongly than the other cones. Red light, for example, can be absorbed by all three types of cones, but those cones most sensitive to red act as red receptors. By comparing the relative responses of the three types of cones, the brain can detect colors of intermediate wavelengths. There will always be a dominant cone sending the electrical color-coded impulse to the brain, but the other two color cones will also be stimulated to some degree, even if it is a faint spark. These various and unlimited combinations are what make possible the millions of shades of color we see.

What are the three different cone pigments?

The three different cone pigments are erythrolabe, chlorolabe, and cyanolabe. Erythrolabe is most sensitive to red light waves; chlorolabe is most sensitive to green light waves; and cyanolabe is most sensitive to blue light waves.

Are the three types of color cones present in equal quantities?

In an individual with normal vision, the cone population consists of 16 percent blue cones, 10 percent green cones, and 74 percent red cones. Although their sensitivities overlap, each type is most sensitive to a specific portion of the visual spectrum.

What causes color blindness?

Color blindness is the inability to perceive one or more colors; this may involve complete or partial loss of color perception. Most forms of color blindness occur more frequently in males as an X-linked genetic disorder. Color blindness is actually a collection of several abnormalities of color vision. The most common form is a red-green blindness, which affects about 8 percent of the male population in the United States. Red blindness is the inability to see red as a distinct color, while green blindness is the inability to see green. A rare form of color blindness is the inability to see the color blue.

How fast do photoreceptors respond?

It takes 0.002 seconds for the brain to recognize an object after its light first enters the eyes.

What are phosphenes?

If the eyes are shut tightly, the "lights" seen are phosphenes, from the Greek words *phos* ("light") and *phainein* ("to show"). The phosphenes can be in the form of multicolored shapes or patterns. Technically, the luminous impressions are due to the excitation of the retina caused by some mechanical event, such as pressure on the eyeball. For example, this can occur when a person puts pressure on his or her closed eye with a finger.

What is the blind spot of the eye?

The area where the optic nerve passes out of the eyeball, the optic disc, is known as the "blind spot" (or scotoma) because it lacks rods and cones. Images falling on the blind spot cannot be perceived. Everyone has a natural blind spot in their vision, and the brain is efficient at using the other eye to fill in the "missing" information. (This blind spot should not be confused with one in a person's field of vision. Such spots may be caused by migraines, glaucoma, macular degeneration, or even a detached retina, and may require a check by an eye care professional.)

What structure of the eye produces tears?

The lacrimal gland, which is part of the larger lacrimal apparatus, produces tears that flow over the anterior surface of the eye. Most of this fluid evaporates, but excess amounts are collected in small ducts in the corner of the eye called, of course, tear ducts. Tears lubricate and cleanse the eye. In addition, tears contain lysozyme, an enzyme that is capable of destroying certain kinds of bacteria and helps fight eye infections.

How often does a person blink?

The rate of blinking varies, but on average the eye blinks once every five seconds (twelve blinks per minute). Assuming the average person sleeps eight hours a day and is awake for sixteen hours, he or she would blink 11,520 times a day, or 4,204,800 times a year. The average blink of the human eye lasts about 0.05 seconds.

What is nearsightedness?

Nearsightedness, or myopia, is the ability to see close objects but not distant ones. It is a defect of the eye in which the focal point is too near the lens and the image is

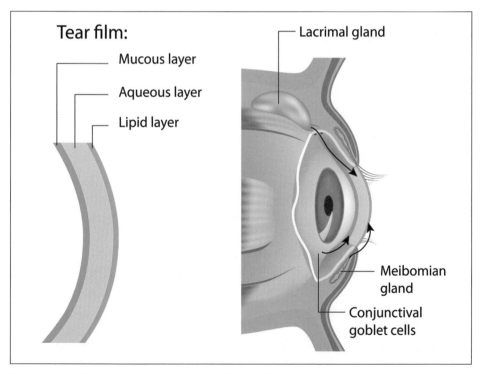

Tear film:
Mucous layer
Aqueous layer
Lipid layer

Lacrimal gland

Meibomian gland

Conjunctival goblet cells

The lacrimal gland in the eye produces tears to cleanse and lubricate the surface.

What does it mean to have 20/20 vision?

Many people think that 20/20 vision equals perfect eyesight, but it actually means that the eye can see clearly at 20 feet (6 meters), which is average for a normal human eye. Some people can see even better, such as 20/15, for example. With 20/15 vision they can view objects from 20 feet (6 meters) away with the same sharpness as a normal-sighted person standing 15 feet (4.5 meters) away.

focused in front of the retina when looking at distant objects. This condition is corrected by concave lenses (eyeglasses or contact lenses) that diffuse the light rays coming to the eyes so that when the light is focused by the eyes, it reaches the proper spot on the retinas.

What is farsightedness?

Farsightedness, or hyperopia, is the ability to see distant objects but not close ones. It is a disorder in which the focal point is too far from the lens, and the image is focused "behind" the retina when looking at a close object. In this condition, the lens must thicken to bring somewhat distant objects into focus. Farsightedness is corrected by a convex lens that causes light rays to converge as they approach the eye to focus on the retina.

What is astigmatism?

Astigmatism is an irregularity in the curvature of the cornea or lens that causes light traveling in different planes to be focused differently. The normal cornea or lens has a spherical curvature like the inside of a ball. In astigmatism, the cornea or lens has an elliptical curvature, like the inside of a spoon. As a result, some portions of an image are in focus on the retina, while other portions are blurred and vision is distorted.

What causes double vision?

Double vision (diplopia) is seeing two images of one object. In order for the brain to receive a clear image, both eyes must move in unison. This condition may result from weakness in one or more of the muscles that control eye movements. When this happens, the good eye focuses on a seen object, but the affected eye is focused elsewhere. Other causes include fatigue, alcohol intoxication, multiple sclerosis, vertigo, or trauma. The sudden and ongoing appearance of double vision may indicate a serious disorder of the brain or nervous system.

What is strabismus?

Strabismus is a lack of coordination between the eyes. As a result, the eyes look in different directions and do not focus simultaneously on a single point. When the two eyes fail to focus on the same image, the brain may learn to ignore the input from one eye. If this continues, the eye that the brain ignores will never see well. This loss of vision is called amblyopia. Treatment for strabismus includes exercises and other strategies to strengthen the weakened eye muscles and realign the eyes. Glasses may also be prescribed. Surgery may be required to realign the eye muscles if strengthening techniques are unsuccessful.

What is the most common cause of blindness in the United States?

A person is considered legally blind in the United States when his or her best cor-

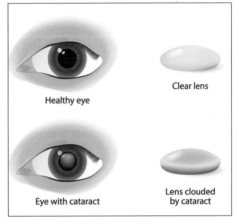

A cataract occurs when the lens of the eye becomes clouded, hampering vision.

rected vision is 20/200 or worse. Cataracts are the most common cause of blindness in the United States. This is a condition in which clouding of the lens occurs as a result of advancing age, infection, trauma, or excessive exposure to ultraviolet (UV) radiation.

COMPARING OTHER ORGANISMS

Do humans or bloodhounds have a keener sense of smell?

Humans smell the world using about twelve million olfactory receptor cells—divided between two odor-detecting patches. Bloodhounds have about four billion such cells and, therefore, a much better sense of smell, although other types of dogs have fewer receptor cells. (It is often said that the receptor patch in a human is the size of a stamp, while that of a bloodhound is the size of a handkerchief.) In fact, the trace of sweat that seeps through shoes and is left in footprints is a million times more powerful than the odor a bloodhound needs to track down someone. Other animals also have keener smell abilities than humans. For example, a rabbit has one hundred million olfactory cells.

Why are some larger birds such efficient hunters?

The main reason larger birds like eagles, hawks, and buzzards are such successful hunters is their keen eyesight. For example, a raptor's eyesight is three to four times sharper than a human's eyesight. A red-tailed hawk scanning the ground at a height of about 10,000 feet can spot a rodent—and as it dives, still keep its prey in focus.

Why do all newborn human babies—and many other baby animals—have blue eyes?

The color of the iris gives the human eye its color. In newborns, the pigment is concentrated in the folds of the iris. When a baby is a few months old, the melanin moves to the surface of the iris and gives the baby his or her permanent eye color.

How wide an angle is covered by the average person's field of vision versus some other animals?

The average field of vision for humans is about 200, but it can range from 160 to 210 degrees. Other animals have different fields of vision. For example, a dragonfly's eyes cover almost the entire head, covering a 360 degree field of vision. Goats, and some other hoofed animals, have horizontal slits in their eyes that can be almost rectangular when dilated. Thus a goat's vision can range from 320 to 340 degrees, allowing the animal to see all around without having to move.

Can cats see in complete darkness?

No, cats cannot see in complete darkness, but almost complete darkness. This is because their unusual eye shape—the slit-shaped pupils—gives them an advantage as nocturnal hunters. According to researchers, a cat has the ability to alter the intensity of light entering the eye and falling on the retina by 135-fold, compared to 10-fold in a human with a circular pupil. The cat's large cornea allows more light into the eye; while the slit pupil dilates more than a round pupil, allowing even more light to enter the eye while hunting prey in the dark.

How do various animals see predators?

The retinas of mammals such as rabbits and squirrels, as well as those of nonmammals like turtles, have a long, horizontal strip of specialized cells called a visual streak, which can detect the fast movement of predators. Primates, as well as some birds, have front projecting eyes allowing binocular vision and thus depth perception; their eyes are specialized for good daylight vision and are able to discriminate color and fine details. Primates and raptors, like eagles and hawks, have a fovea, a tremendously cone-rich spot devoid of rods where images focus.

What is echolocation in bats?

Echolocation is the ability of bats to use high picked pulses of sound, then listening for the return of the sound to determine the distance and direction in their environment. Contrary to popular belief, bats are not blind. It is just that their eyes are less powerful than other predators—and thus these nocturnal hunters rely less on light and more on sound in the search for prey.

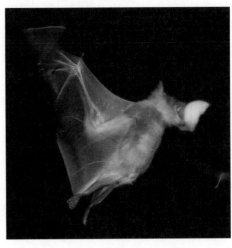

Bats use sound—echolocation—to locate food in dim or no light.

Can animals hear different sound frequencies than humans?

The frequency of a sound is the pitch. Frequency is expressed in Hertz (Hz). Sounds are classified as infrasounds (below the human range of hearing), sonic range (within the range of human hearing), and ultrasound (above the range of human hearing). The following lists the various hearing ranges of animals, including humans:

Animal	Frequency range heard (Hz)
Dog	15 to 50,000
Human	20 to 20,000
Cat	60 to 65,000
Dolphin	150 to 150,000
Bat	1,000 to 120,000

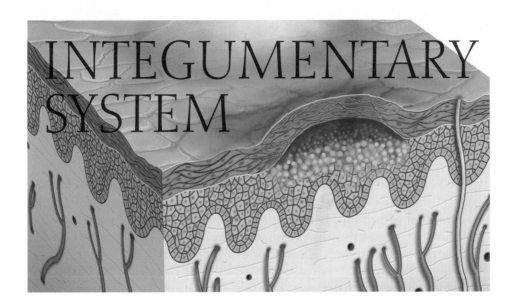

INTEGUMENTARY SYSTEM

INTRODUCTION

What organs are included in the integumentary system?

The integumentary system (from the Latin *integere*, meaning "to cover") includes skin, hair, glands, and nails. The main function of the integumentary system is to provide the body with a protective barrier between the organs inside the body and the changing environment outside.

How much skin does the average person have?

The average person is covered with about 20 square feet (1.9 square meters) of skin, which weighs about 5.6 pounds (2.7 kilograms). The skin is the largest and heaviest organ in the body, representing four percent of the average weight of the human body.

How much skin does a person shed in one year?

An average man or woman sheds about 600,000 particles of skin per hour, which is approximately 1.5 pounds (680 grams) per year. Using this figure, by the age of seventy, a person will have lost an average of 105 pounds (47.6 kilograms) of skin.

What structures are present in an average square inch (6.4 square centimeters) of skin?

The average square inch (6.4 square centimeters) of skin holds 20 feet (6.1 meters) of blood vessels, 77 feet (23.5 meters) of nerves, and more than a thousand nerve endings. In addition to blood vessels and nerves, there are 645 sweat glands, 65 hair follicles, and 97 sebaceous glands per square inch.

How many bacteria are present on skin?

Every square inch (6.4 square centimeters) of skin contains approximately thirty-two million bacteria. Collectively, there are some one trillion bacteria on the average human body. They represent around a thousand species and most of them are harmless.

SKIN STRUCTURE

What are the various layers of the skin?

Skin is a tissue membrane that consists of layers of epithelial and connective tissues. The outer layer of the skin's epithelial tissue is the epidermis and the inner layer of connective tissue is the dermis. A basement membrane that is anchored to the dermis separates these two layers. The epidermis and dermis rest on a supportive layer of connective tissue and fat cells called the hypodermis. This supportive layer is flexible and allows the skin to move and bend while the fat cells cushion against injury and excessive heat loss.

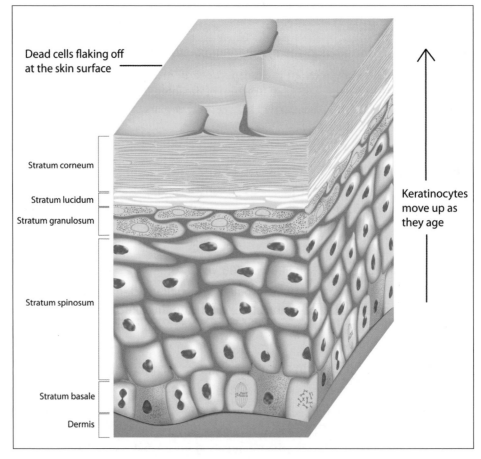

Dead cells flaking off at the skin surface

Stratum corneum

Stratum lucidum

Stratum granulosum

Stratum spinosum

Stratum basale

Dermis

Keratinocytes move up as they age

Anatomy of the epidermis

What causes warts?

Warts, which are noncancerous masses produced by uncontrolled growth of epithelial skin cells, are caused by the human papillomavirus (HPV) on the top layer of the skin. Many strains of the virus exist, most commonly on the hands and feet as benign growths, but such warts are not the same strains of HPV that cause genital warts.

A wart can only be removed by killing the basal cells that harbor the virus. This can be accomplished by cutting away that piece of skin, destroying it by freezing, or killing it with certain compounds (essentially slowly burning off the wart with chemicals, such as salicylic acid). Most of these processes can be done by a physician, but it can often take months before the wart disappears depending on the method. In general, a person should not pick or scratch a wart, as it will make the wart worse and can spread the virus to other parts of the body.

Are there different types of warts?

Yes, there are different types of warts, but each is considered to be caused by the human papillomavirus (HPV), and most are named for where they appear on the body. For example, palmer warts form on the hands. Plantar warts form on the feet; clusters of plantar warts are called mosaic warts.

What are the two types of specialized cells in the epidermis?

The more numerous cells are called keratinocytes, which produce a tough, fibrous, waterproof protein called keratin. Keratinization is the process in which cells form fibrils of keratin and harden. Over most of the body, keratinization is minimal, but the palms of the hands and the soles of the feet normally have a thick, outer layer of dead, keratinized cells. Melanocytes are less numerous than keratinocytes in the epidermis; the average square inch (6.4 square centimeters) of skin contains 60,000 melanocytes. The melanocytes produce a pigment called melanin, which ranges in color from yellow to brown to black and determines skin color.

What determines skin color?

Three factors contribute to skin color: 1) the amount and kind (yellow, reddish brown, or black) of melanin in the epidermis; 2) the amount of carotene (yellow) in the epidermis and subcutaneous tissue; and 3) the amount of oxygen bound to hemoglobin (red blood cell pigment) in the dermal blood cells. Skin color is genetically determined, for the most part. Differences in skin color result not from the number of melanocytes an individual has, but rather from the amount of melanin produced by the melanocytes and the size and distribution of the pigment granules. Although darker-skinned people have slightly more melanocytes than those who are light-skinned, the distribution of melanin in the higher levels of the epidermis contributes to their skin color.

What is albinism?

A genetic trait characterized by the lack of ability to produce melanin causes the condition known as albinism. Individuals with this disorder lack not only pigment in

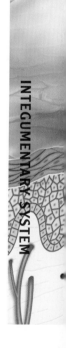

the skin but also in their hair and eyes. (For more about albino animals, see below under "Comparing Other Organisms.")

What are freckles and are they dangerous?

Freckles, those tan or brown spots on the skin, are small areas of increased skin pigment or melanin. There is a genetic tendency to develop freckles, and parents with freckles often pass this trait down to their children. Freckles usually occur on the face, arms, and other parts of the body that are exposed to the sun. Freckles themselves usually pose no health risks, but most individuals who freckle easily are at an increased risk for skin cancer.

What are age spots?

Age spots—also known as sunspots, liver spots, or lentigines—are caused by long-term exposure to the sun. Age spots are flat, irregular, brown discolorations of the skin that usually occur on the hands, neck, and face of people older than forty. They are not harmful and are not cancerous.

How frequently are epidermal cells replaced?

The epidermis is avascular (it has no blood supply of its own). This explains why a man can shave daily and not bleed even though he is cutting off several cell layers each time he shaves. New epidermal cells originate from the deepest layer of the epidermis, the stratum germinativum, and are pushed upward daily to become part of the outermost layer, which flakes off steadily. A totally new epidermis is produced every twenty-five to thirty days.

What are the two regions of the dermis?

The dermis is composed of dense connective tissue and is the site of blood vessels, nerves, and epidermal appendages. The dermis has two regions: the papillary and reticular layers. The papillary layer has ridges, which produce fingerprints and is composed of loose connective tissue. The reticular layer is much thicker and is composed of dense, irregular connective tissue.

When do fingerprints form?

At about thirteen weeks of gestation, the human fetus has developed outer epidermal ridges that will eventually develop into fingerprints. These become more and more defined, and at about twenty-one to twenty-four weeks the ridges resemble their adult form.

Do identical twins have the same fingerprints?

No. Even identical twins have differences in their fingerprints, which, though sub-

Eveyone has a unique set of fingerprints, even identical twins.

> ## What are some newer innovations in identification used by such groups as the FBI?
>
> There are several newer innovations that are being used for identification by such groups as the FBI. In particular, one is called NGI, or Next Generation Identification. For example, an NGI called the Advanced Fingerprint Identification Technology (AFIT) was implemented. This contained a new fingerprint matching algorithm that improved matching accuracy from approximately 92 percent to over 99 percent. And in 2013, an NGI called the National Palm Prints System (NPPS) was expanded and contains millions of palm prints that are now searchable on a nationwide basis.

tle, can be discerned by experts. Research indicates that fingerprints would not be the same even in the clone of an individual.

Who first used fingerprints as a means of identification?

It is generally acknowledged that Francis Galton (1822–1911) was the first to classify fingerprints. However, his basic ideas were further developed by Sir Edward Henry (1850–1931), who devised a system based on the pattern of the thumb print. In 1901 in England, Henry established the first fingerprint bureau with Scotland Yard called the Fingerprint Branch. Today, the Federal Bureau of Investigation (FBI) files house the fingerprints and criminal histories for more than seventy-five million subjects in the criminal master file, along with more than thirty-four million non-criminal prints.

How did John Dillinger and Roscoe Pitts attempt to change their fingerprints?

American gangster John Dillinger (1903–1934) used acid to burn his fingerprints in an attempt to permanently change them by removing the ridge patterns. He failed, and the fingerprints that reappeared were identical to the ones he had tried to change. Another American criminal named Roscoe Pitts (real name Robert Philipps) asked a New Jersey physician (and ex-con) to perform a dramatic surgical procedure to permanently alter fingertips. The physician taped Pitts's cut right hand fingers to the criminal's chest. After three weeks, when the chest skin had grown into the fingertips, Pitts's hand was separated from his chest, resulting in smooth fingertips. Pitts endured the procedure on his left hand for three more weeks. But in the end, his girl-friend told authorities about his fingerprint procedure—and there were enough of the old prints still at the edge of his fingertips, allowing him to be identified.

What is dermatoglyphics?

Dermatoglyphics, the study of fingerprints, recognizes three basic patterns of fingerprints. They are arches, loops, and whorls. The lines or ridges of an arch run from one side of the finger to the other with an upward curve in the center. In a loop, the ridges begin on one side, loop around the center, and return to the same side. The ridges of a whorl form a circular pattern. Dermatoglyphics is of interest in such diverse fields as medicine, anthropology, and criminology.

Can fingerprints be permanently changed or destroyed?

An individual's fingerprints remain the same throughout his or her entire life. Minor cuts or abrasions, and some skin diseases such as eczema or psoriasis, may cause temporary disturbances to the fingerprints, but upon healing the fingerprints will return to their original pattern. More serious injuries to the skin that damage the dermis might leave scars that change or disrupt the ridge pattern of the fingerprints, but examining the skin outside the area of damage will reveal the same fingerprint pattern.

How are fingerprints used for computer security?

Recent technological advances using optical scanners and solid-state readers use software to analyze the geometric pattern of fingerprints and compare it with those of registered, legitimate users of a network system. Less expensive models of these devices have false acceptance rates of less than twenty-five per million and false rejection rates of less than 3 percent. Applications include using fingerprints instead of passwords for computers, linking to individual bank accounts and automated teller machines, and for credit cards and Internet transactions. For example, in 2015, the Apple computer company presented Touch ID, in which a user's fingerprint can be used as a passcode to unlock a personal iPhone.

What is the name of the protein—the most abundant in the human body—that holds our skin together?

The protein that holds the human body skin together is called collagen. It makes up between 25 to 35 percent of the proteins in the human body. Along with humans, it makes up most of the proteins found in all mammals.

How thick is the average person's skin?

The thickness of skin varies, depending on where it is found on the body. Skin averages 0.05 inches (1.3 millimeters) in thickness. The thinnest skin is found in the eyelids and is less than 0.002 inches (0.05 millimeters) thick, while the thickest skin is on the upper back (0.2 inches or 5 millimeters).

What other structures are found in the dermis?

There are several structures in the dermis. The following lists the most common:

Hair—The hair root is in the dermis layer and the shaft is above the surface of the skin.

Oil glands—Also known as sebaceous glands, these glands secrete an oily substance that moistens and softens skin and hair.

Sweat glands—These help regulate body temperature.

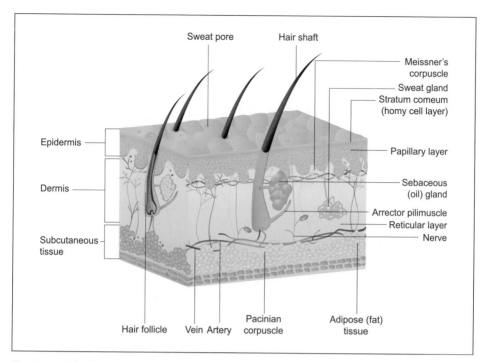

The structure of the skin is far more complex that most people would initially imagine.

Blood vessels—They are responsible for supplying the epidermis and dermis with nutrients and removing wastes.

Nerve endings—These provide information about the external environment.

What is the difference between thick and thin skin?

The terms "thick" and "thin" refer to the thickness of the epidermis. Most of the body is covered by thin skin, which is 0.003 inches (0.08 millimeters) thick. This skin contains hair follicles, sebaceous glands, and arrector pili muscles. The epidermis in thick skin may be six times thicker than the epidermis that covers the general body surface. Thick skin does not have hair, smooth muscles, or sebaceous glands. The skin on the palms of the hands, the fingertips, and soles of the feet may be covered by many layers of keratinized cells that have naturally become thicker.

What causes a blister to the skin?

The epidermis and dermis are usually firmly cemented together by a basement membrane that is anchored to the dermis. However, a burn or friction due to, for example, rubbing by a poorly fitting shoe, may cause the epidermis and dermis to separate, resulting in a blister.

What is a scab?

A scab is made up of the blood clot and dried tissue fluids that form over a wound. It has an important function in keeping the wound bacteria-free while the skin cells underneath divide rapidly to heal the opening. Eventually, the scab will fall off (usually within one or two weeks) and new epithelial tissue will cover the wound.

Which layers of skin are damaged by burns?

Burns may be caused by heat generated by radioactive, chemical, or electrical agents. Two factors affect burn severity: the depth of the burn and the extent of the burned area. There are three categories of burns:

First-degree burns—Burns that are red and painful, but not swollen and blistering, such as from a sunburn, and damage only the epidermis.

Second-degree burns—Burns that are red, painful, and blistering, these burns involve injury to the epidermis and the upper region of the dermis.

Third-degree burns—Burns that are severely painful, giving the skin a white or charred appearance; they destroy all layers of the skin, including blood vessels and nerve endings. Skin damaged by third-degree burns does not regenerate. Damage to the skin affects the body's ability to retain fluids.

What are some differences between cutaneous carcinomas and melanomas?

Cutaneous carcinomas (basal cell and squamous cell) are the most common type of skin cancer. They originate from non-pigmented epithelial cells within the deep layer of the epidermis. These cancers usually appear in light-skinned adults who are regularly exposed to sunlight. Cutaneous carcinomas may be flat or raised and develop from hard, dry growths that have reddish bases. This type of carcinoma is slow growing and can usually be completely cured by surgical removal or treatment with radiation.

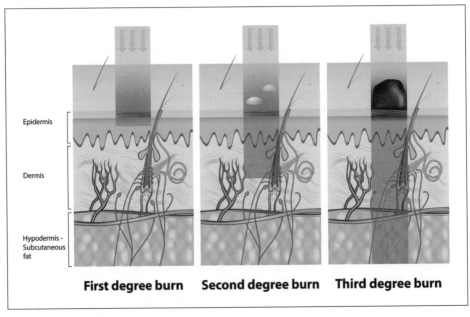

Epidermis

Dermis

Hypodermis - Subcutaneous fat

First degree burn   **Second degree burn**   **Third degree burn**

The degree of a burn depends on how deeply the tissue has been damaged.

Melanomas develop from melanocytes and range in color from brown to black and gray to blue. The outline of a malignant melanoma is irregular, rather than smooth, and is often bumpy. Unlike cutaneous carcinomas, melanoma is generally not associated with continued sun exposure. A cutaneous melanoma may arise from normal-appearing skin or from a mole. The lesion grows horizontally but may thicken and grow vertically into the skin, invading deeper tissues. If the melanoma is removed before it invades the deeper tissues, its growth may be arrested. Once it spreads vertically into deeper tissue layers, it is difficult to treat and the survival rate is very low.

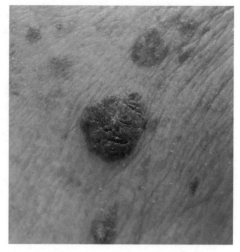
When considering whether a mole might be cancerous, a dermatologist will look at its shape, color, and whether it is getting bigger over time.

How common are moles on the body?

Everyone has moles, which are pigmented, fleshy blemishes of the skin. There are various types of moles, including the congenital moles (those a person is born with) and acquired moles, also called common moles, some of the most abundant on a person's skin. The average person has ten to forty common moles on his or her skin.

What is the "ABCD" rule dermatologists use for recognizing a suspicious mole?

Dermatologists suggest a simple way of judging a suspicious mole on a person's skin. It is called ABCD, and refers to the following:

"A" is for asymmetry, or the two sides of the growth or mole do not match.

"B" is for border irregularity, or the border or outline of the growth is not smooth but shows indentations.

"C" is for color, or the pigmented growth contains areas of different colors such as blacks, browns, tans, grays, blues, and reds.

"D" is for diameter, or the growth is larger than about 0.25 inches (6.35 millimeters) in diameter, or larger than a pencil eraser.

SKIN FUNCTION

What are the functions of the skin?

The skin has several different and important functions. It provides protection from both injury (such as abrasion) and dehydration. Since outer skin cells are dead and keratinized, the skin is waterproof, thereby preventing fluid (water) loss. The skin's waterproofing also prevents water from entering the body when a person is immersed. The skin is a barrier against invasion by bacteria and viruses and is involved in the regulation of body temperature. It is the site for the synthesis of an inactive form of vitamin D. In addition, the skin contains receptors that receive the sensations of touch, vibration, pain, and temperature.

How do skin cells synthesize vitamin D?

Vitamin D is crucial to normal bone growth and development. When ultraviolet (UV) light shines on a lipid present in skin cells, the compound is transformed into vitamin D. People native to equatorial and low-latitude regions of the earth have dark skin pigmentation as a protection against strong, nearly constant exposure to UV radiation. Most people native to countries that exist at higher latitudes—where UV radiation is weaker and less constant—have lighter skin, allowing them to maximize their vitamin D synthesis. During the shorter days of winter, the vitamin D synthesis that occurs in people who live in higher latitudes is limited to small areas of skin exposed to sunlight.

Increased melanin pigmentation, which is present in people native to lower latitudes, reduces the production of vitamin D. Susceptibility to vitamin D deficiency is increased in these populations by the traditional clothing of many cultural groups native to low latitudes, which attempts to cover the body completely to protect the skin from overexposure to UV radiation. Most clothing effectively absorbs irradiation produced by ultraviolet B rays. The dose of ultraviolet light required to stimulate skin synthesis of vitamin D is about six times higher in African Americans than in people of European descent. The presence of darker pigmentation and/or veiling may significantly impair sun-derived vitamin D production, even in sunny regions like Australia.

How do cells become keratinized?

The epithelial layer of the skin is continuously replaced. As the replacement cells move closer to the surface of the epidermis, they produce keratin (from the Greek *keras*, meaning "horn"), a tough protein. The transformation of cells into keratin breaks down the cells' nuclei and organelles until they can no longer be distinguished. When the cells' nuclei have broken down, the cells cannot carry out their metabolic functions. By the time the cells reach the superficial layer of the skin, they are dead and composed mostly of keratin.

Do all cells become keratinized?

No, the nuclei and organelles of stratified squamous cells in a moist environment such as the mouth, esophagus, vagina, and cornea do not break down even as the cells reach the superficial layer. This tissue is known as nonkeratinized stratified squamous epithelium and is much different than the normal skin cells that become keratinized.

How fast do epidermal cells grow in tissue culture?

A piece of epidermis 1.2 square inches (7.7 square centimeters) can be expanded more than 5,000 times within three to four weeks. This process can almost yield enough skin cells to cover the body surface of an adult human.

How is the skin involved in the regulation of body temperature?

The skin is one of several organ systems participating in maintaining a core temperature, meaning the temperature near the center of someone's body. Temperature sensors in the skin and internal organs monitor core temperature and transmit signals to the control center located in the hypothalamus, a region of the brain. When the core temperature falls below its set point, the hypothalamus:

1. Sends more nerve impulses to blood vessels in the skin that cause the vessels to narrow, which restricts blood flow to the skin, reducing heat loss.
2. Stimulates the skeletal muscles, causing brief bursts of muscular contraction, known as shivering, which generates heat.

When the core temperature rises above its set point, the hypothalamus:

1. Sends fewer nerve impulses to blood vessels in the skin, causing them to dilate, which increases blood flow to the skin and promotes heat loss.
2. Activates the sweat glands, and when sweat evaporates off the skin surface it carries a large amount of body heat with it.

What is the purpose of goose bumps?

The puckering of the skin that takes place when goose-flesh is formed is the result of the contraction of smooth, minute muscle fibers in the skin (called arrector pili), which are attached to hair follicles. (When a person is scared, the arrector pili are activated, which is also why hair "stands on end.") Goosebumps have often been compared to the skin of poultry after it has been plucked, thus the term. In humans, it usually occurs if a person is cold, in shock or fear, or even amazed. It is actually muscular activity that produces more heat and raises the temperature of the body.

What skin cells are involved with the immune system?

Keratinocytes, found in the epidermis, assist the immune system by producing hormone-like substances that stimulate the development of certain white blood cells

called T lymphocytes. T lymphocytes defend against infection by pathogenic (disease-causing) bacteria and viruses.

NAILS

Why are nails part of the integumentary system?

Nails, which are actually modifications of the epidermis, are protective coverings on the ends of the fingers and toes. Each nail consists of a nail plate that overlies a surface of skin called the nail bed. The whitish, thickened, half-moon shaped region (lunula) at the base of the nail plate is the most actively growing region. Nails, like hair, are primarily dead keratinized cells.

Do the hair and nails continue to grow after death?

No. Between twelve and eighteen hours after death, the body begins to dry out. That causes the tips of the fingers and the skin of the face to shrink, creating the illusion that the nails and hair have grown.

How fast do fingernails grow?

For the average person, healthy nails grow about 0.12 inches (3 millimeters) each month or 1.4 inches (3.5 centimeters) each year. It takes approximately three months for a whole fingernail to be replaced.

Do all fingernails and toenails grow at the same rate—and are they the same thickness?

Of all the nails on a person's hand, the thumbnail grows the slowest and the middle nail grows the fastest. In general, the longer the finger the faster its nail growth. To compare fingernails and toenails, fingernails tend to grow a little faster than toenails, and toenails are approximately twice as thick as fingernails.

Why do some nails have bumps or ridges?

Bumps or ridges on a finger or toenail often indicate a previous injury to the nail, such as a blow (for example, accidentally hitting the fingernail with a hammer). They can also occur due to the normal aging process, if a person has a certain disease (such as diabetes), or even if a person is malnourished.

Nails are formed from keratin, a fibrous protein.

HAIR

What are the three major types of hairs?

There are three major types of hairs associated with the integumentary system. They are as follows:

Vellus hairs—These are the "peach fuzz" hairs found over much of the body.

Terminal hairs—The thick, more deeply pigmented, and sometimes curly hairs that include the hair on the head, eyelashes, eyebrows, and pubic hair.

Intermediate hairs—Those hairs that change in their distribution and include the hair on the arms and legs.

How many hairs does the average person have on his or her head?

The amount of hair on the head varies from one individual to another. An average person has about 100,000 hairs on their scalp (blonds 140,000, brunettes 155,000, and redheads only 85,000). Most people shed between 50 to 100 hairs daily.

How many hairs are on the human body and where are most located?

On the average human body there are approximately five million hairs. Males have a few hundred thousand more hairs then women. In males and females, hormones are responsible for the development of such hairy regions as the scalp, the axillary

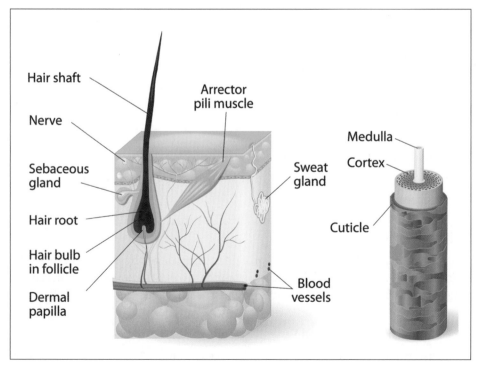

Hair anatomy

(armpit), and pubic areas, and, in men, on the chest. Overall, the various types of hair grow all over the human body except for the soles of the feet, palms of the hand, eyelids, and lips.

How fast do eyelashes grow?

Eyelashes are replaced every three months. The average individual will grow about six hundred complete eyelashes in a lifetime.

Does hair grow faster in summer or winter?

On the average, each hair grows about nine inches (twenty-three centimeters) every year. But during the summertime, human hair grows ten to fifteen percent faster than in the winter. This is because warm weather enhances blood circulation to the skin and scalp, which in turn nourishes hair cells and stimulates growth. In cold weather, when blood is needed to warm internal organs, circulation to the body surface slows and hair cells grow less quickly.

Which hair on the human male body grows the fastest?

A human male's beard grows more rapidly than any other hair on their body. On the average, it grows at a rate of about 5.5 inches (14 centimeters) per year, or about 30 feet (9 meters) in a lifetime.

Does shaving make hair coarser?

No. Uncut body hair is tapered and feels softer at the ends. The bristly feeling of hair that has been shaved is the cut end.

Is a hair living or dead?

Hairs are primarily dead, keratinized protein cells produced by the hair bulb. As the cells are pushed farther and farther away from the growing region, the shaft becomes keratinized and dies. The root is enclosed in a sheath, called the hair follicle.

How is hair color determined?

Genes determine hair color by directing the type and amount of pigment that epidermal melanocytes produce. If these cells produce an abundance of melanin, the hair is dark. If an intermediate quantity of pigment is produced, the hair is blond. If no pigment is produced, the hair appears white. A mixture of pigmented and unpigmented hair is usually gray. Another pigment, trichosiderin, is found only in red hair.

Why does hair turn gray?

The pigment in hair, as well as in the skin, is called melanin. There are two types of melanin: eumelanin, which is dark brown or black, and pheomelanin, which is reddish yellow. Both are made by a type of cell called a melanocyte that resides in the hair bulb and along the bottom of the outer layer of skin, or epidermis. The melanocytes pass this pigment to adjoining epidermal cells called keratinocytes, which produce the protein keratin, or hair's chief component. When the keratinocytes eventually die, they retain the melanin. Thus, the pigment that is visible in the hair and in the skin lies in these dead keratinocyte bodies. Gray hair is simply hair with less melanin, and white hair has no melanin at all.

It remains unclear as to how hair loses its pigment. In the early stages of graying, the melanocytes are still present but inactive. Later, they seem to decrease in number. Genes control this lack of deposition of melanin. In some families, many members' hair turns white in their twenties. Generally speaking, among Caucasians 50 percent are gray by age fifty. There is, however, wide variation. Premature gray hair is hereditary, but it has also been associated with smoking and vitamin deficiencies. Early onset of gray hair (from birth to puberty) is often associated with various medical syndromes, including dyslexia.

What determines the size and shape of hairs?

Hairs are short and stiff in the eyebrows and long and flexible on the head. When the hair shaft is oval, an individual has wavy or kinky hair; when it is flat and ribbon-like,

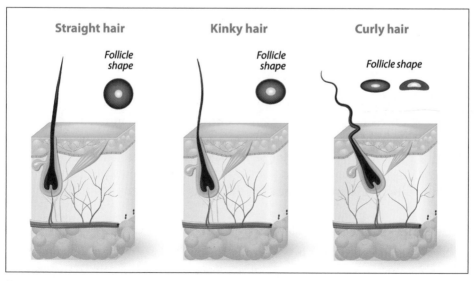

Hair becomes straight, kinky, or curly, depending on the shape of the follicles.

the hair is curly. If the hair shaft is perfectly round, the hair is straight. Humans are born with as many hair follicles as they will ever have, and hairs are among the fastest growing tissues of the body. Each hair follicle on the scalp grows almost 30 feet (9 meters) during an average lifetime.

What is alopecia?

Alopecia is the term used to refer to hair loss, which can have many causes. Male pattern baldness, or androgenic alopecia, is an inherited condition. Alopecia areata is characterized by the sudden onset of patchy hair loss. It is most common among children and young adults and can affect either gender.

What is pattern baldness?

Pattern baldness is an example of sex-influenced inheritance. Sex-influenced genes are expressed in both males and females, but pattern baldness is expressed more often in males than in females. Male pattern baldness is the most frequent reason for hair loss in men. The gene acts as an autosomal dominant in males and an autosomal recessive in females. One out of every five men begins balding rapidly in his twenties. Another one out of five will always keep his hair. The others will slowly bald over time. The level of baldness is often related to the quantity of testosterone in a man.

ACCESSORY GLANDS

What are the two types of cutaneous glands?

The two types of cutaneous glands, both of which are found in the dermis, are sebaceous glands and sweat glands. Sebaceous glands, or oil glands, are found all over the skin, except on the palms of the hands and the soles of the feet. Sebaceous glands produce sebum, a mixture of oily substances and fragmented cells that keeps the skin soft and moist and prevents hair from becoming brittle. A second type of cutaneous gland, the sweat glands, is widely distributed in the skin and is an important part of the body's heat-regulating apparatus.

What is the difference between a whitehead and a blackhead?

The ducts' sebaceous glands usually empty into hair follicles, but some open directly onto the surface of the skin. If the duct of a sebaceous gland becomes blocked by sebum, a whitehead appears on the skin surface. If the accumulated material oxidizes and dries, it darkens, forming a blackhead. If the sebaceous glands become infected, resulting in pimples on the skin, the condition is referred to as acne.

What is the length of sweat glands?

Sweat glands are coiled tubes in the dermis, the middle layer of the skin. If one sweat gland were uncoiled and stretched out, it would be approximately 50 inches (127 centimeters) in length.

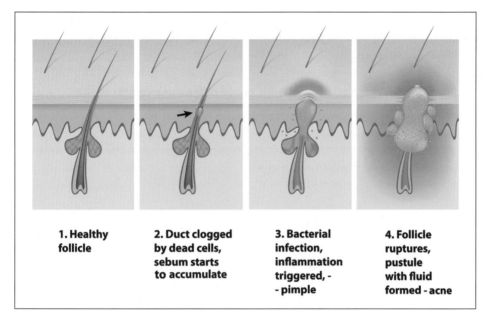

| 1. Healthy follicle | 2. Duct clogged by dead cells, sebum starts to accumulate | 3. Bacterial infection, inflammation triggered, - - pimple | 4. Follicle ruptures, pustule with fluid formed - acne |

How pimples form.

What are the two types of sweat glands?

The two types of sweat glands, also called sudoriferous glands, are eccrine and apocrine glands. Eccrine glands, which are found all over the body and are far more numerous, produce sweat. Sweat is a clear secretion that is largely comprised of water, some salts such as sodium chloride, urea and uric acid (metabolic wastes), and vitamin C. Apocrine sweat glands are found primarily in the axillary (armpit) and genital areas of the body and usually connect to hair follicles. These sweat glands become active when a person is emotionally upset, frightened, or in pain.

How many sweat glands are present in the body?

Sweat glands are present on all regions of the skin. There can be as many as 90 glands per square centimeter on the leg, 400 glands per cubic centimeter on the palms and soles, and an even greater number on the fingertips. Collectively, there are over two million sweat glands in the adult human body.

On a hot day, how much sweat is lost through the skin?

On a typical hot day it is possible to lose up to 7.4 quarts (7 liters) or more of body water in the form of sweat evaporating off the skin surface. Human bodies lose at least 1 pint (0.473 liters) of sweat every day, even when they are relatively inactive.

How do the eccrine sweat glands work?

The eccrine sweat glands are an important and efficient part of the mechanism that regulates body temperature. These glands are supplied with nerve endings that cause them to secrete sweat when the external temperature or body temperature is high.

83

When sweat evaporates off the skin surface, changing from a liquid to a gas, it carries large amounts of body heat with it. The average square inch (6.45 square centimeters) of skin contains 650 eccrine sweat glands.

Why is there an odor associated with perspiration—or sweat?

There are two types of sweat glands that cause perspiration. Both are odorless and sterile and produce two different types of sweat. The sweat produced by a person's armpits and groin are odorless until bacteria naturally found on the skin begin to interact with it, giving it a distinctive smell.

COMPARING OTHER ORGANISMS

What is an integumentary system in an animal?

The integumentary system (IS) in an animal is the skin and fur that cover the animal's body. Like humans, the skin protects the body from the surrounding environment and the internal organs. The animal's fur helps to keep the animal warm or cool, and again, helps to protect the animal from the surrounding environment. There are major differences in the integumentary systems of the numerous animals on our planet. For example, dogs and cats do not sweat through their skin like humans. Instead both animals sweat only through their footpads and nose, along with panting through their mouths to lose water rather than sweating.

How does skin color affect a cat's fur?

The temperature of a cat's skin does translate to the color of a cat's fur. This is more apparent in some breeds than others, such as the Siamese. In general, a cat's fur is lighter in color when the skin is warm and darker when the skin is cooler. This is why a cat's fur seems to be lighter during the warmer summer months and darker during colder winter months. It is also why many Siamese cats have light colored bodies with darker "points," as the skin is cooler at the body's extremities (ears, face, feet, and tail).

Are there any albino animals?

Yes, similar to humans, other animals can be albino. For example, Claude the albino alligator resides at the California Academy of Sciences; and in fiction, Herman Melville's *Moby Dick* portrays the white sperm whale of the same name.

It is not easy for albino animals in the wild, especially after birth and when young. The albino babies are easily seen by predators, and they can also become outcasts within their own species. In addition, many albino animals have health problems, such as bad vision, which, for some animals may mean the odds of survival are slim.

How does human hair compare to the hair of other primates?

In general, humans have as many hair follicles as gorillas, but the type of hair differs. Gorillas are covered with terminal hairs, while humans are mostly covered with vellus hairs.

What animal has almost human-like fingerprints?

Most primates (not including humans) have "fingerprints" that are more like ridges that are used for a better grip. But there is one animal that has fingerprints that are almost human-like: the cuddly koala bear. The fingerprints of these Australian creatures are sometimes difficult to distinguish from a human's, as they are similar in size, shape, and pattern. No other animals have such a distinction, but many animals do have other "prints." For example, animals such as pigs and dogs have hairless snouts and unique nose prints.

How are mammary glands related to the integumentary system in mammals?

Mammary glands are modified sweat glands located within the breasts of mammals. For example, in human females, each breast contains fifteen to twenty-five lobes, which are divided into lobules. Each lobule contains many alveoli, where milk is secreted and enters a milk duct, which leads to the nipple. Milk is produced only after childbirth.

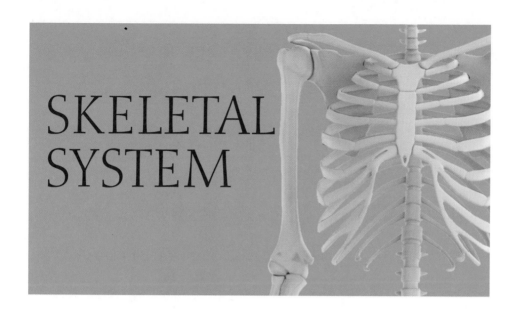

SKELETAL SYSTEM

INTRODUCTION

Who was the first person to study the internal structure of bone?

English scientist Clopton Havers (c. 1655–1702) was the first to study the internal structure of bone using a microscope. Havers's discoveries and observations included finding channels that extend along the shafts of the long bones of the arms and legs. These channels, which allow blood vessels to penetrate dense bone, were named the Haversian canals after their discoverer. He also described cartilage and synovia. In addition, he suggested that the periosteum, which surrounds bones, was sensitive to processes occurring within the bone. This observation was not confirmed for 250 years.

What are the functions of the skeletal system?

The skeletal system has both mechanical and physiological functions. Mechanical functions include support, protection, and movement. The bones of the skeletal system provide the rigid framework that supports the body. Bones also protect internal organs such as the brain, heart, lungs, and organs in the pelvic area. Muscles are anchored to bones and act as levers at the joints, allowing for movement. Physiological functions of the skeletal system include the production of blood cells and the supplying and storing of important minerals.

What are the major divisions of the human skeleton?

The human skeleton has two major divisions: the axial skeleton and the appendicular skeleton. The axial skeleton includes the bones of the center or axis of the body. The appendicular skeleton consists of the bones of the upper and lower extremities.

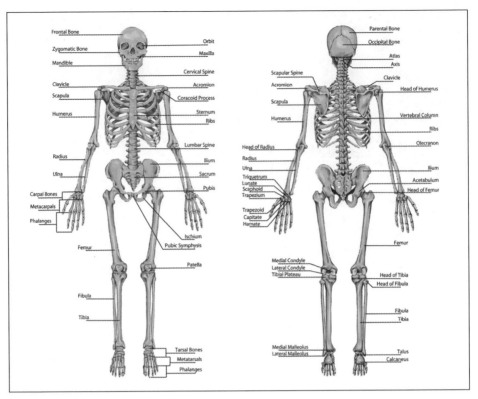

Front and back views of the skeleton

Why is calcium important to the body?

Bones consist mainly of calcium; thus, approximately 99 percent of all the calcium found in the body is stored in bones. Calcium plays an important role as a cofactor for enzyme function, in maintaining cell membranes, in muscle contraction, in nervous system functions, and in blood clotting. When the diet does not provide a sufficient amount of calcium, it is released from the bones, and when there is too much calcium in the body, it is stored in the bones.

How strong is bone?

Bone is one of the strongest materials found in nature. One cubic inch of bone can withstand loads of at least 19,000 pounds (8,626 kilograms), which is approximately the weight of five standard-size pickup trucks. This is roughly four times the strength of concrete. Bone's resistance to load is equal to that of aluminum and light steel. Ounce for ounce, bone is actually stronger than steel and reinforced concrete since steel bars of comparable size would weigh four or five times as much as bone.

Are there differences between the male and female skeletons?

Several general differences exist between the male and female skeletons. The male skeleton is generally larger and heavier than the female skeleton. The bones of the skull are generally more graceful and less angular in the female skeleton. A female also has a wider, shorter breastbone and slimmer wrists. There are significant differences between the pelvis of a female and a male, which are related to pregnancy and childbirth. The female pelvis is wider and shallower than the male pelvis. Females have an enlarged pelvic outlet and a wider, more circular pelvic inlet. The angle between the pubic bones is much sharper in males, resulting in a more circular, narrower, almost heart-shaped pelvis.

BONE BASICS

How many bones are in the human body?

Babies are born with about 300 to 350 bones, but many of these fuse together between birth and maturity to produce an average adult total of 206. Bone counts vary according to the method used to count them, because a structure may be treated as either multiple bones or as a single bone with multiple parts.

Location of Bones	Number
Skull	22
Ears (pair)	6
Vertebrae	26
Sternum	3
Ribs	24
Throat	1
Pectoral girdle	4
Arms (pair)	60
Hip bones	2
Legs (pair)	58
TOTAL	206

What are the major types of bones?

There are four major types of bones: long bones, short bones, flat bones, and irregular bones. The name of each type of bone reflects the shape of the bone. Furthermore, the shape of the bone is indicative of its mechanical function. Bones that do not fall into any of these categories are sesamoid bones and accessory bones.

What are the characteristics of long bones?

The length of a long bone is greater than its width. These bones act as levers that are pulled by contracting muscles. The lever action makes it possible for the body to move. Examples of long bones are the femur (thighbone) and humerus (upper arm bone). Some long bones, such as certain bones in the fingers and toes, are relatively short, but their overall length is still greater than their width.

What is the longest and smallest bone in the human body?

The femur, or thighbone, is the longest bone in the body. The average femur is 18 inches (45.72 centimeters) long. The longest bone ever recorded was 29.9 inches (75.95 centimeters) long. It was from an 8-foot-tall (2.45 meters) German who died in 1902 in Belgium. The stirrup (stapes) in the middle ear is the smallest bone in the body. It weighs about 0.0004 ounces (0.011 grams).

How do short bones compare in size to long bones?

The terms "long" and "short" are not descriptive of the length of the bones. Short bones have approximately the same dimensions in length, width, and thickness as long bones, but they may have an irregular shape. Short bones are almost completely covered with articular surfaces, where one bone moves against another in a joint. The only short bones in the body are the carpal bones in the wrists and tarsal bones in the ankle.

Are flat bones truly "flat"?

Flat bones are generally thin or curved rather than "flat." Examples of flat bones are the bones that form the cranium of the skull, the sternum or breastbone, the ribs, and scapulae (shoulder blades). The curved shape of most flat bones protects internal organs. The scapula, by contrast, is part of the pectoral (or shoulder) girdle.

What are the characteristics of irregular bones?

Irregular bones have complex, irregular shapes and do not fit into any other category of bone. Many irregular bones are short, flat, notched, or ridged, with extensions that protrude from their many bone parts. Examples of irregular bones are the spinal vertebrae, many bones of the face and skull, and the hip bones.

What is unique about sesamoid bones?

Shaped similarly to sesame seeds, sesamoid bones develop inside tendons that pass over a long bone. They are most commonly found in the knees (the patella or kneecap is a sesamoid bone), hands, and feet. Sesamoid bones may form in 26 different locations in the body. However, the number of sesamoid bones varies from individual to individual.

Where are blood cells formed in the skeletal system?

Hematopoiesis (from the Greek *hemato*, meaning "blood" and *poiein*, meaning "to make"), or red blood formation, occurs in the red bone marrow in adults. Adult red marrow, found in the proximal epiphysis (the ends) of the femur and humerus, some short bones, and in the vertebrae, sternum, ribs, hip bones, and cranium, is the site of production of all red blood cells (erythrocytes), platelets, and certain white blood cells.

What are some specialized bone cells?

The four major types of specialized cells in bone are osteogenic cells, osteoblasts, osteocytes, and osteoclasts. The following defines each type:

Osteogenic cells: From the Greek *osteo*, meaning "bone," and *genes*, meaning "born," these are cells that are capable of becoming bone-forming cells (osteoblasts) or bone-destroying cells (osteoclasts).

Osteoblasts: From the Greek *osteo* and *blastos*, meaning "bud or growth," they are the cells that form and build bone. Osteoblasts secrete collagen and other organic components needed to build bone tissue. As they surround themselves with matrix materials, they become trapped in their secretions and become osteocytes.

Osteocytes: From the Greek *osteo* and *cyte*, meaning "cell," osteocytes are the main cells in mature bone tissue.

Osteoclasts: From the Greek *osteo* and *klastes*, meaning "break," these cells are multinuclear, huge cells that are usually found where bone is reabsorbed.

How does compact bone tissue differ from spongy bone tissue?

Bone tissue is classified as compact or spongy according to the size and distribution of the open spaces in the bone tissue. Compact bone tissue is hard or dense with few open spaces. Compact bone tissue provides protection and support. Most long bones consist of compact bone tissue. In contrast, spongy bone tissue is porous with many open spaces. Spongy bone tissue consists of an irregular latticework of thin needle-like threads of bone called trabeculae (from the Latin *trabs*, meaning "beam"). Most flat, short, and irregular-shaped bones are made up of spongy bone tissue.

How do bones grow?

Bones form and develop through a process called ossification. There are two types of ossification: intramembranous ossification and endochondral ossification. Intramembranous ossification is the formation of bone directly on or within the fibrous connective tissue. Examples of bone formed through intramembranous ossification are the flat bones of the skull, mandible (lower jaw), and clavicle (collarbone).

Endochondral ossification, from the Greek *endo*, meaning "within," and *khondros*, meaning "cartilage," is the transformation of the cartilage model into bone. Cartilage cells in the epiphyseal plate grow and move into the metaphysis where they are reabsorbed and replaced by bone tissue. Examples of bone formed through endochondral ossification are the long bones, such as the femur and humerus.

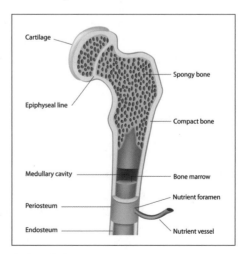

Cartilage
Spongy bone
Epiphyseal line
Compact bone
Medullary cavity
Bone marrow
Nutrient foramen
Periosteum
Endosteum
Nutrient vessel

Anatomy of a bone

What is the average age when bone is completely ossified?

There is variation among the different bones as to when the epiphyseal plates ossify and the bones fuse. The following table indicates the average age of ossification for different bones:

Bone	Chronological Age of Fusion (years)
Scapula	18–20
Clavicle	23–31
Bones of upper extremity	17–20
Os coxae	18–23
Bones of lower extremity	18–22
Vertebrae	25
Sacrum	23–25
Sternum (body)	23
Sternum (manubrium, xiphoid)	After 30

What is remodeling?

Remodeling is the ongoing replacement of old bone tissue by new bone tissue. In order to maintain homeostasis, bone must be replaced or renewed through the selective resorption of old bone and the simultaneous production of new bone. Approximately 5 to 10 percent of the skeleton is remodeled each year. Thus, every seven years the body grows the equivalent of an entirely new skeleton through remodeling.

How does exercise affect bone tissue?

Bone adapts to changing stresses and forces. When muscles increase and become more powerful due to exercise, the corresponding bones also become thicker and stronger through stimulation of osteoblasts. Regular exercise (especially weight-bearing exercise) maintains normal bone structure. Bones that are not subjected to normal stresses, such as an injured leg immobilized in a cast, quickly degenerate. It is estimated that unstressed bones lose up to a third of their mass after a few weeks. The adaptability of bones allows them to rebuild just as quickly when regular, normal weight-bearing activity is resumed.

How serious is osteoporosis?

Osteoporosis (from the Greek *osteo*, meaning "bone," *por*, meaning "passageway," and *osis*, meaning "condition") is a condition that reduces bone mass because the rate of bone resorption is quicker than the rate of bone deposition. The bones become very thin and porous and are easily broken. Osteoporosis is most common in the elderly, who may experience a greater number of broken bones as a result of the mechanical stresses of daily living and not from accidents or other trauma. Generally, osteoporosis is more severe in women, since their bones are thinner and less massive than men's bones. In addition, estrogen helps to maintain bone mass, so the loss of estro-

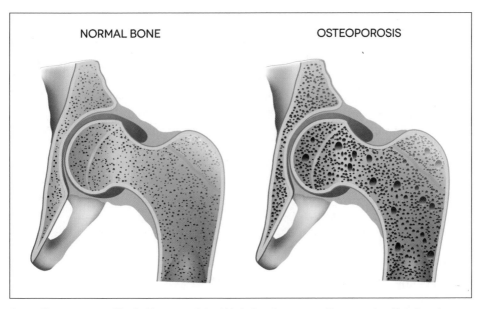

A normal bone versus one afflicted with osteoporosis in which the bone loses mass and becomes vulnerable to fracturing.

gen in women after menopause contributes to more severe osteoporosis. (For more about testing for osteoporosis, see the chapter "Helping Human Anatomy.")

How does osteoporosis differ from osteomalacia?

In osteomalacia the bones are weakened and softened from a loss of calcium and phosphorous. The volume of the bone matrix does not change. In osteoporosis, the volume of the bone matrix is reduced, leaving holes in the bones.

AXIAL SKELETON

How many bones are part of the axial skeleton?

The adult axial skeleton consists of eighty bones, including the bones of the skull, auditory ossicles (ear bones), hyoid bone, vertebral column, and the bones of the thorax (sternum and ribs). The following lists the structure and the number of bones in that structure:

Structure	Number of Bones
Skull	22
Auditory ossicles	6
Hyoid	1
Vertebral column	26
Thorax	25
TOTAL	80

What is TMD?

TMD stands for temporomandibular disorders, which occur when a person has problems with his or her jaw and facial muscles. It is associated with the TMJ, or the temporomandibular joint, a hinge that connects the jaw to the temporal bones of the human skull located at the front of each ear. This joint allows a person to move his or her jaw from side to side, and up and down to chew, talk, and open the mouth wide, as with yawning. The reason (or reasons) for TMD—often wrongly referred to as TMJ—is difficult to determine in most people. It is thought that problems with the jaw and muscles stem from teeth grinding, stress, arthritis of the joint, or misaligned teeth (or even dentures) that put unnecessary pressure on the joint.

What are the two sets of bones of the skull?

The skull consists of two sets of bones: the cranial and the facial bones. The cranial cavity that encloses and protects the brain consists of the eight cranial bones. There are fourteen facial bones that form the framework of the face. The facial bones also provide support and protect the entrances to the digestive and respiratory systems.

Which is the only bone that does not touch another bone?

The hyoid bone is the only bone that does not touch another bone. Located above the larynx, it supports the tongue and provides attachment sites for the muscles of the neck and pharynx used in speaking and swallowing. The hyoid is carefully examined when there is a suspicion of strangulation, since it is often fractured from such trauma.

Where is the vomer bone located?

The vomer bone (*vomer* is a Latin word meaning "plowshare") is part of the nasal septum, which divides the nose into left and right halves. A deviated septum, often occurring when the nasal septum has a bend in it, may cause chronic sinusitis and blockage of the smaller nasal cavity. The deviated septum can usually be corrected or improved by surgery.

Do all of the bones in the face and skull move?

Although the hyoid and ossicles of the ear move, the only bone that can be moved voluntarily and has the greatest range of movement is the mandible. The mandible, the U-shaped bone that forms the jaw and chin, can be raised, lowered, drawn back, pushed forward, and moved from side to side.

What are the functions of the paranasal sinuses?

There are four pairs of paranasal sinuses in the bones of the skull located near the nasal cavity. They are lined with mucus. One of their important functions is to act as resonating chambers to produce unique voice sounds. Since they are air-filled, they lighten the weight of the skull bones.

What are the functions of the vertebral column?

The vertebral column, known also as the spine or backbone, encloses and protects the spinal cord, supports the head, and serves as a point of attachment for the ribs and the muscles of the back. The vertebral column also provides support for the weight of the body, permits movement, and helps the body maintain an erect position.

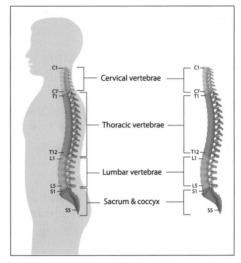

Major regions of the human vertebral column

How are the twenty-six bones of the vertebral column distributed?

The average length of the vertebral column is 28 inches (70 centimeters) in males and 24 inches (60 centimeters) in females. The following lists the five major regions of the vertebral column (cervical, thoracic, lumbar, sacral, and coccygeal) and their locations:

Region	Number of Bones	Location
Cervical	7	Neck region
Thoracic	12	Chest area
Lumbar	5	Small of the back
Sacral	1 fused bone in adults; 5 individual bones at birth	Below lumbar region
Coccygeal	1 fused bone in adults; 3–5 individual bones at birth	Below sacrum

What is the general structure of each vertebra?

Each vertebra (from the Latin *vertere*, meaning "something to turn on") consists of a vertebral body, a vertebral arch, and articular processes. The vertebral body is the thick, disc-shaped, front portion of the vertebra that is weight-bearing. The vertebral arch extends backwards from the body of the vertebra. Each vertebral arch has lateral walls called pedicles (from the Latin *pedicle*, meaning "little feet") and a roof formed by flat layers called laminae (from the Latin *lamina*, meaning "thin plate"). The spinal cord passes through the area between the vertebral arch and the vertebral body. Seven vertebral processes (bony projections) extend from the lamina of a vertebra. Some of the processes are attachment sites for muscles. The other four processes form joints with other vertebrae above or below. Intervertebral discs separate each vertebra.

What causes an intervertebral disc to herniate?

A herniated disc (from the Latin *hernia*, meaning "to bulge" or "stick out") occurs when the soft inner part of an intervertebral disc protrudes through a weakened or torn outer ring and pushes against a spinal nerve. A herniated disc may be the result of an injury or degeneration of the intervertebral joint. Although herniated discs may occur anywhere along the spine, they are most common in the lumbar or sacral regions.

← Normal Disc

← Degenerative Disc

← Bulging Disc

← Herniated Disc

← Thinning Disc

← Disc Degeneration with Osteophyte formation

Different types of spinal disc ailments

Which two cervical vertebrae allow the head to move?

The first two cervical vertebrae, C1 and C2, allow the head to move. The first cervical vertebra, the C1 or atlas, articulates with the occipital bone of the skull and makes it possible for a person to nod his or her head. The second cervical vertebra, C2, known as the axis, forms a pivot point for the atlas to move the skull in a side-to-side rotation.

Why is the first cervical vertebra, C1, called "the atlas"?

The first cervical vertebra is called "the atlas" after the Greek god Atlas, who was condemned to carry the earth and heavens on his shoulders. This vertebra has a ring-like structure with a large central opening and supports the head.

How does the shape of the spine change from birth to adulthood?

The spine of a newborn infant forms a continuous convex curve from top to bottom. At about three months of age, a concave curve develops in the cervical region as a

How long does it take for the "soft spots" on a baby's skull to disappear?

The "soft spots" of a baby's skull are areas of incompletely ossified bones called fontanels. The bones of the skull are connected by fibrous, pliable, connective tissue at birth. The flexibility of these connections allows the skull bones to move and overlap as the infant passes through the birth canal. The fontanels begin to close about two months after birth. The largest of the fontanels, the frontal fontanel located on the top of the skull, does not close until eighteen to twenty-four months of age.

baby learns to hold up his or her head. When the baby learns to stand during the second half of the first year of life, a concave curve appears in the lumbar region. The thoracic and sacral curves remain the same throughout the life of an individual. These are considered primary curves, since they are present in a fetus and retain their original shape.

What is a "dowager's hump"?

The so-called "dowager's hump" is caused by the collapse of thoracic vertebrae (the front edges), resulting in the abnormal curving of the spine. It is often seen in post-menopausal women with osteoporosis of the spine (although other research indicates it may also be caused by improper posture and skeletal alignment). It causes a person to stoop and creates a hump at the upper back.

What is the most common abnormal curvature of the spine?

Scoliosis (from the Greek *scolio*, meaning "crookedness") is the most common abnormal curvature of the spine and occurs in the thoracic region or lumbar region or both (thorocolumbar). Individuals with scoliosis have a lateral bending of the vertebral column. In the thoracic area the curves are usually convex to the right, and in the lumbar region curves are usually convex to the left. In most cases of scoliosis, the cause is unknown. Treatment options for scoliosis include observation, bracing, and surgery, depending on the person's age, how much more he or she is likely to grow, the degree and pattern of the curve, and the type of scoliosis.

What are the causes of kyphosis?

Kyphosis (from the Greek *kypho*, meaning "hunchbacked") is characterized by an exaggeration of the thoracic curve of the vertebral column. In adolescents, kyphosis is often the result of infection or other disturbances of the vertebral epiphysis during growth. In adults, kyphosis may be caused by a degeneration of the intervertebral discs, which results in collapse of the vertebrae. Poor posture may also lead to kyphosis.

Which of the curves of the spine is distorted in lordosis?

Lordosis (from the Greek *lord*, meaning "bent backward") is commonly known as "swayback." It is an exaggerated forward curvature of the spine in the lumbar region. The causes of lordosis include poor posture, rickets, tuberculosis of the spine, and obesity. It is not uncommon during the late stages of pregnancy.

One cause of lordosis is simply bad posture, but it can also result from diseases such as tuberculosis and rickets.

What is the thorax?

The thorax, a Greek word meaning "breastplate," is located at the chest. The sternum, twelve costal cartilages (strips of cartilage that attach ribs to the sternum; see explanation below), twelve pairs of ribs, and the twelve thoracic vertebrae form the thoracic skeleton or thoracic cage. This cage encloses and protects the heart, lungs, and some abdominal organs, and also supports the bones of the shoulder girdle and arm.

How many pairs of ribs does an individual have?

Most individuals have twelve pairs of ribs that form the sides of the thoracic cavity. Approximately five percent of the population is born with at least one extra rib, or about one in two hundred people. This extra rib, also known as a cervical rib, is considered a congenital condition.

How do true ribs differ from false ribs and floating ribs?

The upper seven pairs of ribs are true ribs. These ribs attach directly to the sternum by a strip of hyaline cartilage called costal cartilage. The lower five pairs of ribs are known as false ribs because they either attach indirectly to the sternum or do not attach to the sternum at all. The eighth, ninth, and tenth pairs of ribs attach to each other and then to the cartilage of the seventh pair of ribs. The eleventh and twelfth pairs of ribs are floating ribs because they only attach to the vertebral column and do not attach to the sternum at all.

What is the most frequently fractured bone in the body?

Due to its vulnerable position and its relative thinness, the clavicle is the most frequently fractured bone in the body. Fractured clavicles are caused by either a direct blow or a transmitted force resulting from a fall on the outstretched arm.

APPENDICULAR SKELETON

How many bones are part of the appendicular skeleton?

The adult appendicular skeleton consists of 126 bones. It is composed of the bones of the upper and lower extremities, including the pectoral (shoulder) and pelvic girdles, which attach the upper and lower appendages to the axial skeleton.

Structure	Number of Bones
Pectoral (shoulder) girdles	
Clavicle	2
Scapula	2
Upper extremities	
Humerus	2
Ulna	2

Structure	Number of Bones
Radius	2
Carpals	16
Metacarpals	10
Phalanges	28
Pelvic (hip) girdles	2
Lower extremities	
Femur	2
Fibula	2
Tibia	2
Patella	2
Tarsals	14
Metatarsals	10
Phalanges	28
TOTAL	126

How does the anatomical usage of the word "arm" differ from the common usage?

Anatomically, the word "arm" refers only to the humerus, the long bone between the shoulder and the elbow. Common usage of the word "arm" refers to the entire length of the limb from the shoulder to the wrist.

What is the function of the pectoral girdle?

The pectoral (or shoulder) girdle consists of two bones: the clavicle and scapula. The clavicle articulates with the manubrium of the sternum, providing the only direct connection between the pectoral girdle and the axial skeleton. There is no connection to the vertebral column, allowing for a wide range of movement of the shoulder girdle.

Where is the so-called "funny bone" located?

The "funny bone" is not a bone but rather a part of the ulnar nerve located at the back of the elbow, with the nerve essentially sitting on top of the hard elbow. Because it is the largest unprotected nerve in the body—by cartilage, bone, or muscle—a bump or blow to this area can cause a tingling sensation or produce a temporary numbness and paralysis of muscles on the anterior (outside) surface of the forearm. In other words, a person feels tingling from the elbow, down the arm, and to the pinkie and ring fingers

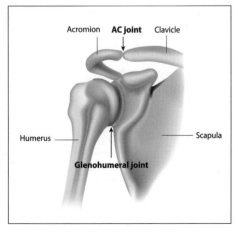

Anatomy of the shoulder

after hitting the funny bone. No one knows where the term "funny bone" originated. But some people speculate the term is associated with the long upper arm bone, the humerus, which runs from the shoulder to the elbow. This was confused with the word "humorous" and could have eventually led to the term "funny bone."

Which structures of the body have more bones for their size than any other part of the body?

The wrist and the hand have more bones in them for their size than any other part of the body. There are eight carpals in the wrist between the forearm and the palm of the hand; five metacarpal bones that form the palm of the hand between the wrist, thumb, and fingers; and fourteen phalanges or finger bones. The presence of many small bones in the wrist and hand with the many movable joints between them makes the human hand highly maneuverable and mobile.

What are the functions of the pelvic girdle?

The pelvic girdle consists of the two hip bones, also called the coxal bones or ossa coxae. The pelvic girdle provides strong, stable support for the vertebral column, protects the organs of the pelvis, and provides a site for the lower limbs to attach to the axial skeleton.

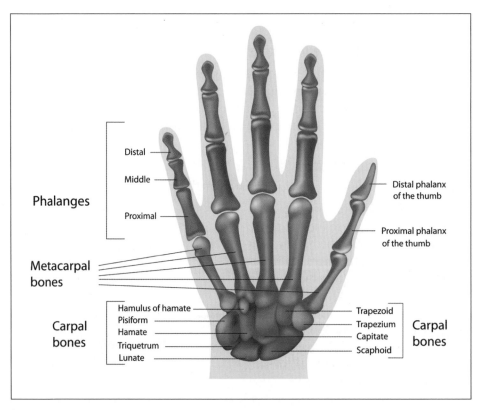

The bones of the human hand and wrist

> **How does wearing high-heeled shoes
> upset the distribution of the body's weight on the feet?**
>
> High-heeled shoes transfer the balanced distribution of the weight of the body from between the calcaneus (the heel bone) and the metatarsals (bones in the ball of the foot) to just the metatarsals. As a result, the arches of the foot do not absorb the force of the body's weight, which may lead to injuries of the soft tissue structures, joints, and bones.

Which is the broadest bone in the body?

The hip bones are the broadest in the body. The hip bones are originally three separate bones in infants: the ilium, ischium, and pubis. These bones fuse together by age twenty-three. The hip bones are united to each other in the front at the pubic symphysis joint and in the back by the sacrum and coccyx.

How does the pelvic brim divide the pelvis?

The pelvic brim divides the pelvis (from the Latin word, meaning "basin") into the false or greater pelvis and the true or lesser pelvis. The part of the pelvis above the pelvic brim is the false pelvis. It does not contain any pelvic organs, except for the urinary bladder, when it is full, and the uterus during pregnancy. The part of the pelvis below the pelvic brim is the true pelvis. The pelvic inlet is the upper opening of the true pelvis, and the pelvic outlet is the lower opening of the true pelvis.

What is the patella?

The patella (from the Latin word, meaning "little plate"), also known as the kneecap, is a small, triangular-shaped, sesamoid bone located within the tendon of the quadriceps femoris, a group of muscles that straighten the knee. The patella protects the knee and gives leverage to the muscles. One of the most common running-related injuries, runner's knee, is essentially an irritation of the cartilage of the kneecap. Normal tracking (gliding) of the kneecap does not occur, since the patella tracks laterally instead of up and down. The increased pressure of abnormal tracking causes pain in this injury.

How is the shape of the foot efficient for supporting weight?

One of the most efficient types of construction for supporting weight is the arch. The bones of the foot are arranged in two arches to support the weight of the body. The longitudinal arch has two parts, medial and lateral, and extends from the front to the back of the foot. The transverse arch extends across the ball of the foot.

SKELETAL SYSTEM

101

JOINTS

What is a joint?

A joint is the place where two adjacent bones, adjacent cartilages, or adjacent bones and cartilages meet. Joints are also called articulations (from the Latin *articulus*, meaning "small joint"). Some joints are very flexible, allowing movement, while others are strong, providing protection of the internal tissues and organs, but do not permit movement.

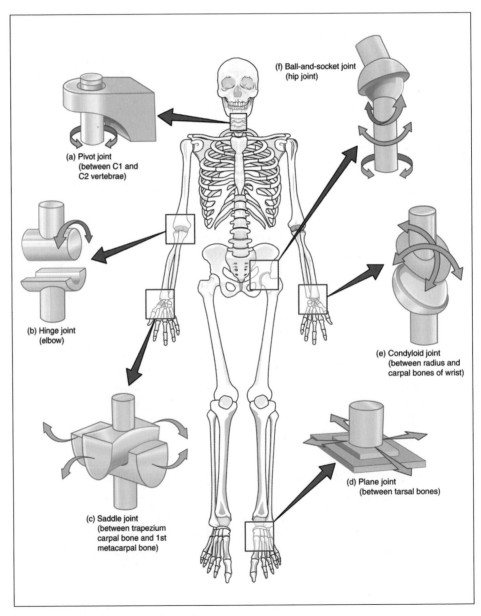

(f) Ball-and-socket joint
(hip joint)

(a) Pivot joint
(between C1 and
C2 vertebrae)

(b) Hinge joint
(elbow)

(e) Condyloid joint
(between radius and
carpal bones of wrist)

(c) Saddle joint
(between trapezium
carpal bone and 1st
metacarpal bone)

(d) Plane joint
(between tarsal bones)

The six types of synovial joints, along with examples

How are joints classified?

There are two classification methods to categorize joints. The structural classification method is based only on the anatomical characteristics of the joint. The functional classification method is based on the type and degree of movement allowed by the joint. Joints are classified both structurally and functionally.

What are the structural classes of joints?

The two main criteria for the structural classification of joints are the presence or absence of a cavity known as the synovial cavity and the type of tissue that binds the bones together. The three types of structural categories are fibrous, cartilaginous, and synovial.

What are the functional classes of joints?

The functional classification of joints is determined by the degree and range of movement the joint allows. The three functional categories for joints are synarthrosis, amphiarthrosis, and diarthrosis. A synarthrosis joint (from the Greek *syn*, meaning "together," and *arthrosis*, meaning "articulation") is immovable. An amphiarthrosis joint (from the Greek *amphi*, meaning "on both sides") is slightly movable. A diarthrosis joint (from the Greek *dia*, meaning "between") is a freely movable joint. The following shows the general classification of joints based on function:

Functional Category	Structural Category	Example
Synarthrosis (immovable joints)	*Fibrous* Suture	Between bones of adult skull
	Gomphosis	Between teeth and jaw
	Cartilaginous Synchondrosis	Epiphyseal cartilages
Amphiarthrosis (little movement)	*Fibrous* Syndesmosis	Between the tibia and fibula
	Cartilaginous Symphysis	Between right and left pubic bones of pelvis; and between adjacent vertebral bodies along vertebral column
Diarthrosis (free movement)	*Synovial*	Elbow, ankle, ribs, wrist, shoulder, hip

What are the three types of fibrous joints?

The fibrous joints are mostly immovable. The three types of fibrous joints are sutures, syndesmoses, and gomphoses. Sutures provide protection for the brain and are only found in the adult skull. They are immovable joints. A syndesmosis joint is a joint where the bones do not touch each other and are held together by fibrous connective tissue. One example of a syndesmosis joint is the distal articulation between the tibia and fibula. A gomphosis joint (from the Greek *gomphos*, meaning "bolt") is composed of a peg and socket. The only gomphosis joints in the human body are the teeth. The roots of the teeth articulate with the sockets of the alveolar processes of the maxillae and mandible.

How do a woman's pubic bones change during pregnancy?

The symphysis pubis, a cartilaginous joint between the two pubic bones, is somewhat relaxed during pregnancy. This allows the mother's hipbones to move in order to accommodate the growing fetus.

Which is the most common type of joint found in the body?

Synovial joints are the most common type of joint in the body, permitting the greatest range of movement. Joints of the hip, shoulder, elbow, ankle, and knee are all examples of synovial joints. The basic structure of a synovial joint consists of a synovial cavity, articular cartilage, a fibrous articular capsule, and ligaments. The synovial cavity (also called joint cavity) is the space between two articulating bones. The articular cartilage covers and protects the bone ends. The articular cartilage also acts as a shock absorber. The articular capsule encloses the joint structure. It consists of an outer layer, the fibrous membrane, and an inner lining, the synovial membrane. Ligaments are fibrous thickenings of the articular capsule that help provide stability.

What is synovial fluid?

Synovial fluid (from the Greek *syn*, meaning "together," and *ovum*, meaning "egg") is a lubricating fluid in joint cavities. It is a thick fluid with a consistency similar to an egg white that reduces the friction between the cartilages of synovial joints as they move.

How many different types of synovial joints are found in the body?

There are six different types of synovial joints, which are classified based on the shape of their articulating surfaces and the types of joint movements those shapes permit. The types of synovial joints are gliding or planar joints, hinge joints, pivot joints, condyloid or ellipsoidal joints, saddle joints, and ball-and-socket joints.

What are some examples of synovial joints and their movements?

There are numerous ways in which synovial joints move. The four basic categories of movements are gliding, angular movements, rotation, and special movements. Each of these groups is defined by the form of motion, the direction of movement, or the relationship of one body part to another during movement. The following lists the type of joint and movement, along with examples of each:

Type of Joint	Type of Movement	Example
Planar (Plane)	Gliding	Joints between carpals and tarsals
Hinge	Flexion and extension	Elbow, knee, and ankle joints

Why is there a "popping" sound when you crack your knuckles, and is it dangerous to crack them?

A number of reasons have been given for the characteristic "popping" sound associated with someone cracking his or her knuckles. One reason is that when a joint is contracted, small ligaments or muscles may pull tight and snap across the bony protuberances of the joint. Another possibility is that when the joint is pulled apart, air can pop out from between the bones, creating a vacuum that produces a popping sound. A third reason, discovered by British scientists in 1971, is that when the pressure of the synovial fluid is reduced by the slow articulation of a joint, tiny gas bubbles in the fluid may burst, producing the popping sound.

Research has not shown any connection between knuckle cracking and arthritis. One study found that knuckle cracking may be the cause of soft tissue damage to the joint capsule and a decrease in grip strength. The rapid, repeated stretching of the ligaments surrounding the joint is most likely the cause of damage to the soft tissue. Some researchers believe that since the bones of the hand are not fully ossified until approximately age eighteen, children and teenagers who crack their knuckles may deform and enlarge the knuckle bones. However, most researchers believe knuckle cracking does not cause serious joint damage.

Type of Joint	Type of Movement	Example
Pivot	Rotation	Atlantoaxial joint (between first and second vertebrae)
Condyloid	Abduction and adduction	Wrist joint
Saddle	Flexion, extension, metacarpal abduction, adduction	Carpometacarpal joint (between bone of thumb and carpal bone of wrist), circumduction
Ball-and-Socket	Rotation, abduction, adduction, circumduction	Shoulder and hip joints

Which type of joint is most easily dislocated?

The ball-and-socket joint is most susceptible to dislocation. Joint dislocation, or luxation (from the Latin *luxare*, meaning "to put out of joint"), occurs when the there is a drastic movement of two bone ends out of their normal end-to-end position. The most frequently dislocated joint is the shoulder joint. Since the socket is shallow and the joint is loose, allowing for the tremendous amount of mobility, there is also the greatest possibility of dislocation.

Which joint is the largest and most complex joint in the body?

The knee joint (tibiofemoral joint) is the most complex in the human body. It is comprised of three different joints: the medial femoral and medial tibial condyles, the lateral femoral and tibial condyles, and the articulation between the patella and the femur. The knee joint is capable of flexion, extension, and medial and lateral rotation to a certain degree. It is also the joint most vulnerable and susceptible to injury. Common knee injuries are tears to the anterior cruciate ligaments (ACL) and tears to the meniscus or cartilage.

What is arthritis?

Arthritis (from the Greek *arthro*, meaning "joint," and *itis*, meaning "inflammation") is a group of diseases that affects synovial joints. Arthritis may originate from an infection, an injury, metabolic problems, or autoimmune disorders. All types of arthritis involve damage to the articular cartilage. Two major categories of arthritis are degenerative diseases and inflammatory diseases.

Which is the most common type of arthritis?

The most common type of arthritis is osteoarthritis. Osteoarthritis is a chronic, degenerative disease most often beginning as part of the aging process. Often referred to as "wear and tear" arthritis because it is the result of life's everyday activities, it is a degradation of the articular cartilage that protects the bones as they move at a joint site. Osteoarthritis usually affects the larger, weight-bearing joints first, such as the hips, knees, and lumbar region of the vertebral column.

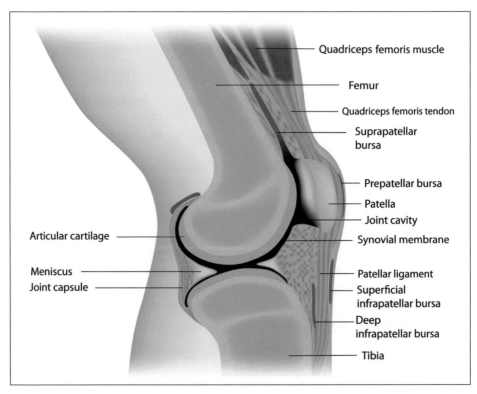

Anatomy of the knee

How does rheumatoid arthritis differ from osteoarthritis?

Rheumatoid arthritis is an inflammatory disease mainly characterized by inflammation of the synovial membrane of the joints. The disease may begin with general symptoms of malaise, such as fatigue, low-grade fever, and anemia, before affecting the joints. Unlike osteoarthritis, rheumatoid arthritis usually affects the small joints first, such as those in the fingers, hands, and feet. In the first stage of the disease there is swelling of the synovial lining, causing pain, warmth, stiffness, redness, and swelling around the joint. This is followed by the rapid division and growth of cells, or pannus, which causes the synovium to thicken. In the next stage of the disease, the inflamed cells release enzymes that may digest bone and cartilage, often causing the involved joint to lose its shape and alignment and leading to more pain and loss of movement. This condition is known as fibrous ankylosis (from the Greek *ankulos*, meaning "bent"). In the final stage of the disease, the fibrous tissue may become calcified and form a solid fusion of bone, making the joint completely nonfunctional (bony ankylosis). (For more about osteoporosis, see this chapter.)

COMPARING OTHER ORGANISMS

How do the number of bones in humans compare to some common animals?

Although they represent only a fraction of the animals on the planet, animals called vertebrates include fish, frogs, reptiles, birds, and mammals. The name is derived from these creatures having a spine (the bony vertebrae) that supports the body and head. Overall, all vertebrates except fish have the same type of skeletons: four limbs with the same basic patterns; a skull that houses and protects the brain; sensory organs; and a rib cage that protects the heart and lungs (in mammals it helps make breathing possible).

What are some of the largest animal bones ever discovered?

To date, the long-extinct dinosaurs seem to be the winners of the category "largest animal bones ever discovered." In 2014, one of the largest thigh bones ever found—around 130 feet (40 meters) long and 65 feet (20 meters) tall—was uncovered in Patagonia, Argentina. The Dreadnoughtus, thought to be a new species of the herbivore titanosaur, is estimated to have weighed 77 tons or heavy as fourteen African elephants, and lived at the end of the Cretaceous period (which ended about 65 million years ago).

What are dinosaurs?

"Dinosaur" comes from the term "dinosauria," which is a combination of the Greek words *deinos* and *sauros*. It means "terrible reptiles" or "terrible lizards." The term was invented by the well-known British anatomist Sir Richard Owen (1804–1892). He coined the term in 1842 to describe the 175-million-year-old fossil remains of two groups of giant reptiles that corresponded to no known living creatures. In 1854, Owen prepared one of the first dinosaur exhibitions for display at the Crystal Palace, a famous museum in London, England.

Are there animals without internal bones?

Yes, there are plenty of animals on Earth without internal bones and are referred to as invertebrates. In fact, more than 98 percent of all animal species are considered to be invertebrates. They include insects, including those with hard outer shells, and those that lack a skeleton or any "hard parts," such as a jellyfish or earthworm. The most common invertebrates include the echinoderms, mollusks, annelids, protozoa, and arthropods, which include insects, crustaceans, and arachnids (spiders).

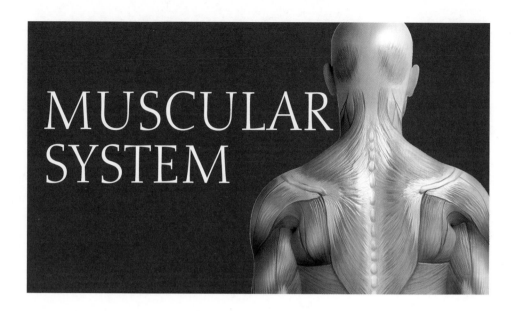

MUSCULAR SYSTEM

INTRODUCTION

What is myology?

Myology is the study of the structure and function of muscles. It also entails the diseases of the muscles, such as inflammation and genetic disorders.

What are the functions of the muscular system?

The major functions of the muscular system are as follows:

1. Body movement due to the contraction of skeletal muscles

2. Maintenance of posture also due to skeletal muscles

3. Respiration due to movements of the muscles of the thorax

4. Production of body heat, which is necessary for the maintenance of body temperature, as a byproduct of muscle contraction

5. Communication, such as speaking and writing, which involve skeletal muscles

6. Constriction of organs and vessels, especially smoother muscles that can move solids and liquids in the digestive tract and other secretions, including urine, from organs

7. Heartbeat caused by the contraction of cardiac muscle that propels blood to all parts of the body

How many muscles are in the human body?

There are about 650 muscles in the body, although some authorities believe there are as many as 850 muscles. No exact figure is available because experts disagree about which are separate muscles and which ones branch off larger ones. Also, there is some variability from one person to another, though the general musculature remains the same.

How important and prominent are muscle cells in the body?

Muscle cells are found in every organ and tissue in the body and participate in every activity that requires some type of movement. Together, all muscles comprise almost 50 percent of body mass. Nearly 40 percent of body weight in males and almost 32 percent in females is skeletal muscle.

If a man weighs 170 pounds (77 kilograms), how many pounds of muscle does he have?

A man who weighs 170 pounds (77 kilograms) has about 81 pounds (37 kilograms) of muscle distributed as 68 pounds (31 kilograms) of skeletal muscle and 13 pounds (6 kilograms) of cardiac and smooth muscle. A woman who weighs 120 pounds (54 kilograms) has a total muscle weight of 45 pounds (20 kilograms).

What is the difference between voluntary and involuntary muscle movements?

Muscle movements that an individual consciously controls are referred to as voluntary. Some examples of voluntary muscle movements would be when an individual walks or picks up an object. Involuntary muscle movements are those that occur without an individual's conscious control. An example of involuntary muscle movement is the pumping action of the heart.

How does strength training affect muscles differently than aerobic training?

Strength training, such as weight lifting, involves doing exercises that strengthen specific muscles by providing some type of resistance that makes them work harder. It builds more myofibrils (contractile fibers within muscle cells) and increases the size of individual muscle cells (hypertrophy). Strength training builds muscle mass and strength, but it does not increase the number of muscle cells (hyperplasia). In

---

Does weightlessness in outer space affect skeletal and cardiac muscles?

The effect of weightlessness on humans results in a loss of muscle strength and volume. Similar to bone deterioration, skeletal muscles atrophy as a result of disuse. In space, actions and movement require considerably less exertion because the force of gravity is practically nonexistent. As a result, astronauts' skeletal muscles become deconditioned. The effects of weightlessness on cardiac (heart) muscles resemble what happens to skeletal muscles in that a weakening can result under such conditions. Just as an athlete will strengthen his or her heart muscles through exercise and make them more efficient as a pump, any reduction in demand will lessen this efficiency.

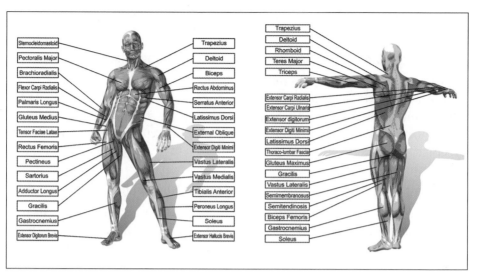

The major muscles of the body

general, the heavier the weight used, the more visible the increase in muscle size. Whereas resistance training strengthens muscles, aerobic training builds endurance. Aerobic training, such as jogging, running, biking, and swimming, involves exercise in which the body increases its oxygen intake to meet the increased demand for oxygen by the muscles. With aerobic training, the number of blood capillaries supplying muscles increases. In addition, the numbers of mitochondria and the amount of myoglobin available to store oxygen both increase.

How quickly does muscle strength increase?

Some muscles gain strength faster than others. In general, large muscles, such as those present in the chest and back, grow faster than smaller ones, including those in the arms and shoulders. Most people can increase their strength between 7 and 40 percent after ten weeks of training each muscle group at least twice a week.

Do creatine supplements improve muscular performance?

Creatine phosphate (CP) is a molecule stored in muscle that yields energy when the creatine splits from the attached phosphate. This energy is used to resynthesize the small amount of ATP (adenosine triphosphate) that is available to the muscle in the initial seconds of high intensity work (think a one hundred-yard dash or a power lift). Because greater amounts of CP in the muscle can potentially allow for those high intensity efforts to be sustained a bit longer or to be performed more effectively, creatine supplementation has become popular within the last fifteen years. Some research indicates that such supplementation can improve performance in the short term and in high intensity activities, but for more sustained activities it has little or no effect because of the ATP's great dependence on aerobic metabolism. The long-term effect of such supplementation on the human body is unknown. (For more about ATP, see this chapter and the chapter "Levels of Organization.")

ORGANIZATION OF MUSCLES

Which are the largest and smallest muscles in the human body?

The largest muscle is the gluteus maximus (buttock muscle), which moves the thigh-bone away from the body and straightens out the hip joint. It is also one of the stronger muscles in the body. The smallest muscle is the stapedius in the middle ear. It is thinner than a thread and 0.05 inches (0.127 centimeters) in length. It activates the stirrup that sends vibrations from the eardrum to the inner ear.

What are the longest and fastest muscles in the human body?

The longest muscle is the sartorius, which runs from the waist to the knee. Its purpose is to flex the hip and knee. The extraocular muscles, which allow you to move your eye, and the laryngeal muscles associated with vocal cords (or folds) are the fastest contracting muscles in the body.

What muscles act on the skin around the eyes and eyebrows?

The occipitofrontalis muscles raise the eyebrows. The orbicularis oculi muscles close the eyelids; they are also responsible for the "crow's feet" wrinkles in the skin at the lateral corners of the eyes.

How many muscles are involved in the movement of each eyeball?

Six skeletal muscles called the extrinsic eye muscles move the eyeball. They include the superior, inferior, medial, and lateral rectus muscles, as well as the superior and inferior oblique muscles.

What muscles are associated with the lips and the area surrounding the mouth?

The orbicularis oris and buccinator, the kissing muscles, pucker the mouth. The buc-cinator also flattens the cheeks, as when one whistles or blows a trumpet, and is

---

How many muscles does it take to produce a smile and a frown?

Not everyone agrees on how many muscles it takes to smile and frown. Some research points to seventeen muscles used in smiling while the average frown uses forty-three muscles; another says ten to smile, six to frown; and still another study says eleven to smile and twelve to frown. All of this is further confused by other variables. For example, some studies take into (or do not take into) consideration variables such as energy used by the muscle, or the variability in each person's facial muscles. Some people also have more developed facial muscles than others. But it appears that most researchers agree that, although the number of muscles is a matter of interpretation, smiling uses more muscles but takes less effort than frowning.

sometimes called the trumpeter's muscle. Smiling is accomplished primarily by the zygomaticus muscles. Sneering is accomplished by the leavator labii superioris, and frowning or pouting is largely caused by the depressor anguli oris.

How many muscles are involved in chewing food?

In humans, there are four pairs of muscles involved in chewing (mastication) and are some of the strongest muscles of the body. The temporalis (triangular-shaped muscles at the temples on the sides of the skull) raises the jaw; the masseter closes the mouth. There are also two pterygoids: the lateral pterygoid actively opens the jaw and the medial pterygoid elevates and closes the jaw. All four of these muscles are not only responsible for chewing, but also for moving the jaw from side to side, helping a person to speak, and grinding the teeth.

How many muscles are involved in swallowing?

There are many muscles involved as a human swallows—from taking in and chewing food or swallowing a liquid, to moving the food or liquid to the stomach. Overall, there are about fifty pairs of muscles (along with many nerves to help the muscles) that contribute to human consumption. In general, there are three stages of swallowing that involve special muscles. The first stage moves the tongue and jaw so the food can be taken in and chewed in the mouth. The second stage occurs when the tongue pushes the food or liquid to the back of the mouth, which triggers the swallowing response (and is also where the gag response is located in most people). At this stage, the larynx closes and the body stops breathing briefly to prevent the food or liquid from being taken into the lungs. The last stage is when the food or liquid enters the esophagus, or the long tube that carries the food or liquid to the stomach. At this stage, too, the muscles in the esophagus contract, pushing the food down the tube (this rhythmic contraction is called peristalsis). Depending on the food, it can

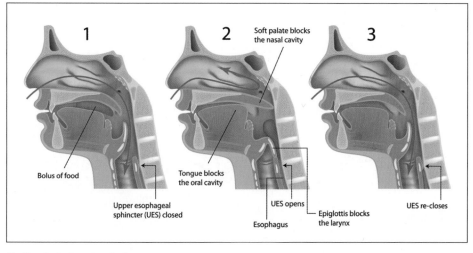

The three basic steps of swallowing

take about three seconds to move the food down the esophagus or longer, for example, if the food is dry or the person is swallowing a pill. (For more about swallowing, see the chapter "Digestive System.")

How many muscles are in the human ear?

There are six muscles in the human ear, but they do not have the same function as other animals. For most animals, such as monkeys, ear muscles allow the animal to move their ears in order to pick up sounds from various directions. But in humans, they do not move, suggesting that the ear muscles have lost their biological function (with the exception of people who can wiggle their ears, but such movements do not enhance a person's hearing ability).

What is the function of the corrugator muscle?

Located on the forehead, the corrugator is the muscle that contracts the forehead into wrinkles and pulls the eyebrows together. The small, narrow, pyramidal muscle is close to the eye and is one of the most important human expressive muscles.

What is the triangle of auscultation?

The triangle of auscultation is a small area of the back where three muscles (trapezius, latissimus dorsi, and rhomboideus major) converge. This area is near the scapula and becomes enlarged when a person leans forward with arms folded across the chest. When a physician places a stethoscope on the triangle of auscultation, the sounds of the respiratory organs can be clearly heard.

What are the hamstring muscles?

There are three hamstring muscles, which are located at the back of the thigh. They flex the leg on the thigh, such as when one kneels. They also extend the hip whenever one, for example, sits in a chair. Hamstring injuries are probably the most common muscle injury among runners. Maintaining flexibility and strengthening the muscle help to prevent injury. Hamstring muscles are also prone to reinjury.

What are some examples of muscles named for size or shape?

Early anatomists often named a muscle based on its size, including length. If a muscle was long, its name would likely include the Latin term *longus*. Muscles that were large would have the term *maximus* (Latin for "largest" or "greatest"), such as *glu-*

Which muscle is the most variable among humans?

The platysma muscle in the side of the neck is probably the most variable. It can cover the whole region in some people, while in others it is straplike. In a few cases it is missing completely.

teus maximus. Other terms related to size include *brevis* (Latin for "short"), *major* (Latin for "larger"), and *minor* (Latin for "smaller"). Some muscles whose names are based on shape include the deltoid (a shoulder muscle that gets its name from the Greek *delt*, meaning "triangle" and *oid*, meaning "like") and the trapezeius (another shoulder muscle whose name derives from the Greek word *trapez*, meaning "table").

What are some muscles named as a result of the action produced by the muscle?

The names of many muscles also include the type of movement or action that they bring about. The flexors of the wrist and fingers are examples of flexion (from the Latin *flex*, meaning "to bend"), that is, decreasing the angle at a joint. The ad-

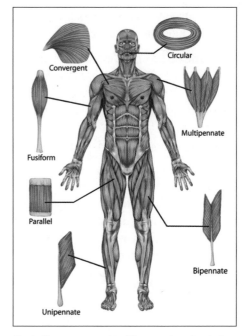

Muscle fibers are arranged in different configurations in the body to support specific movements.

ductors of the thigh pull the limb toward the midline. This type of movement is called adduction (from the Latin *ad*, meaning "to" or "toward" and *duct*, meaning "lead").

How many muscles are in the fingers and thumbs?

There are no muscles in the fingers or thumbs. Even though these structures are among the most frequently used parts of the body, they consist only of tendons. The muscles that control these tendons are located in the hand and forearm.

How are some muscles named for the location of the muscle's origin and insertion on bones?

The first part of the name of some muscles indicates the origin, while the second part indicates the insertion. For example, the muscle that has its origin on the breast bone and clavicle (collarbone) and that inserts on a breast-shaped process of the skull is termed the sternocleidomastoid (from the Greek words *sterno*, meaning "breast bone," *cleido*, meaning "clavicle," and *mastoid*, meaning "breast shape.")

How are some muscles named after the direction of muscle fibers?

When looking at a muscle, one can often see that it appears to have lines running within it. These lines are made of muscle fibers, and the direction that these fibers run in relation to the midline of the body is often used to provide partial names to different muscles. If the fibers of the muscle are running with or parallel to the body's midline, the term *rectus* (from the Latin word for "straight") is often used to describe that muscle. Some examples of muscles that have the term rectus in their name include the rectus femoris and rectus abdominis.

If the fibers of the muscle run at an angle to the body's midline, they are said to run obliquely. The term *oblique* is also of Latin origin. Some examples of muscle that have the term oblique associated with their name include the internal and external oblique muscle of the abdominal area.

Are there other categories of how muscles are named?

There are several other categories of muscle names. One of these is based on the location of the muscle attachment and its association with a particular bone. For example, the temporalis muscle is found covering the temporal bone, while the frontalis muscle is found covering the frontal bone of the skull. Another category deals with the number of origins. Some muscles have multiple origins and the number of origins is often used in the muscle's name. An example is the biceps brachii muscle, which has two heads that attach to two different origins. A final category of muscle names deals with the relation of the muscle to the bone. Not only is a muscle sometimes named because of the bone to which it attached, but the name may be even more detailed to describe where its position is in relation to the bone or body part. Some of the terms and prefixes that describe position are: supra (above or over), infra (below or beneath), medialis (middle), external (outer), and inferior (underneath). The infraspinatus (an arm muscle connected to the humerus bone) is one example of muscles in this category.

MUSCLE STRUCTURE

What are the three types of muscle tissue and their characteristics?

The three types of muscle tissue are skeletal, cardiac, and smooth. The main and most unique characteristic of muscle tissue is its ability to contract (shorten), making some type of movement possible. The following lists the characteristics of the muscle tissues:

Characteristics	Skeletal Muscle	Smooth Muscle	Cardiac Muscle
Location	Attached to tendons that attach to bones	Found in the walls of blood vessels and in the walls of organs of the digestive, respiratory, urinary, and reproductive tracts and the iris of the eye	Found only in the heart
Function	Movement of the body, maintenance of posture	Control of blood vessel contents in hollow organs	Pumping of blood
Cellular characteristics			
Cell shape	Long, cylindrical	Spindle-shaped	Branched

Characteristics	Skeletal Muscle	Smooth Muscle	Cardiac Muscle
Striations	Present	Absent	Present
Nucleus	Many nuclei	Single nucleus	Single nucleus
Special features	Well-developed transverse tubule system	Lacks transverse tubules	Well-developed transverse tubule system, intercalated discs, separate adjacent cells
Mode of control	Voluntary	Involuntary	Involuntary
Initiation of contraction	Only by a nerve cell	Some contraction always maintained, modifiable by nerves	Autorhythmic (pacemaker cells), modifiable by nerves
Speed of contraction	Fast (0.05 seconds)	Slow (1 to 3 seconds)	Moderate (0.15 seconds)
Sustainability of contraction	Not sustainable	Sustainable indefinitely	Not sustainable
Likelihood of fatigue	Varies widely depending on type of skeletal muscle fiber and work load	Low; relaxation between contractions reduces the likelihood of fatigue	Generally does not fatigue

What are the four major characteristics of skeletal muscle?

The four major functional characteristics of skeletal muscle are as follows:

Contractility—The ability to shorten, which causes movement of the structures to which the muscles are attached.

Excitability—The ability to respond or contract in response to chemical and/or electrical signals.

Extensibility—The capacity to stretch to the normal resting length after contracting.

Elasticity—The ability to return to the original resting length after a muscle has been stretched.

Do cardiac muscle cells continue to divide throughout a person's life?

The vast majority of heart muscle cells are thought to stop dividing by the time a person reaches the age of nine. These cells then pump blood for the rest of a healthy person's life. In people stricken by a heart attack, the cardiac muscle cells die and are replaced by scar tissue.

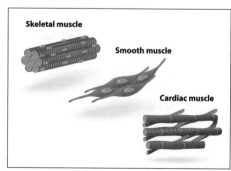

The three types of muscle tissue

What is the basic contractile unit of a muscle cell?

The basic structural and functional unit of a muscle cell is the sarcomere, which consists of thin filaments of the protein actin and thicker filaments of the protein myosin. The repetition of sarcomeres within the muscle fiber gives the muscle its characteristic striated pattern.

What are troponin and tropomyosin?

Troponin and tropomyosin are two proteins that are part of the actin filament. Although they do not directly participate in contractions, they help to regulate them.

What is dystrophin?

A muscle cell is packed with filaments of actin and myosin. Less abundant, but no less important, is a protein called dystrophin. It literally holds skeletal muscle cells together by linking actin in the cell to glycoproteins (called dystrophin-associated glycoproteins, or DAGs) that are part of the cell membrane.

What causes muscular dystrophy?

Often thought of as a single disorder, muscular dystrophy is a group of genetic diseases characterized by progressive weakness and degeneration of the skeletal muscles that control movement. Missing or abnormal dystrophin causes muscular dystrophies. There are thirty types of muscular dystrophies and they are subdivided by mode of inheritance, age of onset, and clinical features. Discovery of the gene that causes the most common forms of muscular dystrophy took many years because dystrophin compromises only 0.002 percent of the protein in skeletal muscle. There is no specific treatment for muscular dystrophies. Ultimately, fat and connective tissue replace muscle. But there are treatments that can help reduce or even prevent problems in the joints and spine. This allows people with the disease to remain relatively mobile for as long as possible.

What muscular dystrophies are inherited?

The Duchenne and Becker muscular dystrophies are inherited as X chromosome-linked, recessive traits, meaning that only males inherit the condition. Duchenne muscular dystrophy is the most common form of muscular dystrophy. The age of onset is between one and five years of age and affects one in every 3,500 males. Progressive muscle weakness continues rapidly, and by twelve years of age affected individuals are typically confined to wheelchairs. Death usually occurs by the age of twenty due to respiratory infection or cardiac failure. The age of onset of Becker muscular dystrophy is between five and twenty-five years, has a slow progression with milder symptoms, and some individuals do have a normal life span.

What is myoglobin?

Myoglobin is a pigment synthesized in muscle cells that stores oxygen and is responsible for the reddish brown color of skeletal muscle tissue. It is considered to be an iron- and oxygen-binding protein and is related to hemoglobin, the pigment that gives red blood cells their color.

What is the difference between the origin and insertion of a muscle?

The skeleton is a complex set of levers that can be pulled in many different directions by contracting or relaxing skeletal muscles. Most muscles extend from one bone to another and cross at least one joint. One end of a skeletal muscle, the origin, attaches to a bone that remains relatively stationary when the muscle contracts. The other end of the muscle, the insertion, attaches to another bone that will undergo the greatest movement when the muscle contracts. When a muscle contracts, its insertion is pulled toward its origin. The origin is generally closer to the midline of the body and the insertion is farther away.

Do skeletal muscles ever have more than one origin or insertion?

The biceps brachii, located in the arm, is an example of a muscle that has two origins. This is reflected in the name biceps, which means "two heads." This muscle, which originates on the scapula, extends along the front surface of the humerus and is inserted by means of a tendon on the radium. When the biceps brachii contracts, its insertion is pulled toward its origin, and the forearm flexes at the elbow.

What are tendons?

Tendons are bundles of white fibrous connective tissue that attach a muscle to a bone. They are similar to ligaments that attach bone to bone. Tendonitis is the result of a tendon becoming painfully inflamed and swollen following injury or the repeated stress of an activity. The tendons most commonly affected are those associated with the joint capsules of the shoulder, elbow, and hip, as well as those that move the wrist, hand, thigh, and foot. Rest, ice, and anti-inflammatory medication will often relieve the inflammation.

How does tennis elbow differ from golfer's elbow—and even a cellphone elbow?

Tennis elbow and golfer's elbow are both injuries caused by overuse of the muscles of the forearm. Tennis elbow, also called lateral epicondylitis, affects the tendons of the forearm, which attach to the bony prominence on the outside of the elbow. Golfer's elbow, also called medial epicondylitis, affects the muscles of the forearm that attach to the inside of the elbow. Another common "elbow" in-

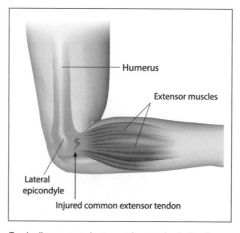

Tennis elbow occurs when a muscle or tendon in the elbow becomes irritated or inflamed.

jury is called the cellphone elbow. It is not caused by a muscular problem, but mainly by compression of nerves in the arm.

What is unusual about the calcaneal tendon?

The gastrocnemius muscle located at the back of the lower leg forms part of the calf. The distal end of the gastrocnemius joins the calcaneal tendon (also known as the Achilles tendon), which descends to the heel. The calcaneal tendon is the thickest and strongest human tendon, but it can be partially or completely torn as a result of strenuous athletic activities, including those that involve quick movements and directional changes.

What are the major types of smooth muscles?

The two major types of smooth muscles are multiunit and visceral. Multiunit smooth muscles are found in the walls of blood vessels and the iris of the eye. These fibers are separate rather than organized into sheets, and they contract only in response to stimulation by nerve impulses or selected hormones. Visceral smooth muscles, which are more common, are composed of sheets of spindle-shaped cells in close contact with one another. They are found in the walls of organs such as the stomach, intestine, uterus, and urinary bladder. Fibers of visceral muscles can stimulate each other and also display rhythmicity or repeated contractions.

What are the characteristics of smooth muscle cells?

Smooth muscle cells are elongated with tapering ends containing filaments of actin and myosin in myofibrils that extend the length of the cells. However, the actin and myosin filaments are organized differently than in skeletal muscle and lack striations.

How do synergistic muscles differ from antagonistic muscles?

Synergistic muscles are groups of muscles that work together to cause the same movement. Muscles that oppose each other are called antagonistic muscles. Antagonist muscles must oppose the action of an agonist muscle so that movement can occur. For example, when the biceps brachii on the front of the upper arm contracts and shortens (agonist), the triceps brachii must relax and lengthen (antagonist) so that the arm can flex.

MUSCLE FUNCTION

Who discovered how muscles work?

British molecular biologist Hugh Huxley (1924–2013) and English physiologist and biophysicist Andrew Huxley (1917–2012) (the scientists were unrelated) researched theories regarding muscle contraction. Hugh Huxley was initially a nuclear physicist who entered the field of biology at the end of World War II. He used both X-ray diffraction and electron microscopy to study muscle contraction. Andrew Huxley was a muscle biochemist who obtained data similar to Hugh's, indicating that the contractile proteins thought to be present in muscles are not contractile at all, but rather slide past each other to shorten a muscle. This theory is called the sliding filament theory of muscle contraction.

How do muscle cells work?

Muscle cells—whether the skeletal muscles in the arms or legs, the smooth muscles that line the digestive tract and other organs, or the cardiac muscle cells in the heart—work by contracting. Skeletal muscle cells are comprised of thousands of contracting units known as sarcomeres. The proteins actin (a thin filament) and myosin (a thick filament) are the major components of the sarcomere. These units perform work by moving structures closer together through space. Sarcomeres in the skeletal muscles pull parts of the body through space relative to each other (for example, walking or swinging the arms).

What are the four steps in the contraction and relaxation of a skeletal muscle?

The four key steps in contraction and relaxation of a skeletal muscle are as follows:

1. A skeletal muscle must be activated by a nerve, which releases a neurotransmitting chemical.

2. Nerve activation increases the concentration of calcium in the vicinity of actin and myosin, the contractile proteins.

3. The presence of calcium permits muscle contraction.

4. When a muscle cell is no longer stimulated by a nerve, contraction ends.

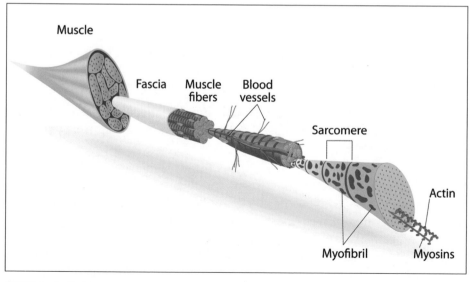

Skeletal muscle structure

What causes muscle cramps?

Normally, a muscle at work contracts, tightening to exert a pulling force, then stretches out when the movement is finished or when another muscle exerts force in the opposite direction. But sometimes a muscle contracts with great intensity and stays contracted, refusing to stretch out again. This is a muscle cramp. Muscles contract or lengthen in response to electrical signals from nerves. Minerals such as sodium, calcium, and magnesium, which surround and permeate muscles cells, play a key role in the transmission of these signals. Imbalances in these minerals, as well as certain hormones, body fluids, and chemicals, or a malfunction in the nervous system itself, can disrupt the flow of electrical signals and cause a muscle to cramp. Fatigued and cold muscles are more likely to cramp.

What is a muscle spasm?

A muscle spasm is a sudden, strong, and painful involuntary contraction. When a muscle is in spasm it feels tight and is described as being in a knot. Muscle spasms occur more frequently in muscles that are overworked or injured. Rest and time resolve most muscle spasms. A "charley horse" is a common name for a muscle spasm in the leg.

What are neuromuscular junctions and motor units?

Each skeletal muscle fiber connects to an axon from a nerve cell called a motor neuron. The connection between the motor neuron and muscle fiber is called a neuromuscular junction. A motor unit is all of the muscle cells (approximately 1,500 muscle fibers of skeletal muscle) that are controlled by a single motor neuron.

How does acetylcholine interact with muscle cells?

Acetylcholine is a neurotransmitter, a chemical substance released by nerve cells that either excites or inhibits another excitable cell such as another nerve cell or a

What are writer's and gamer's cramps?

Writer's cramp is actually a localized muscle spasm called focal dystonia. It is caused by holding a pen or pencil too long, especially too tightly, and occurs only during handwriting. Relaxing the hand periodically, exercising the hand, holding the pen more loosely, and taking frequent breaks from writing usually solve the problem.

Gamer's cramp is more a symptom of today's technology world. It is the overuse of such technology as video games, cellphone text messaging, and even tapping a stylus on a computer pad. This result in such problems as tendonitis, bursitis, and carpal tunnel syndrome, all brought on by stress on tendons, ligaments, and nerves in the hands and upper arms. There is also a condition called Gamer's Thumb, a repetitive stress injury (RSI) that causes swelling at the base of the thumb due to overuse of video games and handheld devices.

What is Botox?

Botox is the trade name for botulinum toxin type A, a protein produced by the bacterium *Clostridium botulinum*, the same toxin that causes botulism. In the past decade or so, Botox has been associated with anti-aging serums. This purified and sterilized botulinum toxin is converted to a form that can be injected under the skin primarily to smooth out wrinkled skin. It was first approved by the Food and Drug Administration (FDA) in 1989 to treat two eye muscle disorders (blepharospasm, or uncontrollable blinking, and strabismuis, or misaligned eyes). It has only been recently that Botox has been used for cosmetic purposes. In small doses, Botox blocks nerve cells from releasing a chemical called acetylcholine, a compound that signals muscle contractions. By interfering with certain muscles' abilities to contract, a person's wrinkles appear to smooth out at the site and surrounding skin. The biggest problem with the toxin as an anti-aging serum is a possible allergic reaction.

muscle cell. In the case of skeletal muscle, acetylcholine excites or activates the muscle cells.

What is botulinum toxin and what symptoms does it produce?

The bacterium *Clostridium botulinum* produces a poison called botulinum toxin that can prevent the release of acetylcholine from motor neuron axons at neuromuscular junctions, causing botulism, a very serious form of food poisoning. This condition is most likely the result of eating home-processed food that has not been heated enough to kill the bacteria in it or to deactivate the toxin. The endospores of this bacterium are very heat resistant and can withstand several hours of boiling at 212°F (100°C) and ten minutes at 248°F (120°C). Botulinum toxin blocks stimulation of muscle fibers, paralyzing muscles, including those responsible for breathing. Without prompt medical treatment, the fatality rate for botulism is very high.

Do all muscle cells work the same way?

Although all muscles work by contracting, not all muscle types have sarcomeres, the muscle contraction units. Cardiac muscle cells have sarcomeres but use different support structures during contraction than those found in skeletal muscles. Smooth muscle cells do not use sarcomeres at all.

How do muscle cells use calcium?

Calcium ions are stored inside muscle cells. The calcium ions are released from storage when a muscle cell gets a signal to induce contraction, which initiates the movement of the contractile proteins within muscles. When calcium concentrations fall, muscle contractions stop.

What is rigor mortis?

Dead bodies are at first limp. Several hours after death, the skeletal muscles undergo a partial contraction that fixes the joints. This condition, known as rigor mortis, may continue for seventy-two hours or more. When neurons signal living muscle fibers to contract, they do so with a neurotransmitter that is received at the surface of the muscle fiber. The signal makes the fiber open calcium ion channels, and it is the calcium that causes the contraction. The muscle then removes the calcium in two ways: it stores some in its mitochondria, and it pumps out the rest. When a body dies, stored calcium leaks and calcium pumps no longer function. The excess calcium causes the actin and myosin filaments of the muscle fibers to remain linked, stiffening the whole body until the muscles begin to decompose.

What sources do muscle cells use for energy?

Muscle cells use a variety of energy sources to power their contractions. For quick energy, the cells utilize their stores of ATP (adenosine triphosphate) and creatine phosphate, which is another phosphate-containing compound. These stored molecules are usually depleted within the first twenty seconds of activity. The cells then switch to other sources, most notably glycogen, a carbohydrate that is made of glucose molecules strung together in long-branching chains.

What energy sources are available for muscles to contract and relax?

Muscle contraction requires significant amounts of energy. Like most cells, muscle cells use ATP as the energy source. In the presence of calcium ions, myosin acts as an enzyme, splitting ATP into ADP (adenosine diphosphate) and inorganic phosphate and releasing energy to do work. Muscle cells store only enough ATP for about ten seconds worth of activity. Once this is used up, the cells must produce more ATP from other energy sources, including creatine phosphate, glycogen, glucose, and fatty acids.

How is shivering related to the muscular system?

Shivering is the body's natural way of keeping warm and can actually serve as a life saver in extreme cold. Shivering produces heat by forcing skeletal muscles to contract and relax rapidly. Heat is produced as a byproduct when muscles metabolize ATP

Why does a runner continue to breathe heavily after completing a ten-mile run with a sprint at the end?

During a ten-mile run, aerobic metabolism is the primary source of ATP production for muscle contraction. Anaerobic metabolism provides the short (fifteen to twenty seconds) burst of energy for the sprint to the finish. After the race, aerobic metabolism is elevated for some time to repay the oxygen debt, causing the heavy breathing after the race.

for contractions. Approximately 80 percent of the muscle energy used in this process is turned into body heat. One study has shown that warmth from external sources such as blankets and hot water bottles can actually be harmful in some cases of hypothermia because the shivering reflex is shut down.

What is the all-or-none response in muscle cells?

According to the all-or-none response, muscle cells are completely under the control of their motor neuron. Muscle cells never contract on their own. A skeletal muscle does not contract partially. If it contracts, it contracts fully.

What is muscle fatigue?

Muscle fatigue results from strenuously exercising a muscle for a prolonged period of time. The muscle may lose its ability to contract due to interruption in the muscle's blood supply (and therefore an interruption in the oxygen supply) or the lack of acetylcholine in motor neuron axons. However, muscle fatigue is most commonly associated with the accumulation of lactic acid in the muscle as a result of anaerobic respiration. During vigorous exercise, the circulatory system cannot supply oxygen to muscle fibers quickly enough. In the absence of oxygen, the muscle cells begin to produce lactic acid, which accumulates in the muscle. The lactic acid buildup lowers pH, and as a result muscle fibers no longer respond to stimulation.

Are pulled muscles a result of muscle fatigue?

Muscle soreness that develops approximately twenty-four hours after exercise is the result of microtrauma to the muscle fibers. Pulled muscles, frequently called torn muscles, result from stretching a muscle too far, causing some of the fibers to tear apart. Internal bleeding, swelling, and pain often accompany a pulled muscle.

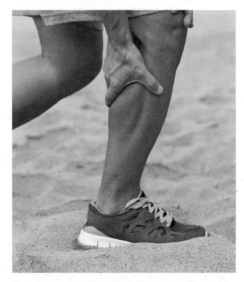

What is oxygen debt?

During rest or moderate exercise, muscles receive enough oxygen to respire aerobically. During strenuous exercise, oxygen deficiency may cause lactic acid to accumulate. Oxygen debt is the amount of oxygen required to convert accumulated lactic acid to glucose and to restore supplies of ATP and creatine phosphate.

Does muscle regularly convert to fat when a person stops exercising?

When a person stops exercising regularly, muscles begin to atrophy and fat

Pulled or torn muscles can result from overexercising and overstretching muscle fibers, sometimes causing internal bleeding and swelling.

cells may begin to expand. This process gives the appearance of muscle converting to fat, but the number of muscle cells remains the same.

Are smooth muscle contractions the same as skeletal muscle contractions?

There are similarities, as well as differences, in comparing smooth and skeletal muscle contractions. Both types of muscles include reactions involving actin and myosin, both are triggered by membrane impulses and an increase in intracellular calcium ions, and both use energy from ATP. One difference between smooth and skeletal muscle contractions is that smooth muscle is slower to contract and to relax than skeletal muscle. Smooth muscle can maintain a forceful contraction longer with a set amount of ATP. In addition, smooth muscle fibers can change length without changing tautness (as when the stomach is full), while this does not occur in skeletal muscles.

What is fibrosis?

Fibrosis is a process in which increasing amounts of fibrous connective tissue develop in skeletal muscle. This makes the muscle less flexible and the collagen fibers can restrict movement and circulation.

What is the effect of aging on the muscular system?

As the body ages, there is a general reduction in the size and power of all muscle tissues. In general, skeletal muscle fibers become smaller in diameter, reflecting a decrease in the number of myofibrils as well as less ATP, glycogen reserves, and myoglobin. In addition, skeletal muscles become smaller and less elastic. The tolerance for exercise decreases as does the ability to recover from muscular injuries. Much of the decrease in muscle strength associated with aging is due to decreased activity. Strength training among older adults can help to slow such losses.

COMPARING OTHER ORGANISMS

Do humans have dark and white muscles similar to those of a chicken?

A chicken has white wing meat and dark leg meat, and humans are much the same in having dark leg muscles and white arm muscles. These differences in color are due to the use of and demands on the limbs. Dark muscle is specialized for endurance and its color comes from a rich blood supply and high myoglobin content. Endurance in dark muscle is at the expense of speed. Your legs can carry you all day, but they cannot move with the speed of a magician's hand. White muscle specializes in very fast contractions and movements, such as wildly clapping hands or swinging a tennis racquet. White muscle tires quickly because it is less well supplied with blood.

What are some animals with the most muscular strength?

There are several animals that have amazing muscular strength. For example, in the oceans, the blue whale is one of the strongest marine animals. In fact, some research indicates that the blue whale is the strongest animal on the planet. On land, the elephant is considered to be one of the strongest animals (and the strongest mammal). In the insect world (considered to be under the animal kingdom), the dung beetle is the world's strongest animal compared to body weight, and the strongest insect. It

Elephant trunks are remarkably dextrous and sensitive limbs. There are over 40,000 muscles in a trunk, compared to only 639 muscles in the entire human body!

is estimated that the dung beetle can pull about 1,141 times its own body weight. To compare, this is equal to a human pulling six double-decker buses filled with people.

Why are trunks so important to elephants?

Elephants are the largest mammals on the planet. The male African elephant (bull) can weigh up to 13,225 pounds (6,000 kilograms). Asian elephants are slightly smaller and lighter, weighing between 10,000 and 12,000 pounds (4,535 and 5443 kilograms). Both types of elephants are known for their huge, muscular trunks. This extension of the animal's nose also contains more than 40,000 muscles, allowing certain elephants to lift close to 800 pounds (350 kilograms).

This remarkable ability involves an intricate balance of muscles and precise contractions within the trunk. The trunks are referred to as muscular hydrostats, or body parts composed of almost all muscle tissue that uses water pressure to move (the human tongue is also called a muscular hydrostat). This allows the animal to twist and turn the trunk, collecting food and water, and allowing the animals to wash (the trunk can hold up to two gallons of water), sand bathe, and communicate with other elephants. The trunk is not only used for major physical tasks, but even delicate ones, such as cracking open a peanut shell. This is because of muscular "fingers" at the end of the trunk—Asian elephants have one, African elephants have two—that can pick up smaller objects.

What animal has the strongest jaw muscles or bite?

Although it is often debated, most researchers believe the saltwater crocodile has the strongest jaw muscles in the animal kingdom. One study measured the bite of a saltwater crocodile at around 7,700 pounds per square inch (PSI). To compare, an adult human male has an average PSI of 150. In the ocean world, it is thought that sharks—such as the bull shark—have the strongest bite. But not all sharks have a high PSI. For example, the great white shark may be one of the largest sharks, but they do not have a high PSI (670). Instead, they use their teeth to kill and cut their prey.

What animals leap to great heights?

Many animals use their muscular strength to leap great heights, mainly to obtain food or escape predators. For example, kangaroo rats can jump forty-five times their own body length, making them the longest jumpers (compared to body size) in the mammal world. There is much debate over which creature is the highest jumper in the entire animal kingdom. One contender is the copepod, small crustaceans found in the oceans and freshwater. In 2010, researchers found that they may be the world's most powerful animal jumpers, leaping with greater force per muscle unit than any other animal.

Do other primates besides humans have muscles with which to smile?

Humans are born with the ability to smile, so it's not something a baby learns (researchers believe this because even blind babies smile). Other primates also have the muscular ability to smile, especially the ape. In particular, they have the same muscles as humans and also use these muscles to express certain emotions and non-verbal cues to other apes. Chimpanzees also use certain facial muscles to communicate information and remain organized while hunting in groups.

Why can some deep diving mammals, such as seals, stay underwater for long periods?

There are many animals that have a high amount of myoglobin, the oxygen-carrying pigment of muscle tissues, especially those that hold their breath for long periods. For example, some deep-sea diving seals and whales have a high amount of myoglobin in their muscles.

The reason for this ability was often debated. In 2013, researchers found that the increase in the oxygen-carrying muscles were apparently due to small changes in an animal's sequence of amino acids—the building blocks for proteins that also give the muscular myoglobin a "positive" charge. Because all the positive charges in the muscles repel each other, this allows the myoglobin to be evenly distributed in the muscle cells. In turn, this allows the oxygen to spread out throughout the muscles, which means a longer time underwater.

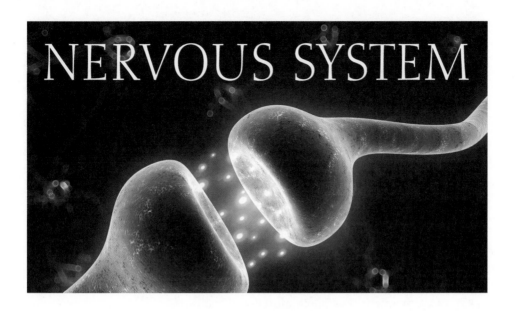

NERVOUS SYSTEM

INTRODUCTION

What are the functions of the nervous system?

The nervous system is one of the major regulatory systems of the body maintaining homeostasis. Its functions are to: 1) monitor the body's internal and external environments; 2) integrate sensory information; and 3) direct or coordinate the responses of other organ systems to the sensory input.

What are the two subsystems of the nervous system?

The nervous system is divided into the central nervous system and the peripheral nervous system. The central nervous system consists of the brain and spinal cord, while the peripheral nervous system consists of all the nerve tissue in the body, excluding the brain and spinal cord. Communication between the central nervous system and the rest of the body is via the peripheral nervous system. Specialized cells of the peripheral nervous system allow communication between the two systems.

What are the cells in the nervous system?

There are approximately twenty billion nerve cells—called neurons—in the nervous system. Neurons have a very limited capacity for regeneration. In general, they neither replicate themselves nor repair themselves. Although highly debated, some scientists suggest that there are a few small concentrations of neuronal stem cells that remain in adults that can produce a limited number of new neurons. They are also working on possible future treatments that will allow adult human neurons to regenerate, but such discoveries are still in their infancy.

What are the two types of cells found in the peripheral nervous system?

The peripheral nervous system consists of afferent or sensory neurons and efferent or motor neurons. The afferent nerve cells (from the Latin *ad*, meaning "toward," and *ferre*, meaning "to bring") carry sensory information from the peripheral to the central nervous system. They have their cell bodies in ganglia and send a process into the central nervous system. The efferent nerve cells (from the Latin *ex*, meaning "away from," and *ferre*, meaning "to bring") carry information away from the central nervous system to the effectors (muscles and tissues). They have cell bodies in the central nervous system and send axons into the periphery.

What are the divisions of the peripheral nervous system?

The peripheral nervous system is divided into the somatic nervous system and autonomic nervous system. The somatic nervous system has afferent and efferent divisions to receive and process sensory input from the skin, voluntary skeletal (striated) muscles, tendons, joints, eyes, tongue, nose, and ears. The autonomic, or visceral, nervous system innervates smooth muscle and glands.

What are the divisions of the autonomic nervous system?

The autonomic nervous system is divided into three parts: 1) the sympathetic nervous system; 2) the parasympathetic nervous system; and 3) the enteric nervous system. The parasympathetic and sympathetic nervous systems usually have opposing actions. For example, while the sympathetic nervous system controls the "fight or flight" responses, which increase the heart rate under stress, the parasympathetic nervous system will slow the heart rate. The enteric nervous system consists of nerve cells in the gastrointestinal tract.

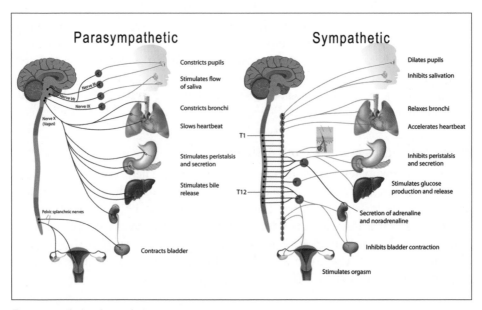

The parasympathetic and sympathetic nervous systems.

What is amyotrophic lateral sclerosis (ALS)?

Amyotrophic lateral sclerosis (ALS)—also called Lou Gehrig's disease after the New York Yankees first baseman who retired from baseball in 1939 after being diagnosed with ALS—is a fatal neurological disease that attacks the nerve cells (motor neurons) responsible for controlling voluntary muscles. Motor neurons serve as controlling units and vital communication links between the nervous system and the voluntary muscles of the body. Messages from motor neurons in the brain (upper motor neurons) are transmitted to motor neurons in the spinal cord (lower motor neurons) and from them to particular muscles. In individuals with ALS, both the upper and lower motor neurons degenerate or die and cease to send messages to muscles. Eventually, all muscles under voluntary control are affected and patients lose the strength and ability to move their arms, legs, and perform other body functions. In the end, even the ability to breathe is affected. The disease does not impair a person's mind, personality, intelligence, or memory.

NEURON FUNCTION

How do neurons transmit information to other neurons?

Most neurons communicate with other neurons or muscle by releasing chemicals called neurotransmitters. These transmitters influence receptors on other neurons. In a few specialized places, neurons communicate directly with other neurons via pores called "gap junctions."

What is a resting membrane potential?

All cells (including neurons) have resting membrane potentials. The ionic environment inside a cell differs from the ionic environment outside the cell. This difference is maintained by special ion pumps that are embedded in the cell membrane. Because the ions have a charge (cations are positively charged and anions are negatively charged), this difference in ionic content sets up an electrical potential difference between the interior and the exterior of the cell. Excitable cells (for example, neurons, cardiac muscle cells, and striated muscle cells) also have other ion channels across the cell membrane that can be activated (or gated) by different conditions. For a neuron, this potential difference produced by the ion pumps when the cell is "at rest" is called the "resting membrane potential."

The resting membrane potential of an average neuron is approximately –70 millivolts with respect to the exterior of the cell. This means that the electrical charge on the inside of the plasma membrane measures 0.07 volts less than that on the outside of the plasma membrane.

What is an action potential?

An action potential is a series of rapidly occurring events that locally decrease and reverse the membrane potential and then eventually restore it to the resting state. The two phases of an action potential are the depolarizing phase and the repolarizing phase. During the depolarizing phase, the inside of the neuron becomes more positive than the outside, reaching thirty millivolts. During the repolarizing phase, the membrane polarization is restored to its resting state of –70 millivolts. The depolarizing and repolarizing phases of an action potential last approximately one millisecond.

The production of the action potential by the opening of a special ion channel in the cell membrane was first described by Alan Lloyd Hodgkin (1914–1998) and Andrew Fielding Huxley (1917–2012) in the 1940s. They shared the Nobel Prize in Physiology or Medicine in 1963, along with Sir John Eccles (1903–1997), for their discoveries concerning the ionic mechanisms of the action potential and the excitation and inhibition of the nerve cell membrane.

What is a synapse?

A synapse is the location of intercellular communication. Every synapse has components associated with two cells: the presynaptic neuron and the postsynaptic neuron. The presynaptic neuron is the cell that sends the message, while the postsynaptic neuron is the cell that receives the message.

When is a nerve impulse generated?

A nerve cell receives many synapses from other neurons (and sometimes from itself). Each time one of these axons conducts an action potential, the presynaptic terminal releases neurotransmitters that can open "chemically gated" ion channels on the postsynaptic neuron (the neuron that receives the terminal). The opening of the ion channels produce local, graded changes in the resting potential of the neuron. If it depolarizes the cell (reduces the potential difference between the inside and outside of the neuron), the small change is called an excitatory postsynaptic potential (EPSP). If it hyperpolarizes the cell (makes the cell's internal potential more negative with respect to its exterior), then it is called an inhibitory postsynaptic potential (IPSP). The sum of the EPSPs and IPSPs change the membrane potential. A nerve impulse is generated

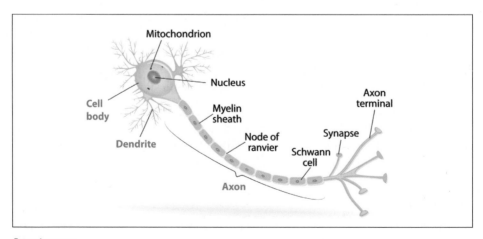

What is myelin?

Myelin is a white, fatty substance that forms an insulating wrapping around large nerve axons. In the peripheral nervous system, myelin is formed by Schwann cells (a type of supporting cell) that wrap repeatedly around the axon. In the central nervous system, myelin is formed by repeated wrappings of processes of oligodendrocytes (a different type of supporting cell). The process of each cell forms part of the myelin sheath.

when the membrane potential reaches a critical threshold. This threshold is typically about fifty-five millivolts. If the neuron does not reach this critical threshold, it does not fire an action potential. (Each action potential of a nerve cell is the same for that cell. For this reason, an action potential is called an "all-or-none" response.)

How quickly do nerve impulses travel?

Nerve impulses travel at an average of 160 feet/second (50 meters/second). The slowest nerve impulses travel at 2.5 feet/second (0.7 meters/second) in small unmyelinated (uninsulated) fibers. Nerve impulses in large myelinated (insulated) fibers can travel at 395 feet/second (120 meters/second) or faster.

What was the first neurotransmitter to be discovered?

The concept of chemical neurotransmission is generally attributed to British physician and physiologist Thomas Renton Elliott (1877–1961). As early as 1904, Elliott had published a theory emphasizing the similarity between adrenaline and sympathetic nerve stimulation. It was not until 1921, however, that German-born pharmacologist and psychobiologist Otto Loewi (1873–1961) demonstrated experimentally that the transmitter substance at the parasympathetic nerve endings (*Vagusstoff*) is acetylcholine and that a substance closely related to adrenaline played a corresponding role at the sympathetic nerve endings. Loewi shared the Nobel Prize in Physiology or Medicine in 1936 with English pharmacologist and physiologist Sir Henry Hallett Dale (1875–1968) for their discoveries concerning the chemical transmission of nerve impulses.

What are some major neurotransmitters?

Scientists have identified at least fifty neurotransmitters in the nervous system, and there may be several dozen

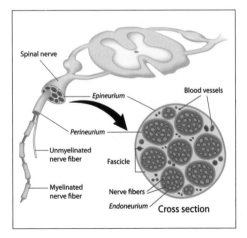

Anatomy of a nerve

NERVOUS SYSTEM

133

more. There are four groups of neurotransmitters: 1) acetylcholine, 2) amino acids, 3) monoamines, and 4) neuropeptides.

Acetylcholine, perhaps one of the best-known neurotransmitters, is the most important neurotransmitter between motor neurons and voluntary muscle contraction. It has an inhibitory effect on heart muscle and excitatory effect on smooth muscles, through the effects on different types of acetylcholine receptors.

Amino acid neurotransmitters include glutamate and asparate. These neurotransmitters are some of the most potent excitatory neurotransmitters in the central nervous system. They are found in the brain.

There are two important groups of *monoamines*: catecholamines and indoleamines. Catecholamines include norepinephrine and dopamine. Serotonin, believed to be involved in sleep, mood, appetite, and pain, is an indoleamine.

Neuropeptides include somatostatin, endorphins, and enkephalins. Somatostatin is a growth-hormone inhibiting hormone. Endorphins and enkephalins suppress synaptic activity leading to pain sensation.

What are some drugs and toxins that affect acetylcholine (ACh) activity at synapses?

The chart below explains the various drugs and toxins and their effects on acetylcholine (ACh) activity.

Drug or Toxin	Mechanism	Effects	Examples
Botulinus toxin (produced by *Clostridium botulinum*, a bacteria)	Inhibits and blocks ACh release	Paralyzes voluntary skeletal muscles	Used therapeutically, such as with Botox, in small doses to remove wrinkles
d-tubocurarine	Prevents ACh binding to postsynaptic receptor sites	Paralyzes voluntary muscles	Known as curare; used by certain South American tribes to paralyze their prey
Atropine	Prevents ACh binding to muscarinic postsynaptic receptor sites	Reduces heart rate and smooth muscle activity; decreases salivation; dilation of pupils; high doses produce skeletal weakness	Used therapeutically by ophthalmologists to dilate pupils; may be used therapeutically to counteract the effects of anticholinesterase poisoning
Nicotine	Binds to nicotinic ACh receptor sites and stimulates the postsynaptic membrane	Low doses facilitate voluntary muscles; high doses cause paralysis	Active ingredient in cigarette smoke
Black widow spider venom	Release of ACh	Produces intense muscular cramps and spasms	

Drug or Toxin	Mechanism	Effects	Examples
Neostigmine or phyostigmine	Prevents ACh inactivation by the enzyme cholinesterase	Extreme sustained contraction of skeletal muscles; effects on cardiac muscle, smooth muscle, and glands	Military nerve gases; used as insecticides (malthion); used therapeutically to treat myasthenia gravis by inhibiting acetylcholinesterase, thereby increasing the usable amount of ACh; counteracts overdoses of tubocurarine

How do local anesthetics block the sensation of pain?

Local anesthetics, such as Novocain and lidocaine, reduce the permeability of the membrane to sodium. Nerve impulses cannot pass through the membrane, and so the stimulation of sensory neurons is prevented. Pain signals do not reach the central nervous system.

What is epilepsy?

Epilepsy is a brain disorder in which clusters of neurons in the brain sometimes signal abnormally. During an epileptic seizure, neurons may fire as many as five hundred times a second, much faster than the normal rate of about eighty times a second. When the normal pattern of neuronal activity becomes disturbed, an individual may experience strange sensations, emotions, and behavior; convulsions; muscle spasms; and loss of consciousness.

What are the causes of epilepsy?

Epilepsy may develop because of an abnormality in brain wiring, an imbalance of neurotransmitters, or some combination of these factors. Researchers believe that some people with epilepsy have an abnormally high level of excitatory neurotransmitters that increase neuronal activity, while others have an abnormally low level of inhibitory neurotransmitters that decrease neuronal activity in the brain. Either situation can result in too much neuronal activity and cause epilepsy.

What are the different types of seizures?

There are more than thirty types of seizures, which are categorized as either focal seizures or generalized seizures. Focal seizures, also called partial seizures, occur in just one part of the brain. They are frequently described by the area of the brain in which they originate (for example, focal frontal lobe seizures). Two examples of focal seizures are simple focal seizures and complex focal seizures. In simple focal seizures, the person will remain conscious but experience sudden and unusual feelings or sensations, such as unexplainable feelings of joy, anger, sadness, or nausea. He or she also may hear, smell, taste, see, or feel things that are not real. In complex focal seizures, the person has a change in or loss of consciousness. People having a complex focal seizure may display strange, repetitive behaviors such as blinks, twitches, mouth movements, or even walking in a circle. These repetitive movements are called automatisms. Some people with focal seizures may experience seeing auras. These seizures usually last just a few seconds.

Generalized seizures are a result of abnormal neuronal activity on both sides of the brain. These seizures may cause loss of consciousness, falls, or massive muscle spasms. There are many kinds of generalized seizures. Two of the better-known generalized seizures are absence seizures and tonic-clonic seizures. In absence seizures, formerly called petit mal seizures, the person may appear to be staring into space and/or have jerking or twitching muscles. Tonic-clonic seizures, formerly called grand mal seizures, cause a mixture of symptoms, including stiffening of the body and repeated jerks of the arms and/or legs, as well as loss of consciousness.

Which neurotransmitter is depleted in Parkinson's disease?

Parkinson's disease was first formally described by James Parkinson (1755–1824), a London physician, in "An Essay on the Shaking Palsy," published in 1817. Parkinson's disease results from a deficiency of the neurotransmitter dopamine in certain brain neurons that regulate motor activity. Parkinson's disease is characterized by stiff posture, tremors, slowness of movement, postural instability, and reduced spontaneity of facial expressions. There is no cure for Parkinson's disease, but certain medications provide relief from the symptoms by increasing the amount of dopamine in the brain. Patients are usually given levodopa combined with carbidopa. Carbidopa delays the conversion of levodopa into dopamine until it reaches the brain. Nerve cells can use levodopa to make dopamine and replenish the brain's dwindling supply.

CENTRAL NERVOUS SYSTEM

What are the characteristics of the central nervous system?

The central nervous system (brain and spinal cord) is protected by a bony covering. The skull surrounds the brain and the vertebral column protects the spinal cord. The central nervous system is responsible for integrating the sensory information it receives from the surrounding environment and allows the body to respond accordingly.

Which membranes cover and protect the brain and spinal cord?

The meninges (from the Greek *meninx*, meaning "membrane") cover and protect the brain and spinal cord. The meninges have three layers: 1) the dura mater, 2) the arachnoid, and 3) the pia mater. The dura mater is the outermost layer covering the central nervous system. The arachnoid (from the Greek *arachne*, meaning "spider") is a weblike network of collagen and elastic fibers. The innermost layer of the meninges is the pia mater. The pia mater is firmly attached to the neural tissue of the spinal cord and brain. Cerebrospinal fluid fills the space between the pia mater and the arachnoid membrane. Most of the blood vessels that supply blood to the central nervous system are in the pia mater.

What is meningitis?

Meningitis is an infection or inflammation of the meninges, or the tissue that covers the brain and the spinal cord. Meningitis is most often caused by a bacterial or viral infection, although certain fungal infections and tumors may also cause meningitis. The usual symptoms and signs of meningitis are sudden fever, severe headache, and a stiff neck. In more severe cases, neurological symptoms may include nausea and vomiting, confusion and disorientation, drowsiness, sensitivity to bright light, and poor appetite. Early treatment of bacterial meningitis with antibiotics is important to reduce the risk of dying from the disease.

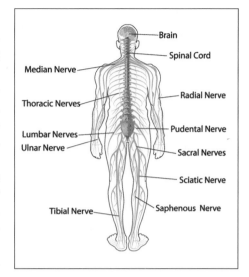

The central nervous system

Can meningitis be prevented?

The introduction and widespread use of *Haemophilus influenzae* type b and *Streptococcus pneumoniae* conjugated vaccines has dramatically reduced the incidence of meningitis caused by these bacteria. The Centers for Disease Control and Prevention recommend routine vaccination of adolescents and college freshmen with the meningococcal vaccine, which prevents four types of meningococcal disease caused by the bacteria *Neisseria meningitides*. But such vaccines against bacteria do not help viral meningitis, the most common type of this disease. In most cases, viral meningitis is less severe than bacterial meningitis, and most people get better on their own without treatment.

What are gray and white matters?

Gray matter consists of neurons and unmyelinated dendrites and axons. In the spinal cord, the gray matter is shaped like an "H" around the very small, narrow central

What is the difference between meningococcal disease and meningitis?

According to the Centers for Disease Control and Prevention, if a person has meningitis it does not mean they have meningococcal disease, and vice versa. Meningococcal disease is any infection caused by the bacterium *Neisseria meningitidis*, or *meningococcus*, and any infection caused by the bacteria is known as meningococcal disease. One serious infection it can cause is meningococcal meningitis.

canal. It has a grayish appearance in autopsy specimens. White matter consists of myelinated nerve tissue. Since myelin is white the tissue appears whitish.

What are demyelinating diseases?

Demyelinating diseases involve damage to the myelin sheath of neurons in either the peripheral or central nervous system. Multiple sclerosis (MS) is a chronic, potentially debilitating disease that affects the myelin sheath of the central nervous system. The illness is believed to be an autoimmune disease. In MS the body directs antibodies and white blood cells against proteins in the myelin sheath surrounding nerves in the brain and spinal cord. This causes inflammation and injury to the myelin sheath.

Demyelination is the term used for a loss of myelin, a substance in the white matter that insulates nerve endings. Myelin helps the nerves receive and interpret messages from the brain at maximum speed. When nerve endings lose this substance, they cannot function properly, leading to patches of scarring, or "sclerosis." The result may be multiple areas of sclerosis. The damage slows or blocks muscle coordination, visual sensation, and other functions that rely on nerve signals.

How is the myelin sheath affected in Guillain-Barré syndrome?

In the autoimmune disorder known as Guillain-Barré syndrome, the body's immune system attacks part of the peripheral nervous system. The immune system starts to destroy the myelin sheath that surrounds the axons of many peripheral nerves, or even the axons themselves. The loss of the myelin sheath surrounding the axons slows down the transmission of nerve signals. In diseases such as Guillain-Barré, in which the peripheral nerves' myelin sheaths are injured or degraded, the nerves cannot transmit signals efficiently. Consequently, muscles begin to lose their ability to respond to the brain's commands, commands that must be carried through the nerve network. The brain also receives fewer sensory signals from the rest of the body, resulting in an inability to feel textures, heat, pain, and other sensations. Alternatively, the brain may receive inappropriate signals that result in tingling, "crawling-skin," or painful sensations. Because the signals to and from the arms and legs must travel the longest distances, these extremities are most vulnerable to interruption.

What are the symptoms of Guillain-Barré syndrome?

The first symptoms of this disorder include varying degrees of weakness or tingling sensations in the legs. In many instances the weakness and abnormal sensations spread to the arms and upper body. In severe cases the patient may be almost totally paralyzed since the muscles cannot be used at all. In these cases the disorder is life threatening because it potentially interferes with breathing and, at times, with blood pressure or heart rate. Such a patient is often put on a respirator to assist with breathing and is watched closely for problems such as an abnormal heartbeat, infections, blood clots, and high or low blood pressure. Most patients, however, recover from even the most severe cases of Guillain-Barré syndrome, although some continue to have a certain degree of weakness.

How much cerebrospinal fluid is in the central nervous system?

The entire central nervous system contains between 3 to 5 ounces (80 to 150 milliliters) of cerebrospinal fluid, a clear, colorless liquid. The choroid plexus produces nearly 17 ounces (500 milliliters) of cerebrospinal fluid per day, effectively replacing the cerebrospinal fluid every eight hours (three times per day). Normally, cerebrospinal fluid flows through the ventricles, exits into cisterns (closed spaces that serve as reservoirs) at the base of the brain, bathes the surfaces of the brain and spinal cord, and then is absorbed into the bloodstream.

What are the functions of cerebrospinal fluid?

Cerebrospinal fluid has three important, life-sustaining functions: 1) to keep the brain tissue buoyant, acting as a cushion or "shock absorber"; 2) to act as the vehicle for delivering nutrients to the brain and removing waste; and 3) to flow between the cranium and spine to compensate for changes in intracranial blood volume (the amount of blood within the brain).

THE BRAIN

How large is the brain?

The brain weighs about 3 pounds (1.4 kilograms). The average brain has a volume of 71 cubic inches (1,200 cubic centimeters). In general, the brain of males averages about 10 percent larger than the brain of females due to overall differences in average body size. The brain contains approximately 100 billion neurons and 1 trillion neuroglia (or glial, types of cells that support neurons).

How is the weight of the brain reduced in cerebrospinal fluid?

Since the brain is buoyant and floats in the cerebrospinal fluid, its weight of approximately 3 pounds (1.4 kilograms) is reduced to about 14 percent of its unsupported weight, less than 2 ounces (50 grams).

What is the ailment that was once called "water on the brain"?

Obstructive hydrocephalus, commonly called "water on the brain," results from an imbalance of production, circulation, and reabsorption of cerebrospinal fluid. Since cerebrospinal fluid is being produced continually, once the balance is disrupted the volume of cerebrospinal fluid within the brain will continue to increase. The increased volume of fluid leads to compression and distortion of the brain. Left untreated, the intracranial pressure increases, often causing brain function to deteriorate. In infants, treatment often includes the installation of a shunt to either avoid the site of the blockage or drain the excess cerebrospinal fluid.

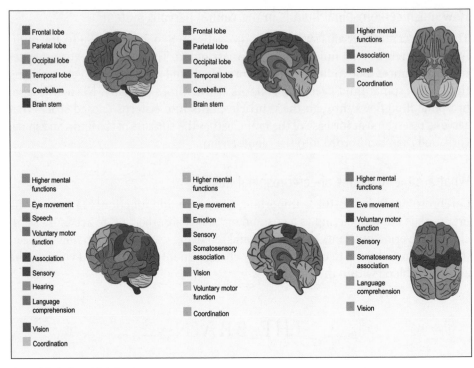

Areas of the brain and their functions

What is the paleomammalian brain—or the limbic system?

The paleomammalian brain, also called the limbic system, is a set of brain structures located on both sides of the thalamus, right under the cerebrum. This collection of structures supports many "automatic" functions, such as emotion, behavior, adrenaline flow, motivation, olifaction, and long-term memory. It is thought to have evolved early in mammalian evolution.

Is brain size an indication of intelligence?

There is no correlation between brain size and intelligence. Individuals with the smallest brains (as small as 46 cubic inches [750 cubic centimeters]) and the largest brains (as large as 128 cubic inches [2,100 cubic centimeters]) have the same functional intelligence.

How does the size of the brain change from birth to adulthood?

Brain cells grow in size and degree of myelination as a child grows from birth to adulthood. Although the number of neurons does not increase after infancy, the number of glial cells does increase. An adult brain is approximately three times as heavy as it was at birth. Between ages twenty and sixty, the brain loses approximately 0.033 to 0.10 ounces (1 to 3 grams) a year as neurons die and are not replaced. After age sixty the annual rate of shrinkage increases to 0.10 to 0.143 ounces (3 to 4 grams) per year.

What are the major divisions of the brain and their function?

The brain has four major divisions: 1) brainstem, including the medulla oblongata, pons, and midbrain; 2) cerebellum; 3) cerebrum; and 4) diencephalon. The diencephalon is further divided into the thalamus, hypothalamus, epithalamus, and ventral thalamus or subthalamus. Each area of the brain has a specific function, as seen in the table below:

Brain Area	General Functions
Brainstem	
Medulla oblongata	Relays messages between spinal cord and brain and to cerebrum; center for control and regulation of cardiac, respiratory, and digestive activities
Pons	Relays information from medulla and other areas of the brain; controls certain respiratory functions
Midbrain	Involved with the processing of visual information, including visual reflexes, movement of eyes, focusing of lens, and dilation of pupils
Cerebellum	Processing center involved with coordination of movements, balance and equilibrium, and posture; processes sensory information used by motor systems
Cerebrum	Center for conscious thought processes and intellectual functions, memory, sensory perception, and emotions
Diencephalon	
Thalamus	Relay and processing center for sensory information
Hypothalamus	Regulates body temperature, water balance, sleep-wake cycles, appetite, emotions, and hormone production

Which structure connects the spinal cord to the brain?

The medulla oblongata connects the spinal cord to the brain. The medulla oblongata regulates autonomic functions, such as heart rate, blood pressure, and digestion, and automatic functions, such as respiratory rhythm. It relays sensory information to the thalamus and to other portions of the brain stem.

Who proposed that the left and right sides of the brain have different functions?

Scientists have been aware for at least one hundred years that each side of the brain controls actions on the opposite side of the body. However, it was not until the 1950s that American neuropsychologist and neurobiologist Roger Sperry (1913–1994) conducted the pioneering research to determine the different functions of the left side and right side of the brain. His experiments contributed to the "split-brain" theory. Sperry received the Nobel Prize in Physiology or Medicine in 1981 for his work.

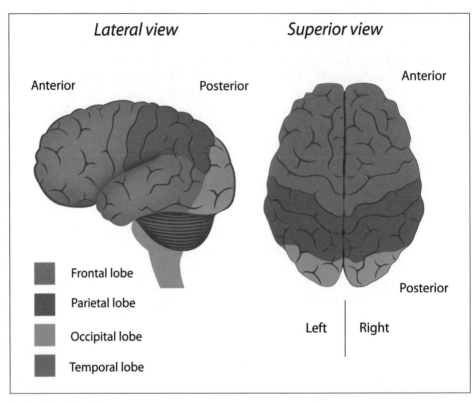

The four lobes of the brain

Which is the largest part of the brain?

The largest part of the brain is the cerebrum. The outer surface of the cerebrum is covered with a series of elevated ridges called gyri, and grooves or shallow depressions called sulci. The deepest sulci are called fissures. The cerebrum is divided into the right and left hemispheres. The corpus callosum connects the two halves at their lower midportion. Each hemisphere is divided into four sections called lobes, which have been named for the bones of the skull that cover them. The lobes are identified as the frontal lobe, the parietal lobe, the temporal lobe, and the occipital lobe.

Which skills are controlled by the left cerebral hemisphere and which are controlled by the right cerebral hemisphere?

The left side of the brain controls the right side of the body, as well as spoken and written language, logic, reasoning, and scientific and mathematical abilities. In contrast, the right side of the brain controls the left side of the body and is associated with imagination, spatial perception, recognition of faces, and artistic and musical abilities.

What is a concussion?

A concussion is an injury to the brain caused by a blow or jolt to the head that disrupts the normal functioning of the brain. Concussions are usually not life-threatening. Since the brain is very complex, there is great variation in the signs and symptoms of a concussion. Some people lose consciousness; others never lose consciousness. Some symptoms may appear immediately, while others do not appear for several days or even weeks. Symptoms include the following:

- Headaches or neck pain that will not go away
- Difficulty with mental tasks such as remembering, concentrating, or making decisions
- Slowness in thinking, speaking, acting, or reading
- Getting lost or easily confused
- Feeling tired all of the time, having no energy or motivation
- Mood changes (feeling sad or angry for no reason)
- Changes in sleep patterns (sleeping a lot more or having a hard time sleeping)
- Light-headedness, dizziness, or loss of balance
- Urge to vomit (nausea)
- Increased sensitivity to lights, sounds, or distractions
- Blurred vision or eyes that tire easily
- Loss of sense of smell or taste
- Ringing in the ears

What are the two forms of stroke?

There are two forms of stroke: ischemic and hemorrhagic. Ischemic stroke is the blockage of a blood vessel that supplies blood to the brain. Ischemic strokes account for 80 percent of all strokes. Hemorrhagic stroke is bleeding into or around the brain. Hemorrhagic strokes account for 20 percent of all strokes.

What are the symptoms of stroke?

The symptoms of stroke appear suddenly and include numbness or weakness, especially on one side of the body; confusion or trouble speaking or understanding speech; trouble seeing in one or both eyes; trouble walking, dizziness, or loss of balance or coordination; or severe headache with no known cause. Often more than one of these symptoms will be present, but they all appear suddenly.

How does a mini-stroke differ from a regular stroke?

A mini-stroke, technically called a transient ischemic attack (TIA), begins like a stroke but then resolves itself, leaving no noticeable symptoms or deficits. The average duration of a TIA is a few minutes. For almost all TIAs, the symptoms go away within an hour. A person who experiences a TIA should consider it a warning, since approximately one-third of the

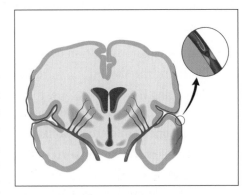

An ischemic stroke is caused by a blockage of a blood vessel supplying blood to the brain.

143

fifty thousand Americans who have a TIA have an acute stroke sometime in the future. Since all stroke symptoms appear suddenly and it is not possible to determine whether it is a TIA or full stroke, medical treatment should be sought immediately.

What are two of the most common forms of dementia?

The term "dementia" describes a group of symptoms that are caused by changes in brain function. The two most common forms of dementia in older people are Alzheimer's disease and multi-infarct dementia (sometimes called vascular dementia). There is no cure for these types of dementia. In Alzheimer's disease, nerve-cell changes in certain parts of the brain result in the death of a large number of cells. Some researchers believe there is a genetic origin to Alzheimer's disease. The symptoms of Alzheimer's disease range from mild forgetfulness to serious impairments in thinking, judgment, and the ability to perform daily activities.

In multi-infarct dementia, a series of small strokes or changes in the brain's blood supply may result in the death of brain tissue. The location in the brain where the small strokes occur determines the seriousness of the problem and the symptoms that arise. Symptoms that begin suddenly may be a sign of this kind of dementia. People with multi-infarct dementia are likely to show signs of improvement or remain stable for long periods of time, then quickly develop new symptoms if more strokes occur. In many people with multi-infarct dementia, high blood pressure is to blame.

How many people in America are estimated to have Alzheimer's disease?

According to the Alzheimer's Association, it is estimated that more than five million Americans are living with Alzheimer's disease. It is also the sixth leading cause of death in the United States. Researchers estimate that three percent of people over the aged of sixty-five have Alzheimer's disease and nearly half of all people over 85 may have the disease. Around two-thirds of Americans with Alzheimer's are women, and for women over sixty, it is estimated that they have a one in six chance of developing the disease.

What are the seven warning signs of Alzheimer's disease?

In general, the seven warning signs of Alzheimer's disease are as follows:

1. Asking the same question over and over again
2. Repeating the same story, word for word, again and again
3. Forgetting how to cook, how to make repairs, how to play cards, or any other activities that were previously done with ease and regularity
4. Losing one's ability to pay bills or balance one's checkbook
5. Getting lost in familiar surroundings, or misplacing household objects
6. Neglecting to bathe, or wearing the same clothes over and over again, while insisting that one has taken a bath or that clothes are still clean

7. Relying on someone else, such as a spouse, to make decisions

It is important to understand that even if someone has several or even most of these symptoms, it does not mean they definitely have Alzheimer's disease. It does mean they should be thoroughly examined by a medical specialist trained in evaluating memory disorders, such as a neurologist or a psychiatrist, or by a comprehensive memory disorder clinic with an entire team of experts knowledgeable about memory problems.

SPINAL CORD

Where is the spinal cord located?

The spinal cord lies inside the vertebral column. It extends from the occipital bone of the skull to the level of the first or second lumbar vertebra. In adults, the spinal cord is 16 to 18 inches (42 to 45 centimeters) long and 0.5 inches (1.27 centimeters) in diameter. The spinal cord is the connecting link between the brain and the rest of the body. It is the site for integration of the spinal cord reflexes. (For more about tests on the spine, see the chapter "Helping Human Anatomy.")

What are spinal cord tracts?

Tracts are bundles of axons that are relatively uniform with respect to diameter, myelination, and conduction speed. All the axons within a tract relay the same type of information in the same direction.

How do ascending tracts differ from descending tracts?

Ascending tracts consist of sensory fibers that carry information up the spinal cord to the brain. Descending tracts consist of motor fibers that carry information from the brain to the spinal cord.

What is a reflex?

A reflex is a predictable, involuntary response to a stimulus. It was given this name in the eighteenth century because it appeared that the stimulus was reflected off of the spinal cord to generate the response, just as light is reflected by a mirror. Reflexes allow the body to respond quickly to internal and external changes in the environment in order to maintain homeostasis. Reflexes that involve the skeletal muscles are called somatic reflexes. Reflexes that involve responses of smooth muscle, cardiac muscle, or a gland are called visceral or autonomic reflexes.

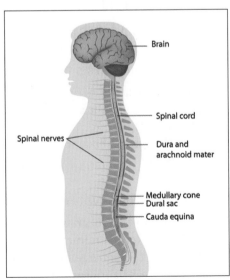

The spinal cord

How is trauma to the spinal cord classified?

Injury to the spinal cord produces a period of sensory and motor paralysis termed spinal shock. The severity of the injury will determine how long the paralysis will last and whether there will be permanent damage.

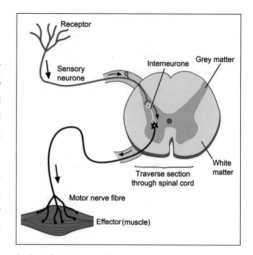

In the reflex arc mechanism, a stimulus is registered through the nerves of the spinal cord, leading to a reaction in a muscle without the involvement of the brain.

- *Spinal concussion* results in no visible damage to the spinal cord. The resulting spinal shock is temporary and may last for as short as a couple of hours.

- *Spinal contusion* involves the white matter of the spinal cord being injured (bruised). Recovery is more gradual and may involve permanent damage.

- *Spinal laceration*, caused by vertebral fragments or other foreign bodies penetrating the spinal cord, often requires a longer recovery and is less complete.

- *Spinal compression* occurs when the spinal cord is squeezed or distorted within the vertebral canal. Relieving the pressure usually relieves the symptoms.

- *Spinal transection* is the complete severing of the spinal cord. Surgical procedures cannot repair a severed spinal cord.

PERIPHERAL NERVOUS SYSTEM: SOMATIC NERVOUS SYSTEM

What are the components of the peripheral nervous system?

The cranial and spinal nerves constitute the somatic peripheral nervous system. These nerves connect the brain and spinal cord to peripheral structures such as the skin surface and the skeletal muscles. Overall, the total length of the peripheral nerves in the human body measure approximately 93,000 miles (150,000 kilometers) in length.

How many cranial nerves are in the human body?

There are twelve pairs of cranial nerves in the human body. The cranial nerves are designated by Roman numerals and names. The Roman numerals indicate the order in which they emerge from the brain. The name indicates an anatomical feature of function. The trigeminal nerve is the largest, although not the longest, cranial nerve.

What are the cranial nerves and their functions?

The cranial nerves and their functions are as follows:

Name of Cranial Nerve	Function
I Olfactory	Smell
II Optic	Vision
III Oculomotor	Movement of eyeball and eyelid; constricts pupil; focuses lens
IV Trochlear	Movement of eyeball (down and out)
V Trigeminal	Sensations to the face, including scalp, forehead, cheeks, upper lip, palate, tongue, and lower jaw; chewing
VI Abducens	Lateral movement of eye
VII Facial	Facial expressions, taste, secretion of tears, saliva
VIII Vestibulocochlear	Hearing and equilibrium (balance)
IX Glossopharyngeal	Taste and other sensations of the tongue; swallowing and secretion of saliva
X Vagus	Swallowing, coughing, voice production; monitors blood pressure and oxygen and carbon dioxide levels in blood
XI Accessory (also called spinal accessory)	Voice production; skeletal muscles of palate, pharynx, and larynx; movement of head and shoulders
XII Hypoglossal	Movement of tongue during speech and swallowing

Which cranial nerve is involved in *tic douloureux*?

Tic douloureux is caused by compression or degeneration of the fifth cranial nerve, the V trigeminal. Individuals afflicted with this condition have sudden, severe, stabbing pain on one side of the face, along the jaw or cheek. The pain may last several seconds and may be experienced repeatedly over several hours, days, weeks, or even months. The episodes may subside as rapidly as they began with no incidents of pain for months or even years.

Which cranial nerve is responsible for Bell's palsy?

Bell's palsy, a form of temporary facial paralysis, is the result of damage or trauma to the seventh cranial nerve, the VII facial. The nerve may be swollen, inflamed, or compressed, resulting in an interruption of messages from the brain to the facial muscles. Individuals with Bell's palsy may exhibit twitching, weakness, or paralysis on one or both sides of the face; drooping of the eyelid and corner of the mouth; drooling; dryness of the eye or mouth; impairment of taste; and excessive tearing in one eye. Although the symptoms appear suddenly, individuals begin to recover within two weeks and return to normal function within three to six months.

How many pairs of spinal nerves are in the human body—and which is the longest?

There are thirty-one pairs of spinal nerves in the human body. The spinal nerves are grouped according to where they leave the vertebral column. There are eight pairs of cervical nerves (C_1–C_8), twelve pairs of thoracic nerves (T_1–T_{12}), five pairs of lumbar nerves (L_1–L_5), five pairs of sacral nerves (S_1–S_5) and one pair of coccygeal nerves (Co_1). The longest spinal nerve is the tibial nerve, which averages twenty inches (fifty centimeters) long.

How does the location of an injury along the spinal cord cause a certain paralysis?

Injuries or damage to the spinal cord at or above the fifth cervical vertebra eliminates sensation and motor control of the upper and lower limbs, as well as any part of the body below the level of the injury. The paralysis after a high spinal injury is termed quadriplegia, and the person is unable to feel or move both arms and legs. Damage that occurs in the thoracic region of the spinal cord effects motor control of the lower limbs only. This paralysis is called paraplegia, and the person is unable to move or feel both lower limbs.

How are spinal nerves attached to the spinal cord?

Spinal nerves divide in the vertebral canal into two branches: the dorsal root and the ventral root. The dorsal root, which is the posterior branch, contains the axons of sensory neurons that bring information to the spinal cord. The ventral root, which is the anterior branch, contains the axons of motor neurons that carry commands to muscles or glands. Therefore, each spinal nerve is considered a mixed nerve with both sensory and motor neurons.

What is a plexus?

A plexus (from the Latin *plectere*, meaning "braid") is an interwoven network of spinal nerves. There are four major plexuses on each side of the body: 1) the cervical plexus innervates the muscles of the neck, the skin of the neck, the back of the head, and the diaphragm muscle; 2) the brachial plexus innervates the shoulder and upper limb; 3) the lumbar plexus innervates the muscles and skin of the abdominal wall; and 4) the sacral plexus innervates the buttocks and lower limbs. The nerves then divide into smaller branches.

What causes sciatica?

Sciatica is caused by compression of the sciatic nerve, such as from a herniated disc or even from sitting for extended periods of time with a wallet in the back pocket. The pain usually subsides after a few weeks, although over-the-counter pain relievers may be helpful.

What are dermatomes and what infection affects the skin of a single dermatome?

Dermatomes (from the Greek *derma*, meaning "skin," and *tomos*, meaning "cutting") are areas on the skin surface supplied by an individual spinal nerve. Shingles, or herpes zoster, appears as a painful rash on the skin that most often corresponds

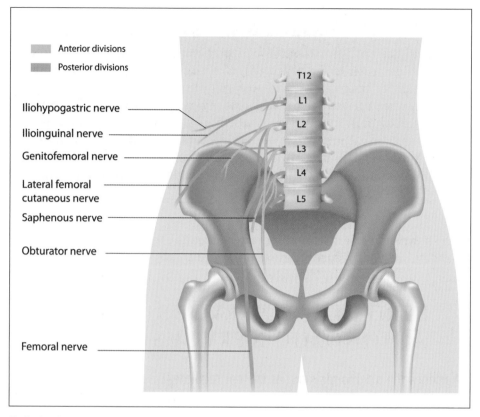

Anterior divisions
Posterior divisions

T12
L1
L2
L3
L4
L5

Iliohypogastric nerve

Ilioinguinal nerve

Genitofemoral nerve

Lateral femoral
cutaneous nerve

Saphenous nerve

Obturator nerve

Femoral nerve

The lumbar plexus

to the sensory nerve in the area of a single dermatome. The virus is the same one that causes chicken pox. If someone has chicken pox as a child, the virus may lie dormant in the nerve roots of the spinal nerves for decades. If reactivated for some reason (usually stress is implicated), the virus will present itself as shingles.

Which nerve is responsible for carpal tunnel syndrome?

The median nerve controls sensations to the palm side of the thumb and fingers (although not the little finger), as well as impulses to some small muscles in the hand that allow the fingers and thumb to move. Carpal tunnel syndrome occurs when the median nerve, which runs from the forearm into the hand, becomes pressed or squeezed at the wrist. The carpal tunnel, a narrow, rigid passageway of ligament and bones at the base of the hand, houses the median nerve and tendons. At times thickening from irritated tendons or other swelling narrows the tunnel and causes the median nerve to be compressed. Carpal tunnel syndrome is characterized by pain, weakness, or numbness in the hand and wrist, often radiating up the arm.

According to the National Institute of Neurological Disorders and Stroke, initial treatment of carpal tunnel syndrome includes resting the affected hand and wrist for a minimum of two weeks. Nonsteroidal, anti-inflammatory drugs may be used to ease the pain. Ice and corticosteroids may relieve the swelling and pressure on the

149

nerve. If symptoms persist, surgery may be required to sever the band of tissue around the wrist and reduce pressure on the median nerve.

PERIPHERAL NERVOUS SYSTEM: AUTONOMIC NERVOUS SYSTEM

What does the autonomic nervous system regulate?

The autonomic nervous system regulates "involuntary" activity, which is not controlled on a conscious level. Specifically, the autonomic nervous system innervates the activity of smooth muscle, cardiac muscle, and glands of the body.

What causes the sensation of "pins and needles" when your foot "falls asleep"?

Local pressure, such as crossing or sitting on your legs, may temporarily compress a nerve, removing sensory and motor function in your foot. When the local pressure is removed, the familiar feeling of "pins and needles" is felt as the nerve endings become reactivated.

How is the autonomic nervous system organized?

The autonomic nervous system consists of two divisions: the sympathetic nervous system and the parasympathetic nervous system. The sympathetic division is often called the "fight or flight" system because it usually stimulates tissue metabolism, increases alertness, and generally prepares the body to deal with emergencies. The parasympathetic division is considered the "rest and repose" division because it conserves energy and promotes sedentary activities, such as digestion. In general, both the sympathetic and parasympathetic divisions innervate the target cells.

How does the somatic nervous system differ from the autonomic nervous system?

The table below explains the main differences between the somatic and autonomic nervous systems:

	Somatic Nervous System	Autonomic Nervous System
Effectors	Skeletal muscles	Cardiac muscle, smooth muscle, and glands
Type of control	Voluntary	Involuntary
Neural pathway	One motor neuron extends from the central nervous system and synapses directly with a skeletal fiber	One motor neuron (preganglion neuron) extends from the central nervous system and synapses with another motor neuron in a ganglion; the second motor neuron (postganglion neuron) synapses with a visceral effector
Neurotransmitter	Acetylcholine	Acetylcholine or norephinephrine
Action of neurotransmitter on effecor	Always excitatory (causing contraction of skeletal muscle)	May be excitatory (causing contraction of smooth muscle, increased heart rate, increased force of heart contraction, or increased secretions from glands) or inhibitory (causing relaxation of smooth muscle, decreased heart rate, or decreased secretions from glands)

LEARNING AND MEMORY

Which parts of the brain are involved in higher order functions, such as learning and memory?

Higher order functions, such as learning and memory, involve complex interactions among areas of the cerebral cortex and between the cortex and other areas of the brain. Information is processed both consciously and unconsciously. Since higher order functions are not part of the programmed "wiring" of the brain, the functions are subject to modification and adjustment over time.

What are the areas of the cerebral cortex and their functions?

The cerebral cortex is divided into three functional areas: 1) sensory areas, 2) motor areas, and 3) association areas. The sensory areas receive and interpret sensory impulses. The motor areas control muscular movement. The association areas are involved in integrative functions such as memory, emotions, reasoning, will, judgment, personality traits, and intelligence.

Which areas of the brain are responsible for specific functions?

Researchers know that certain areas of the brain are responsible for certain general functions. In 1909, German physician and researcher Korbinian Brodmann (1868–1918) published *Vergleichende Lokalisationslehre der Grosshirnrinde in ihren Prinzipien dargestellt auf Grund des Zellenbaues*. This treatise included maps of the localization of functions in the cerebral cortex. Brodmann's maps are still used to depict the areas of cerebral cortex that are responsible for specific functions.

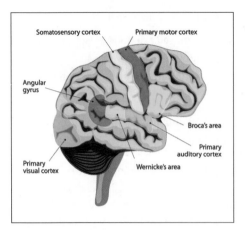
Broca's area of the brain is responsible for speech production, while Wernicke's area allows one to understand and interpret speech.

Who discovered which area of the brain is responsible for speech and language?

French physician Pierre Paul Broca (1824–1880) identified an area of the brain responsible for speech production in 1861. Having observed a patient who could not speak except for the meaningless utterance of "tan, tan," Broca examined the brain of the individual upon his death. Broca determined the patient was missing a section of the frontal lobe from the cerebral hemisphere. Broca continued to examine the brains of individuals with a lack of speech and found they all lacked the same area of the brain.

How do Broca's area and Wernicke's area differ?

Broca's area and Wernicke's (named after German physician, anatomist, psychiatrist, and neuropathologist Karl [or Carl] Wernicke, 1848–1905) area are both associated with speech. Broca's area is associated with the production of speech. It controls the flow of words from brain to mouth. Wernicke's area is associated with the interpretation and understanding of speech.

What are the causes of aphasia?

Aphasia is a language disorder that results from damage to portions of the brain that are responsible for language. Strokes are the most common cause of aphasia, although aphasia can also result from a brain tumor, infection, head injury, or dementia that damages the brain. Individuals with aphasia have difficulty speaking—both in producing words and complete sentence structure—or understanding speech, or both. Depending on the severity of the aphasia (and the degree of permanent brain damage),

What did scientists recently discover about Broca's area?

In 2015, scientists knew that the Broca's area of the brain is the command center for human speech (and vocalization), but they have now found out that the area switches off when we talk out loud. They also believe that the area may remain active during conversation as the person plans future words and full sentences. Overall, the researchers hope that this leads to diagnoses and treatments of stroke, epilepsy, and brain injuries that often affect and/or impair language skills of the affected person.

some patients regain their speech capabilities with little or no rehabilitation. In most cases, however, speech therapy is necessary to regain language capabilities.

What is intelligence?

There is no clear, standard definition of intelligence. Psychologists identify intelligence as an individual's adaptation to the environment as fundamental to understanding what intelligence is and what it does. Most researchers agree that intelligence is a person's ability to comprehend his or her environment, evaluate it rationally, and form appropriate responses.

Is it possible to measure intelligence?

The earliest test created to measure intelligence was developed by French physiologist Alfred Binet (1857–1911) in 1905. The purpose of the test was to measure skills such as judgment, comprehension, and reasoning in order to place children in the appropriate classes in school. The test was brought to the United States by Stanford University psychologist Lewis Terman (1877–1956) in 1916 and renamed the Stanford-Binet test. Since then, other intelligence tests, such as the Wechsler Adult Intelligence Scale and the Wechsler Intelligence Scale for Children have been developed. These tests have produced a score referred to as an intelligence quotient (IQ).

How is IQ calculated?

IQ, or the intelligence quotient, was originally computed as the ratio of a person's mental age to his or her chronological age, multiplied by 100. Following this method, a child of ten years old who performed on the test at the level of an average twelve year old (mental age of twelve), was assigned an IQ of 12/10 x 100 = 120. More recently, the concept of "mental age" has fallen into disrepute and IQ is computed on the basis of the statistical percentage of people who are expected to have a certain IQ. An IQ of 100 is considered average. An IQ of 70 or below indicates mental retardation, and an IQ of 130 or above indicates gifted abilities.

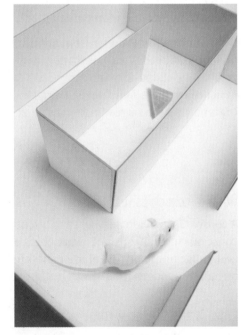

What is memory?

Memory is the ability to recall information and experiences. Memory and learning are related because in order to be able to remember something it must first be "learned." Memories may be facts or skills. Memory "traces" have been described traditionally as concrete things

Rats have been shown to have very good memories when it comes to finding food sources.

153

that are formed during learning and imprinted on the brain when neurons record and store information. However, the way that memories are formed and represented in the brain is not well understood.

How does short-term memory differ from long-term memory?

Short-term memory, also called primary memory, refers to small bits of information that can be recalled immediately. The recalled information has no permanent importance, such as a name or telephone number that is only used once. Long-term memory is the process by which information that for some reason is interpreted as being important is remembered for a much longer period. Short-term memories may be converted to long-term memories.

Which areas of the brain are involved in memory?

Several areas of the brain are associated with memory, including the association cortex of the frontal, parietal, occipital, and temporal lobes, the hippocampus, and the diencephalon. Damage to the hippocampus results in an inability to convert short-term memories to long-term memories. Memory loss may be the result of trauma or injury, disease, lifestyle choices, such as alcoholism and drug use, and aging.

What is amnesia?

Amnesia refers to the loss of memory from disease or trauma. The extent and type of memory loss is dependent on the area of the brain that is damaged. Individuals with retrograde amnesia suffer from memory losses of past events. This is a common occurrence when head injury is involved. Oftentimes, an individual will not be able to recall the events and moments immediately preceding an accident or fall.

Individuals who suffer from anterograde amnesia are unable to store additional memories, but their earlier memories are intact and accessible. They have difficulty creating new long-term memories. As a consequence, every experience is a new experience for these individuals, even if they have experienced it earlier, such as meeting a new person or reading a new book.

SLEEP AND DREAMS

What is consciousness?

A conscious individual is alert and attentive to his or her surroundings, while an unconscious individual is not aware of his or her surroundings. Conscious states, however, range from normal consciousness to the conscious yet unresponsive state, while unconscious states range from being asleep to being in a coma.

What is the Glasgow coma scale?

The Glasgow coma scale is a system of classifying the severity of head injuries or other neurologic diseases. It rates three areas of response, involving eye, verbal, and motor responses, and then tallies a total score. It is based on a fifteen-point score and is the most common scoring system used to describe the level of consciousness in a person following a traumatic brain injury.

What are the stages of sleep?

Data collected from EEGs (electroencephalograms) of brain activity during sleep have shown at least four separate stages of sleep. During stage 1, heart and breathing rates decrease slightly, the eyes roll slowly from side to side, and an individual experiences a floating sensation.

- *Stage 1* sleep is not usually classified as "true" sleep. This stage generally lasts only five minutes. Individuals awakened during stage 1 sleep will often insist that they were not sleeping, but merely "resting their eyes."

- *Stage 2* sleep is characterized by the appearance of short bursts of waves known as "sleep spindles" along with "K complexes," which are high-voltage bursts that occur before and after a sleep spindle. Eyes are generally still and heart and breathing rates decrease only slightly. Sleep is not deep.

- *Stage 3* sleep is intermediate sleep and is characterized by steady, slow breathing, a slow pulse rate, and a decline in temperature and blood pressure. Only a loud noise awakens sleepers in stage 3 sleep.

- *Stage 4* sleep, known as oblivious sleep, is the deepest stage. It usually does not begin until about an hour after falling asleep. Brain waves become even slower, and heart and breathing rates drop to 20 or 30 percent below those in the waking state. The sleeping individual in stage 4 sleep is not awakened by external stimuli, such as noise, although an EEG will indicate that the brain acknowledges such stimuli. Stage 4 sleep continues for close to an hour, after which the sleeper will gradually drift back into stage 3 sleep, followed by stages 2 and then 1, before the cycle begins again.

Why do people need sleep?

Scientists do not know exactly why people need sleep, but studies show that sleep is necessary for survival. Sleep appears to be necessary for the nervous system to work properly. While too little sleep one night may leave us feeling drowsy and unable to concentrate the next day, a prolonged period of too little sleep leads to impaired memory and physical performance. Hallucinations and mood swings may develop if sleep deprivation continues.

What is REM sleep?

REM sleep is rapid eye movement sleep. It is characterized by faster breathing and heart rates than NREM (nonrapid eye movement) sleep. The only people who do not have REM sleep are those who have been blind from birth. REM sleep usually occurs in four to five periods, varying from five minutes to about an hour, growing progressively longer as sleep continues.

What is the sleep cycle?

Typically, there are several cycles of sleep each night. Each cycle begins with a period of REM sleep. Earlier in the night there will be periods of stage 3 and stage 4 sleep, but these diminish towards morning, when there are longer periods of REM sleep and less deep sleep.

When does dreaming occur during the sleep cycle?

Almost all dreams occur during REM sleep. Scientists do not understand why dreaming is important. One theory is that the brain is either cataloging the information it acquired during the day and discarding the data it does not want or is creating scenarios to work through situations causing emotional distress. Regardless of its function, most people who are deprived of sleep or dreams become disoriented, unable to concentrate, and may even have hallucinations.

Dreams occur during REM sleep, it has been discovered, yet still no one understands for certain *why* we dream.

Why is it difficult to remember dreams?

The average person has three or four dreams each night, with each dream lasting ten minutes or more. But it appears that the content of dreams is stored in short-term memory and cannot be transferred into long-term memory unless they are somehow articulated. Sleep studies show that when individuals who believe they never dream are awakened when they are in the middle of a dream, they are more likely to recall the dream.

How much sleep does an individual need?

In 2015, the National Sleep Foundation published a study that took more than two years of research to complete. It was an update of its most-cited guidelines on how much sleep a person needs depending on his or her age. A new range labeled "may be appropriate" has been added to acknowledge the individual variability in appropriate sleep durations. The recommendations now define times as either (a) recommended; (b) may be appropriate for some individuals; or (c) not recommended. Though the research cannot pinpoint the exact amount of sleep needed by people at different ages, the new chart features minimum and maximum ranges for health as well as "recommended" windows. Nevertheless, as most health care professionals point out, it is important for an individual to pay attention to his or her own needs by assessing how he or she feels after different amounts of sleep. The following lists the new chart, along with comparisons of the old chart:

> ## How much time does a human spend sleeping in life?
>
> **B**y the time someone is twenty years old, he or she will have spent approximately eight years of his or her life asleep. By the age of sixty, he or she will have spent about twenty years sleeping.

- Newborns (0–3 months)—Sleep range narrowed to 14–17 hours each day (previously it was 12–18)
- Infants (4–11 months)—Sleep range widened two hours to 12–15 hours (previously it was 14–15)
- Toddlers (1–2 years)—Sleep range widened by one hour to 11–14 hours (previously it was 12–14)
- Preschoolers (3–5)—Sleep range widened by one hour to 10–13 hours (previously it was 11–13)
- School age children (6–13)—Sleep range widened by one hour to 9–11 hours (previously it was 10–11)
- Teenagers (14–17)—Sleep range widened by one hour to 8–10 hours (previously it was 8.5–9.5)
- Younger adults (18–25)—Sleep range is 7–9 hours (new age category)
- Adults (26–64)—Sleep range did not change and remains 7–9 hours
- Older adults (65+)—Sleep range is 7–8 hours (new age category)

How long can a person survive without sleep?

A total lack of sleep will cause death quicker than starvation. A person can survive a few weeks without food, but only ten days without sleep. Sleep-deprived individuals experience extreme psychological discomfort after a few days, followed by hallucinations and psychotic behavior.

What are some common sleep disorders?

The most common sleep disorder is insomnia. Insomnia is ongoing difficulty in falling asleep, staying asleep, or restless sleep. Technically, insomnia is a symptom of other sleep disorders. Consequently, treatment for insomnia depends on the primary cause of insomnia, which may be stress, depression, or too much caffeine or alcohol. Another sleep disorder is hypersomnia, or extreme sleepiness during the day even with adequate sleep the night before. Hypersomnia has been mistakenly blamed on depression, laziness, boredom, or other negative personality traits. Another is narcolepsy, which is characterized by falling asleep at inappropriate times. The sleep may last only a few minutes and is often preceded by a period of muscular weakness. Emotional events may trigger an episode of narcolepsy. Some individuals with narcolepsy experience a state called sleep paralysis. They wake up to find their body is paralyzed except for breathing and eye movement. In other words, the brain is awake but the body is still asleep.

What is sleep apnea and how does it affect sleep?

Sleep apnea is a breathing disorder in which an individual briefly wakes up because breathing has been interrupted and may even stop for a brief period of time. Obstructive sleep apnea (OSA) is the most common form of sleep apnea. It occurs when air cannot flow into or out of the person's nose or mouth as he or she breathes. According to the National Sleep Foundation, sleep apnea often occurs in conjunction with snoring. And although snoring may be harmless for most people, it can be an indication of apnea, especially if it is accompanied by severe daytime sleepiness. Men seem to have apnea more than women, as do people who are obese, but it can occur in anyone at any age for various other reasons, such as having a large neck or small airways (in the nose, throat, or mouth). A person who has apnea awakens frequently during the night gasping for breath. This pause in breathing can reduce the blood oxygen in the body, putting a strain on the cardiovascular system (including the heart), which is why many people with sleep apnea are considered at risk for cardiovascular disease.

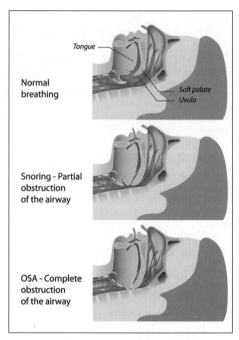

Snoring is caused when the airway becomes obstructed; sleep apnea is a more extreme case when the blockage becomes so bad that breathing can stop, disturbing a person's sleep.

When does sleepwalking occur during the sleep cycle?

Sleepwalking generally occurs during deep sleep but may also be present during periods of NREM (non-rapid eye movement) sleep. It is most common in children, although the National Sleep Foundation estimates that 1 percent to 15 percent of the population may be sleepwalkers. Sleepwalkers generally remain asleep and do not remember leaving their beds. Contrary to popular myth, sleepwalkers should be awakened, although they may be confused when awakened.

What are circadian rhythms?

Circadian (from the Latin *circa*, meaning "about," and *dies*, meaning "day") are the regular, internal body rhythms. Circadian rhythm is a roughly twenty-four-hour cycle in the physiological processes of every living organism, including plants, animals, fungi, protists, viruses, and bacteria. Depending on the organism, these rhythms can be associated with feeding, hormone secretions, brainwave activity, cell regeneration, and/or sleeping patterns.

Although human lives revolve around a twenty-four-hour day, researchers have found that normal circadian rhythms are more on a twenty-five-hour cycle. Many physiological processes, including the sleep/wake cycle, body temperature, gastric secretion, and kidney function, follow a set pattern. For example, body temperature peaks in the late afternoon/early evening and is lowest between 2:00 A.M. and 5:00 A.M. Blood pressure, heartbeat, and respiration follow rhythmical cycles. The production of urine drops at night, allowing for uninterrupted rest. Circadian rhythm disturbances occur when sleep/wake cycles are interrupted. They often affect shift workers whose biological clocks are disrupted by conflicting sleep and work schedules. "Jet lag" is another form of circadian rhythm disturbance.

COMPARING OTHER ORGANISMS

Do all animals have a nervous system?

No, not all animals have a nervous system like that of humans. Instead they have a nerve net, an interconnection of sensory neurons that allow them to respond to physical contact (especially to help find food) and to their surrounding environment. For example, hydra are simple animals with nerve nets.

What are some animal brain sizes compared to humans?

On the average, the human brain weighs 2.7 pounds (1.2 kilograms), representing about 2 percent of a person's body weight (although on the average, the male brain is about 0.3 pounds [100 grams] heavier than the average female brain). The largest brain belongs to the sperm whale and measures about 17.5 pounds (7 kilograms) in weight. The largest land animal brain belongs to the elephant, weighing an average 10.5 pounds (4.78 kilograms). Mountain gorillas have a brain that is 0.95 pounds (430 grams), while the smallest known primate brain belongs to the mouse lemur (*Microcebus murinus*) in Madagascar, weighing 0.004 pounds (2 grams).

How do human and other primate brains differ?

The human brain—with an average weight of about three pounds—is very large for body size. To compare, humans' closest relative, the chimpanzee, has relatively the same body size, but its brain is about one-third the size of a human. In addition, the human brain's cortex

Apes, such as the gorilla seen here, possess an intelligence that is impressive, but their brains are only about one third the size of a human's.

has more neurons and a greater number of fibers connecting certain brain regions, especially those responsible for language, reasoning, and toolmaking.

What are thought to be the most intelligent invertebrates?

Cephalopods (squid, octopi, and nautilus) are unique among the invertebrates because of their intelligence. For example, octopi can be taught to associate geometric shapes with either punishment (a mild electric shock) or reward (food). This can then be used to train them to avoid one type of food and reach for another. Research has indicated that octopi are also tool users; with their flexible arms and suckers, octopi are able to manipulate their environment, as in building a simple home. After an octopus has selected a home site, it will narrow the entrance size by moving small rocks.

Besides humans, what vertebrates are the most intelligent?

The answer to this question is highly debated—even saying "besides humans" is bound to result in a heated discussion. Many lists of "the most intelligent" animals exist, and some of them include such vertebrates as rats and sheep. But according to the famous American behavioral biologist Edward O. Wilson (1929–), the ten most intelligent animals are as follows: 1) chimpanzee (two species); 2) gorilla; 3) orangutan; 4) baboon (seven species, including drill and mandrill); 5) gibbon (seven species); 6) monkey (many species, especially macaques, the patas, and the Celebes black ape); 7) smaller toothed whale (several species, especially killer whale); 8) dolphin (many of the approximately eighty species); 9) elephant (two species); 10) pig. Still another list based on a scientific study includes 1) dolphins; 2) chimps and orangutans; 3) elephants; 4) parrots; and 5) crows. Like all similar scientific studies, the list of animals not only changes as new information is learned, but also varies depending on the researcher's criteria for "intelligent."

Do all land animals sleep?

All land animals sleep, but some "sleep" in different ways. For example giraffes can go without sleep for weeks; rats seem to have the same sleep needs as humans, as they need night sleep to hunt during the day. Large land animals sleep less than four hours on the average per day, such as elephants and cows (because they spend so much time during the day grazing and eating). And some birds sleep with one eye open to keep track of predators.

Do aquatic animals sleep?

Certain marine animals have unique sleep habits. For example, when most species of marine mammals are asleep, one hemisphere of their brain is still awake. Dolphins are known to reach the surface to breathe while "sleeping." It was once thought that sharks never sleep, but now researchers say that most of these animals either keep their brain active enough to respond to some degree—similar to what happens when a human walks in their sleep—or tend to have periods of inactivity, essentially taking a "rest" on the ocean floor.

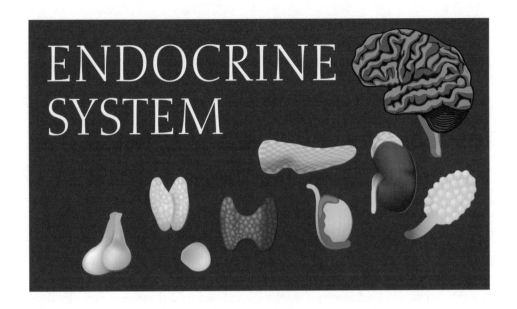

ENDOCRINE SYSTEM

INTRODUCTION

What are the functions of the endocrine system?

The endocrine system, together with the nervous system, controls and coordinates the functions of all of the human body systems. The endocrine system helps to maintain homeostasis and metabolic functions, allows the body to react to stress, and regulates growth and development, including sexual development.

What are the similarities between the nervous system and the endocrine system?

Both the nervous system and endocrine system are devoted to maintaining homeostasis by coordinating and regulating the activities of other cells, tissues, organs, and systems. Both systems are regulated by negative feedback mechanisms. Chemical messengers are important in both systems, although their method of transmission and release differs in the two systems.

How does the endocrine system differ from the nervous system?

Both the endocrine and nervous systems are regulatory systems that permit communication between cells, tissues, and organs. A major difference between the endocrine system and nervous system is the rate of response to a stimulus. In general, the nervous system responds to a stimulus very rapidly, often within a few milliseconds, while it may take the endocrine system seconds and sometimes hours or even days to offer a response. Furthermore, the chemical signals released by the nervous system typically act over very short distances (a synapse), while hormones in the endocrine system are generally carried by the blood to target organs. Finally, the effects of the nervous system generally last only a brief amount of time, while those of the en-

163

docrine system are longer lasting. Examples of endocrine control are growth and reproductive ability.

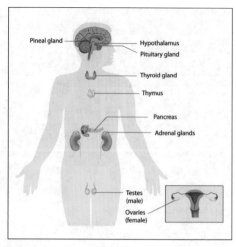
The endocrine system.

What are the organs of the endocrine system?

The endocrine system consists of glands and other hormone-producing tissues. Glands are specialized cells that secrete hormones into the interstitial fluid. Hormones are then transported to the capillaries and circulated via the blood. The major endocrine glands are the pituitary, thyroid, parathyroid, pineal, and adrenal glands. Other hormone-secreting organs are the central nervous system (hypothalamus), kidneys, heart, pancreas, thymus, ovaries, and testes. Some organs, such as the pancreas, secrete hormones as an endocrine function but have other functions also.

HORMONES

What are hormones?

Hormones are chemical messengers that are secreted by the endocrine glands into the blood. Hormones are transported via the bloodstream to reach specific cells, called target cells, in other tissues. They produce a specific effect on the activity of cells that are remotely located from their point of origin.

What are target cells?

Target cells are specific cells that respond to a specific hormone. Target cells have special receptors on their outer membranes that allow the individual hormones to bind to the cell. The hormones and receptors fit together much like a lock and key.

How are hormones classified?

Scientists classify hormones broadly into two classes: those that are soluble in water (hydrophilic) and those that are not soluble in water (hydrophobic) but are soluble in lipids. The chemical structure of hormones determines whether they are water-soluble or lipid-soluble. Water-soluble hormones include amine, peptide, and protein hormones. The steroid hormones are lipid-soluble.

What are the major groups of hormones?

The major groups of hormones are amine hormones, peptide and protein hormones, and steroid hormones. Amine hormones are relatively small molecules that are structurally similar to amino acids. Epinephrine and norepinephrine, serotonin, dopamine, the thyroid hormones, and melatonin are examples of amine hormones.

Peptide hormones and protein hormones are chains of amino acids. The peptide hormones have three to forty-nine amino acids, while the protein hormones are larger with chains of fifty to two hundred or more amino acids. Examples of peptide hormones are antidiuretic hormone and oxytocin. The larger thyroid-stimulating hormone and follicle-stimulating hormone are examples of protein hormones.

Steroid hormones are derived from cholesterol. Cortisol and the reproductive hormones (androgens in males and estrogens in females) are examples of steroid hormones.

Do hormones affect behavior?

Endocrine functions and hormones interact with every other organ system in the human body. Individuals whose hormone levels are abnormal, either due to oversecretion or undersecretion of a particular hormone, will show signs of abnormal behavior and illness. Children whose sex hormones are produced at an early age, for example, may demonstrate aggressive and assertive behavior in addition to the physical characteristics of maturation. In adults, changes in hormonal levels may have significant effects on intellectual capabilities, memory, learning, and emotional states.

How do paracrine hormones differ from circulating hormones?

Local hormones become active without first entering the bloodstream. They act locally on the same cell that secreted them or on neighboring cells. Local secretion of pro-inflammatory factors increases extravasation from blood vessels to produce local edema and flare responses. Circulating hormones are more prevalent than local hormones. Once secreted, they enter the bloodstream to be transported to their target cells.

Which endocrine glands produce which hormones?

Each endocrine gland produces specific hormones. The following table explains the gland and the hormone or hormones produced:

Gland	Hormone(s) Produced
Anterior pituitary	Thyroid-stimulating hormone (TSH); adrenocorticotropic hormone (ACTH); follicle-stimulating hormone (FSH); luteinizing hormone (LH); prolactin (PRL); growth hormone (GH); melanocyte-stimulating hormone (MSH)
Posterior pituitary	Antidiuretic hormone (ADH); oxytocin
Thyroid	Thyroxine (T4); triiodothyronine (T3); calcitonin (CT)
Parathyroid	Parathyroid hormone (PTH)
Pineal	Melatonin
Adrenal (cortex)	Mineralocorticoids, primarily aldosterone; Glucocorticoids, mainly cortisol (hydrocortisone); corticosterone; cortisone
Adrenal (medulla)	Epinephrine (E); norepinephrine (NE)
Pancreas	Insulin; glucagon
Thymus	Thymosins
Ovaries (female)	Estrogens; progesterone
Testes (male)	Androgens, mainly testosterone

How long does a hormone remain active once it is released into the circulatory system?

Hormones that circulate freely in the blood remain functional for less than one hour. Some hormones are functional for as little as two minutes. A hormone becomes inactivated when it diffuses out of the bloodstream and binds to receptors in target tissues or is absorbed and broken down by cells of the liver or kidneys. Enzymes in the plasma or interstitial fluids that break down hormones also cause them to become inactivated. Other hormones (for example, renin) are activated by enzymes that cleave the active portion from a larger circulating precursor molecule.

What is the hormonal response to stress?

The stress response, also known as the general adaptation syndrome (GAS), has three basic phases: 1) the alarm phase, 2) the resistance phase, and 3) the exhaustion phase. The alarm phase is an immediate reaction to stress. Epinephrine is the dominant hormone of the alarm phase. It is released in conjunction with the sympathetic nervous system and produces the "fight or flight" response. Nonessential body functions such as digestive, urinary, and reproductive activities are inhibited.

The resistance phase follows the alarm phase if the stress lasts more than several hours. Glucocorticoids are the dominant hormones of the resistance phase. Endocrine secretions coordinate three integrated actions to maintain adequate levels of glucose in the blood. They are: 1) the mobilization of lipid and protein reserves, 2) the conservation of glucose for neural tissues, and 3) the synthesis and release of glucose by the liver. If the body does not overcome the stress during the resistance phase, the exhaustion phase begins. Prolonged exposure to high levels of hormones involved in the resistance phase leads to the collapse of vital organ systems. Unless there is successful intervention and it can be reversed, the failure of an organ system will be fatal.

Where are hormone receptors located within a cell?

Hormone receptors are located either on the surface of the cell membrane or inside the cell. Water-soluble hormones are not able to diffuse through the plasma membrane easily. Therefore, the receptors for these hormones are located on the surface of the cell. In contrast, lipid-soluble hormones are able to easily penetrate the cell membrane. The receptors for the lipid-soluble hormones are often located inside the cell.

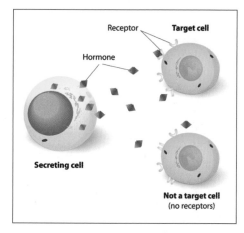

Hormones can be thought of as chemical messages that are targeted to specific cells in the body. Non-target cells can't be influenced by the wrong hormones for them because they do not "fit" properly.

How does aging affect the endocrine system?

Most endocrine glands continue to function and secrete hormones throughout an individual's lifetime. The most noticeable change in hormonal output is in the reproductive hormones. The ovaries decrease in size and no longer respond to FSH and LH, resulting in a decrease in the output of estrogens.

Although the hormonal levels of other hormones may not change with aging and remain within normal limits, some endocrine tissues become less sensitive to stimulation. For example, elderly people may not produce as much insulin after a carbohydrate-rich meal is eaten. It has been suggested that the decrease in function of the immune system is a result of the reduced size of the thymus gland.

PITUITARY GLAND

Where is the pituitary gland located?

The pituitary gland is located at the base of the brain directly below the hypothalamus in the sella turcica ("Turkish saddle"), a depression in the sphenoid bone. It is protected on three sides by the bones of the skull and on the top by a tough membrane called the diaphragma sellae.

How large is the pituitary gland?

The pituitary gland is about the size of a plump lima bean, or, as some say, the size of a pea. It measures 0.39 inches (1 centimeter) long, 0.39 to 0.59 inches (1 to 1.5 centimeters) wide, and 0.12 inches (0.5 centimeters) thick.

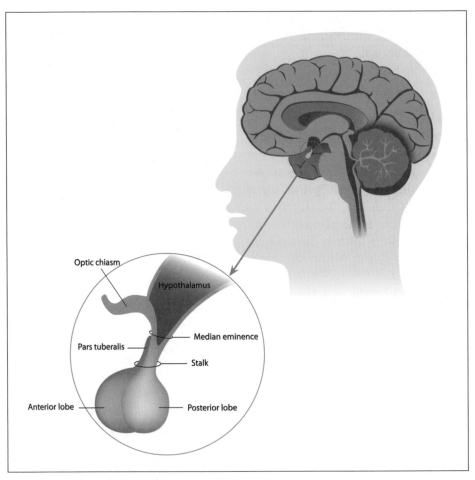

The pituitary gland is a small, but very important, gland at the base of the brain that secretes vital hormones.

What are the differences between the two regions of the pituitary gland?

The pituitary gland is divided into an anterior lobe (or adenohypophysis) and a posterior lobe. The anterior lobe is the larger section of the pituitary, accounting for 75 percent of the total weight of the gland. The anterior lobe contains endocrine secretory cells, which produce and secrete hormones directly into the circulatory system via an extensive capillary network that surrounds the region. The posterior lobe (or neurohypophysis) does not manufacture any hormones. It contains the axons from two different groups of hypothalamic neurons. Hormones produced in the hypothalamus are transported from the hypothalamus to the posterior pituitary within the axons.

How did the pituitary gland receive its name?

The term pituitary comes from the Latin *pituitarius*, or "of phlegm." The name was given to this particular gland because of the erroneous belief around the sixteenth and seventeenth centuries that the gland channeled mucus (phlegm) from the brain into the nose through the palate.

How many different hormones are secreted by the pituitary gland?

The pituitary gland is often called the "master gland," because it is responsible for the release of so many hormones. In all, there are nine different peptide hormones released by the pituitary gland. Seven are produced by the anterior pituitary gland and two are secreted by the posterior pituitary gland. The hormones of the anterior pituitary gland are thyroid-stimulating hormone (TSH), adrenocorticotropic hormone (ACTH), follicle-stimulating hormone (FSH), luteinizing hormone (LH), prolactin (PRL), growth hormone (GH), and melanocyte-stimulating hormone (MSH). The hormones of the posterior pituitary gland are antidiuretic hormone (ADH) and oxytocin.

Which pituitary gland hormones are trophic hormones?

Trophic (from the Greek *trophikos*, meaning "turning toward" or "to change") hormones are hormones that regulate the production of other hormones by different endocrine glands. These hormones "turn on," or activate, the target endocrine glands. The trophic hormones are thyroid-stimulating hormone (TSH), adrenocorticotropic hormone (ACTH), follicle-stimulating hormone (FSH), and luteinizing hormone (LH).

What are the targets of the trophic hormones?

The target of each of the trophic hormones is another endocrine gland. The following lists the pituitary tropic hormones and their targets:

Hormone	Target
Thyroid-stimulating hormone (TSH)	Thyroid gland
Adrenocorticotropic hormone (ACTH)	Adrenal glands
Follicle-stimulating hormone (FSH)	Gonads
Luteinizing hormone (LH)	Gonads

Which medical condition is caused by low production of gonadotropins?

Hypogonadism is caused by abnormally low production of gonadotropins. Children with this condition will not undergo sexual maturation. Adults with hypogonadism cannot produce functional sperm or oocytes (eggs).

How do follicle-stimulating hormone (FSH) and luteinizing hormone (LH) have different actions in males and females?

Both follicle-stimulating hormone and luteinizing hormone are gonadotropic hormones. In females, FSH promotes the growth and development of follicle cells in ovaries. Follicle cells surround a developing oocyte. In response to FSH they grow and develop to the point that one ruptures and expels an ovum to be fertilized. In males, FSH stimulates the production of sperm in the testes.

Luteinizing hormone (LH) induces ovulation, the release of an egg by the ovary, in females. It also stimulates the secretion of estrogen and the progestins, such as

169

progesterone. In males, LH stimulates the production and secretion of androgens, the male sex hormones, including testosterone.

What are the functions of prolactin in females?

Prolactin has two major functions in females. First, it works together with other hormones to stimulate the development of the ducts in the mammary glands. Secondly, it stimulates the production of milk after childbirth. Most researchers believe that prolactin has no effect in males, while some believe it may help regulate androgen production.

Which cells and tissues are most affected by human growth hormone?

Human growth hormone, sometimes called just growth hormone (GH), affects all parts of the body associated with growth. The skeletal muscles and cartilage cells are especially sensitive to the levels of growth hormone. One of the most direct effects of growth hormone is to maintain the epiphyseal plates of the long bones, where growth takes place.

Which conditions result from disorders of the human growth hormone?

A deficiency of human growth hormone in children causes growth at a slower than normal rate during puberty. Slow epiphyseal growth results in short stature and larger than normal adipose tissue reserves. In contrast, if there is no decrease in the secretion of GH towards the end of adolescence, the individual will continue to grow to seven or even eight feet tall, resulting in gigantism. When GH is overproduced after normal growth has ceased, a condition called acromegaly (from the Greek *akros*, meaning "extremity," and *megas*, meaning "great" or "big") occurs. Although the epiphyseal discs cartilages have closed, the small bones in the head, hands, and feet continue to grow, thickening rather than lengthening.

What is the function of antidiuretic hormone (ADH)?

The primary function of antidiuretic hormone (ADH), or vasopressin, is to decrease the amount of urine excreted and increase the amount of water absorbed by the kidneys. It plays a critical role in regulating the balance of fluids in the body. Secretion of ADH increases in response to fluid loss, such as dehydration. Hemorrhaging causes an increase in ADH secretion in order to maintain the body's fluid balance. And strenu-

Why does consumption of alcoholic beverages increase urination?

Alcohol inhibits the secretion of ADH. When the secretion of ADH is decreased, the amount of urine excreted increases. Excess alcohol consumption can lead to dehydration, a major symptom associated with a hangover, as a result of increased urination.

ous exercise, emotional or physical stress, and drugs such as nicotine or barbiturates all increase the secretion of ADH in order to decrease the amount of urine excreted.

What are the functions of oxytocin?

Oxytocin (from the Greek *oxy*, meaning "quick," and *tokos*, meaning "childbirth") stimulates contractions of the smooth muscle tissue in the wall of the uterus during childbirth. Prior to the late stages of pregnancy, the uterus is relatively insensitive to oxytocin. As the time of delivery approaches, the muscles become sensitive to increased secretion of oxytocin.

After delivery, oxytocin stimulates the ejection of milk from the mammary glands. The suckling of an infant stimulates the nerve cells in the brain (the hypothalamus) to release oxytocin. Once oxytocin is secreted into the circulatory system, special cells contract and release milk into collecting chambers from which the milk is released. This reflex is known as the milk let-down reflex.

What external factors can influence the female milk let-down reflex?

The milk let-down reflex may be controlled by a factor that affects the hypothalamus. Anxiety and stress can prevent the flow of milk. Some mothers learn to associate a baby's crying with suckling. These women may experience the milk let-down reflex as soon as they hear their baby crying.

THYROID AND PARATHYROID GLANDS

What are the physical characteristics of the thyroid gland?

The thyroid gland is located in the neck, anterior to the trachea, just below the larynx (the voice box). It has two lobes connected by a slender bridge of tissue called the isthmus. The average weight of the thyroid gland is 1.2 ounces (34 grams). An extensive, complex blood supply gives the thyroid a deep red color.

What are the types of cells in the thyroid gland?

The main types of cells in the thyroid are follicular cells. These cells produce T4 (thyroxine) and T3 (triiodothyronine). Less numerous are the parafollicular cells (or C cells), which are also found between the follicles in the thyroid gland. They produce the hormone calcitonin and control how the body uses calcium. Other, less common cells include immune system cells (lymphocytes) and supportive (stromal) cells.

Where are the thyroid hormones stored in the thyroid gland?

Thyroid hormones are stored in the spherical sacs called follicles. The thyroid follicles, microscopic spherical sacs, are composed of a single layer of cuboidal epithelium tissue. The thyroid hormones are stored in a gelatinous colloid.

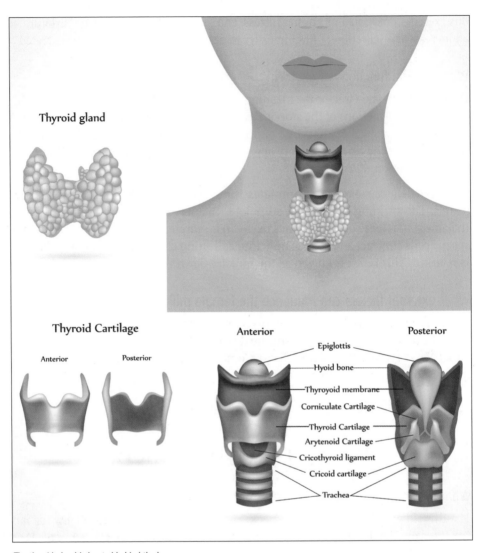

Thyroid gland

Thyroid Cartilage

Anterior

Posterior

Anterior

Posterior

Epiglottis

Hyoid bone

Thyroyoid membrane

Corniculate Cartilage

Thyroid Cartilage

Arytenoid Cartilage

Cricothyroid ligament

Cricoid cartilage

Trachea

The thyroid gland is located behind the larynx.

What is a unique characteristic of the thyroid gland?

The thyroid gland is the only gland that is able to store its secretions outside its principal cells. In addition, the stored form of the hormones is different from the actual hormone that is secreted into the blood system. Enzymes break down the stored chemical prior to its release into the blood.

How does triiodothyronine (T3), differ from thyroxine (T4)?

Thyroxine, or T4, also called tetraiodothyronine, contains four atoms of iodine. Triiodothyronine, or T3, contains only three atoms of iodine. The more common hormone is T4, which accounts for nearly 90 percent of the secretions from the thyroid. The amount of T3 in the body is concentrated and very effective. Both hormones have similar functions. Enzymes in the liver can convert T4 to T3.

172

What are the functions of the thyroid hormones?

Thyroid hormones affect almost every cell in the body. Some important effects of thyroid hormones on various cells and organ systems are as follows:

- Increases body metabolism by increasing the rate at which cells use oxygen and food to produce energy
- Causes the cardiovascular system to be more sensitive to sympathetic nervous activity
- Increases heart rate and force of contraction of heart muscle
- Maintains normal sensitivity of respiratory centers to changes in oxygen and carbon dioxide concentrations
- Stimulates the formation of red blood cells to enhance oxygen delivery
- Stimulates the activity of other endocrine tissues
- Ensures proper skeletal development in children

Why is it important to include iodized salt in one's diet?

Iodized salt provides an adequate supply of iodine in the diet, a necessary element for the production of thyroid hormones. A quarter teaspoon of salt (1.5 grams) provides 67 micrograms of iodine, which is about half of the U.S. Recommended Daily Allowance of 150 micrograms per day for adults. Most adults get this amount in the foods they eat during the day, for example, from breads and cereals that contain a certain amount of iodinized salt.

What medical condition is caused by an overactive thyroid gland?

Hyperthyroidism is the clinical term for an overactive thyroid gland. Patients with hyperthyroidism produce excessive quantities of the thyroid hormones thyroxine and triiodothyronine, which speed up the metabolic processes and functions of the body. Symptoms of hyperthyroidism include sudden weight loss without dieting, rapid and racing heartbeat, nervousness, irritability, tremors, increased perspiration, more frequent bowel movements, and, for women, changes in menstrual patterns.

What is the most common cause of hyperthyroidism?

The most common cause of hyperthyroidism is Graves' disease. Graves' disease is an autoimmune disorder in which antibodies produced by the immune system stimulate the thyroid to produce too much thyroxine. In Graves' disease, antibodies mistakenly attack the thyroid gland and occasionally the tissue behind the eyes and the skin of the lower legs. Some people with Graves' disease have exophthalmos, a bulging of the eyes. Treatment options include antithyroid drugs, radioactive iodine, or surgery. Radioactive iodine is the most common treatment for hyperthyroidism. Some thyroid cells will absorb the radioactive iodine. After a period of several weeks, the cells that took up the radioactive iodine will shrink and thyroid hormone levels will return to normal.

What is thyroid cancer?

Cancer of the thyroid often appears as an enlargement on one side of the thyroid. It is much less common than the benign thyroid nodules—abnormal growths of thyroid tissue—that can form on the gland. Four main types of thyroid cancer are papillary, follicular, medullary, and anaplastic thyroid cancer, and are based on how the cancer cells look under a microscope.

What is a goiter?

A goiter is an enlargement of the thyroid gland. Goiters are often associated with an underactive thyroid (hypothyroidism), although other conditions may also produce a goiter. For example, one or many thyroid nodules can disrupt the secretion of the thyroid hormone, causing a goiter. It can also be the result of Hasiomoto's disease (see below).

What are some common symptoms of hypothyroidism?

Hypothyroidism is caused by a lack of thyroid hormone. In general, the metabolism is slower and fatigue is common. Other symptoms include feeling cold, having dry skin and hair, constipation, weight gain, muscle cramps, and, for women, an increased menstrual flow. These symptoms may be cured or at least controlled by administering a synthetic form of the thyroid hormone thyroxine.

What is Hasiomoto's disease?

Like Graves' disease—which causes hyperthyroidism—Hashimoto's disease is an autoimmune disorder. The disease damages the thyroid, causing it to produce very little thyroid hormone. This causes the pituitary gland to produce more TSH to stimulate the thyroid, and in turn, the gland increases in size and often creates a goiter.

Where are the parathyroid glands located?

The parathyroid glands are embedded in the posterior surface of the thyroid gland. There are usually four parathyroid glands—two pairs—on each side of the thyroid. The parathyroid glands are tiny, pea-shaped glands weighing only a total of 0.06 ounces (1.6 grams) together. Each measures approximately 0.1 to 0.3 inches (3 to 8 millimeters) in length, 0.07 to 0.2 inches (2 to 5 millimeters) in width, and 0.05 inches (1.5 millimeters) in depth.

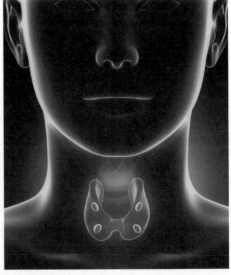

Parathyroid glands secrete a hormone to regulate levels of calcium and phosphorous in the blood.

When were the parathyroid glands discovered?

The parathyroid glands were discovered in 1880 by Swedish medical student Ivar Victor Sandström (1852–1889). They were the last major organ to be discovered in humans.

What is the function of the parathyroid glands?

The parathyroid glands secrete parathyroid hormone (PTH). The main function of PTH is to regulate the levels of calcium and phosphate in the blood.

How does parathyroid hormone increase the level of calcium in the blood?

There are four ways that parathyroid hormone (PTH) increases the level of calcium in the blood.

1. PTH stimulates osteoclasts to break down bone tissue and release calcium ions from the bone.

2. PTH inhibits osteoblasts to reduce the rate of calcium deposition in bone.

3. PTH enhances the absorption of calcium and phosphate from the small intestine in conjunction with the secretion of calcitriol by the kidneys.

4. PTH promotes the reabsorption of calcium at the kidneys, reducing the amount of calcium excreted in urine.

What is the relationship between parathyroid hormone (PTH) and calcitonin?

The thyroid gland secretes calcitonin when the calcium level in the blood is elevated. When the blood calcium level drops, the parathyroid glands increase the secretion of parathyroid hormone until the blood calcium level increases to the normal value. Homeostasis of blood calcium levels is maintained through the interaction of calcitonin and PTH.

Are all four parathyroid glands necessary to maintain homeostasis?

No, not all of the parathyroid glands are needed to maintain homeostasis. The secretion of even a portion of one gland can maintain normal calcium concentrations. When calcium concentration levels are abnormally high, though, a condition called

hyperparathyroidism results. This is typically caused by a tumor, and surgical removal of the overactive tissue will often correct the imbalance.

What are the most common causes of hypoparathyroidism?

Hypoparathyroidism is most often the result of damage to or removal of the parathyroid glands during surgery or when a tumor is on the parathyroid gland. It is suspected that prior to the discovery of the parathyroid glands, many patients died following thyroid surgery since the parathyroid glands were accidentally removed, resulting in abnormally low levels of calcium.

ADRENAL GLANDS

What are the physical characteristics of the adrenal glands?

The adrenal (from the Latin, meaning "upon the kidneys") glands sit on the superior tip of each kidney. Each adrenal gland weighs approximately 0.19 ounces (7.5 grams). The glands are yellow in color and have a pyramid shape. Each adrenal gland has two sections that may almost be considered as separate glands. The inner portion is the adrenal medulla (from the Latin *marrow*, meaning "inside"). The outer portion, which surrounds the adrenal medulla, is the adrenal cortex (from the Latin, meaning "bark," because its appearance is similar to the outer covering of a tree). The adrenal cortex is the larger part of the adrenal glands, accounting for nearly 90 percent of the gland by weight.

How many different types of hormones are secreted by the adrenal cortex?

The adrenal cortex secretes more than two dozen different steroid hormones called the adrenocortical steroids, or simply corticosteroids. The adrenal cortex is divided into three major zones or regions, each of which secretes a different type of corticosteroid. The outer region is the zona glomerulosa, which produces mineralocorticoids. The middle zone is the zona fasciculata, which accounts for the bulk of the cortical volume and produces glucocorticoids. The innermost, and smallest region, is the zona reticularis, which produces small quantities of the sex hormones.

What are the functions of the corticosteroids?

The corticosteroids are vital for life and well-being. Each of the corticosteroids serves a unique purpose. The following lists the corticosteroids and their functions:

Hormone	Target	Effects
Mineralocorticoids	Kidneys	Increases reabsorption of sodium ions and water from the urine; stimulates loss of potassium ions through excretion of urine

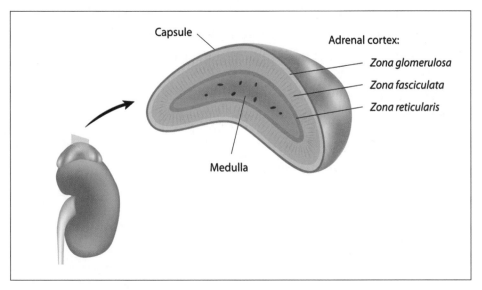

The adrenal glands are located atop the kidneys.

Hormone	Target	Effects
Glucocorticoids	Most cells	Releases amino acids from skeletal muscles, lipids from adipose tissues; promotes liver glycogen and glucose formation; promotes peripheral utilization of lipids; anti-inflammatory effects
Androgens		Promotes growth of pubic hair in boys and girls; in adult women, promotes muscle mass, blood cell formation, and supports the libido; in adult men, adrenal androgens are less significant because androgens are released primarily from the gonads

Which is the main mineralocorticoid hormone?

Mineralocorticoids are responsible for regulating the concentration of minerals in the body fluids. The main mineralocorticoid is aldosterone. It increases reabsorption of sodium from the urine into the blood and it stimulates excretion of potassium into the urine. Aldosterone is released in response to a drop in the blood sodium concentration or blood pressure or to a rise in the blood potassium concentration.

What are the glucocorticoid hormones?

Cortisol, corticosterone, and cortisone are the three most important glucocorticoid hormones. Cortisol, also called hydrocortisone, is the most abundant glucocorticoid produced, accounting for nearly 95 percent of the activity of the glucocorticoids.

What are the effects of glucocorticoids on the body?

Glucocorticoids have many varying effects on the body. The following lists those effects:

177

- The most critical effect of the glucocorticoids is the stimulation of glucose synthesis and glycogen formation, especially within the liver.
- They stimulate the release of fatty acids from adipose tissue, which can be used as an energy source.
- They decrease the effects of physical and emotional stress, such as fright, bleeding, and infection, since the additional supply of glucose from the liver provides tissues with a ready source of ATP. (For more about ATP, see the chapter "Anatomy and Biology Basics.")
- They suppress allergic and inflammatory reactions.
- They decrease and suppress the activities of white blood cells and other components of the immune system.

What two disorders are associated with abnormal glucocorticoid production?

Addison's disease and Cushing's syndrome are both disorders caused by abnormal glucocorticoid production. Addison's disease occurs when the adrenal glands do not produce enough of the hormone cortisol and, in some cases, the hormone aldosterone. Common symptoms of Addison's disease are chronic, worsening fatigue, muscle weakness, loss of appetite, and weight loss. Treatment of Addison's disease involves replacing, or substituting, the hormones that the adrenal glands are not making.

Cushing's syndrome is caused by prolonged exposure of the body's tissues to high levels of the hormone cortisol. The symptoms vary, but most people have upper body obesity, a characteristic rounded "moon" face, increased fat around the neck, and thinning arms and legs. The skin, which becomes fragile and thin, bruises easily and heals poorly. The symptoms of Cushing's syndrome may appear with prolonged use of prescribed glucocorticoid hormones, including prednisone.

Which are the two main hormones secreted by the adrenal medulla?

The adrenal medulla secretes epinephrine (also called adrenaline) and norepinephrine (also called noradrenaline). Epinephrine makes up 75 to 80 percent of the secretions from the adrenal medulla, with norepinephrine accounting for the remainder. These hormones are similar to those released by the sympathetic nervous system, but their effects last longer because they remain in the blood for longer periods of time. While the adrenal medulla assists in the "fight or flight" response, it is not necessary for survival.

PANCREAS

Where is the pancreas located?

The pancreas (from the Greek, meaning "all flesh") is located in the abdominopelvic cavity between the stomach and the small intestine. It is an elongated organ about 6 inches (12 to 15 centimeters) long.

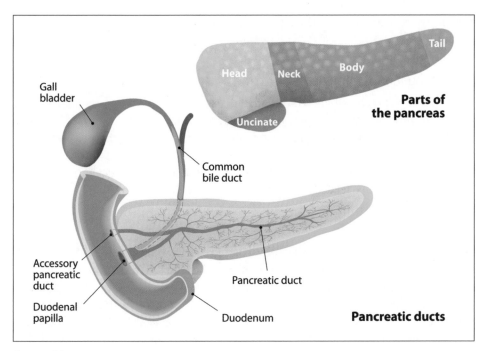

Anatomy of the pancreas

Why is the pancreas called a mixed gland?

The pancreas is a mixed gland because it has both endocrine and exocrine functions. As an endocrine gland, it secretes hormones into the bloodstream. Only 1 percent of the weight of the pancreas serves as an endocrine gland. The remaining 99 percent of the gland has exocrine functions. (For more on the function of the pancreas as an exocrine gland, see the chapter "Digestive System.")

Which cells of the pancreas secrete hormones?

The pancreatic islets (islet of Langerhans) are cluster cells that secrete hormones. There are between 200,000 and 2,000,000 pancreatic islets scattered throughout the adult pancreas.

Who first described the pancreatic islets?

German pathologist and biologist Paul Langerhans (1847–1888) was the first to provide a detailed description of microscopic pancreatic structures in the late 1860s. He noticed unique polygonal cells in the pancreas. It was not until 1893 that French pathologist Gustave E. Laguesse (1861–1927) discovered that the polygon-shaped cells were the endocrine cells of the pancreas that secreted insulin.

How many different types of cells are found in the islets of Langerhans?

There are four different types of cells in each of the islets of Langerhans. The four groups of cells are alpha cells, beta cells, delta cells, and F cells. The two most important types of cells are alpha cells, which produce glucagon, and beta cells, which produce insulin.

What is the function of glucagon?

Glucagon is secreted when the blood glucose levels fall below normal values. Glucagon stimulates the liver to convert glycogen to glucose, which causes the blood glucose level to rise. Glucagon also stimulates the production of glucose from amino acids and lactic acid in the liver. Glucagon stimulates the release of fatty acids from adipose tissue. When blood glucose levels rise, the secretion of glucagons decreases as part of the negative feedback system.

Who discovered insulin?

Insulin was discovered by Canadian physician Frederick Banting (1891–1941), Scottish biochemist and physiologist John James R. Macleod (1876–1935), and Canadian medical scientist Charles Best (1899–1978). Although earlier researchers had suspected that the pancreas secreted a substance that controlled the metabolism of sugar, it was not proved until 1922, when Banting, Macleod, and Best announced their discovery. The Nobel Prize in Physiology or Medicine in 1923 was awarded to Banting and Macleod. Banting shared his half of the prize with Best, while Macleod shared his half of the prize with Canadian biochemist James Bertram Collip (1892–1965), who had also collaborated with the team.

What is the function of insulin?

Insulin is secreted when blood glucose levels rise above normal values. One of the most important effects of insulin is to facilitate the transport of glucose across plasma membranes, allowing the diffusion of glucose from blood into most body cells. It also stimulates the production of glycogen from glucose. The glucose is then stored in the liver to be released when blood glucose levels drop.

When was the structure of insulin determined?

The full structure of insulin, a peptide hormone, was determined in 1955 by British biochemist Frederick Sanger (1918–2013). It was the first protein to have its full structure determined. Sanger won the Nobel Prize in Chemistry in 1958 for his re-

When was human insulin first synthesized?

Human insulin was first synthesized in 1978 by Arthur Riggs (c. 1939–) and Keiichi Itakura (1942–) using *E. coli* bacteria with recombinant DNA technology. The first human insulin, Humulin, was not marketed until 1982. Although called "human insulin," most insulin available today is synthetic insulin. It is almost identical to the insulin produced by the human pancreas.

search (he also won a Nobel Prize in Chemistry in 1980, only one of two people to have ever done so in the same category).

Which medical condition is caused by the body's inability to produce or use insulin?

Diabetes mellitus is a disorder of the metabolism caused when the pancreas either produces little or no insulin, or when the cells do not respond appropriately to the insulin that is produced. Glucose builds up in the blood, overflows into the urine, and passes out of the body. As a result, the body does not benefit from glucose as a source of energy.

How long has diabetes been known as a disease?

Diabetes mellitus as a disease has been known since around the first century. The first reference is from the celebrated physician Aretaeus the Cappadocia (c. 1 C.E.–?), who called it *diabainein*, from the Greek *dia* ("through") and *bainein* ("to go"), referring to the excessive urination associated with the disease. The term "diabetes" was first noted in 1425 (as *diabete*), from the Latin, which in turn, comes from the ancient Greek word meaning "siphon." The word *mellitus* was added around 1675, from the Latin for "like honey" to reflect the sweet smell and taste of the patient's urine.

What is the difference between Type I and Type II diabetes?

Diabetes mellitus is a hormonal disease that occurs when the body's cells are unable to absorb glucose from the blood. Type I is insulin-dependent diabetes mellitus (IDDM), and Type II is noninsulin-dependent diabetes mellitus (NIDDM). Insulin is completely deficient in Type I diabetes. In Type II diabetes, insulin secretion may be normal, but the target cells for insulin are less responsive than normal.

How does diabetes insipidus differ from diabetes mellitus?

Diabetes mellitus results from an inability to produce insulin, while diabetes insipidus is the result of the pituitary not releasing sufficient quantities of antidiuretic hormone (ADH). Water conservation at the kidneys is impaired and excessive amounts of urine are excreted.

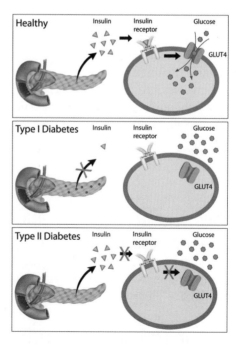

In Type I diabetes the pancrease does not make enough insulin, while in Type II it does, but the target cells can't receive the insulin as they are supposed to.

PINEAL GLAND

What is the pineal gland?

The pineal gland (from the Latin *pinea*, meaning "pinecone") is a small gland located in the midbrain at the posterior end of the third ventricle. The physiological functions of the pineal are unclear. It secretes the hormone melatonin, which appears to be associated with circadian rhythms and setting the biological clock.

What is melatonin and when was it first discovered?

Melatonin is a hormone that is mostly released at night; its secretion diminishes during the day when it is light. The release of melatonin can be influenced by artificially mimicking day and night, such as with indoor lighting; also, its effects on sleep explain why it is used in some medications to induce sleep. It was discovered by American physician and researcher Aaron B. Lerner (1920–2007) in 1958. American neuroscientist Richard J. Wurtman (1936–) did much of the pioneering research on the benefits of melatonin.

What is seasonal affective disorder (SAD)?

Seasonal affective disorder (SAD) is a type of depression that affects some individuals during the winter months when there is less sunlight. One hypothesis is that since there are fewer hours of daylight during the winter months, the production of melatonin is affected, resulting in physical ailments such as drowsiness and lethargy. Additional symptoms of SAD include a craving for carbohydrates, increased appetite, weight gain, mood swings, and anxiety. While for severe cases antidepressants may be recommended for a SAD patient, many researchers believe light therapy is also an effective treatment for a moderate case of SAD. Light therapy, also called phototherapy, involves sitting near a specially designed light box that produces a strong light. Most light boxes emit a light of 2,500 to 10,000 lux, which is between the average living room lighting of 100 lux and a bright sunny day of about 100,000 lux.

REPRODUCTIVE ORGANS

Which reproductive organs secrete hormones?

The gonads (from the Greek *gonos*, meaning "offspring") release hormones in both males and females. In males, the testes secrete hormones, while in females the ovaries are responsible for secreting hormones. (For more about reproduction, see the chapter "Reproductive System.")

What are the hormones of the reproductive glands?

The androgen hormone testosterone is the most important male hormone secreted by the testes. The testes also produce inhibin, which inhibits the secretion of FSH (follicle-stimulating hormone). The three major hormones secreted by the ovaries are estrogens, progestins, and relaxin.

Do males have estrogen and females have testosterone in their respective systems?

Many people erroneously believe that estrogen and testosterone are gender-specific, and that males do not have estrogen and females do not have testosterone in their respective systems. In reality, both hormones are found in both sexes. As a man ages, testosterone is converted to estrogen. The estrogen is created by the aromatase reaction; aromatase is an important enzyme in the formation of estrogen, and it is found in fat cells. Thus in general, the more body fat the more estrogen a male will have in his system. In males, an excess in estrogen as they age can lead to several health issues, including an increased risk of heart attacks and strokes. (In females, it is the opposite: after menopause, as estrogen levels *drop*, women are more at risk for heart attacks, strokes, and osteoporosis.)

In females the ovaries produce the sex hormones, testosterone, and estrogen, but only a fraction of the amount each day a male produces. In addition to the ovaries, the adrenal glands also release small amounts of testosterone into the bloodstream. The two hormones are secreted in short bursts, and the amounts can vary hourly and daily and can depend on the day in the woman's menstrual cycle. Women with more testosterone than average may have deeper voices, have more hair, experience irregular or no menstrual cycles, and experience increased muscle mass. As a woman ages, the testosterone levels change, and in particular, after menopause, a woman's testosterone level declines.

What are the functions of the sex hormones?

Testosterone is stimulated by luteinizing hormone (LH) from the pituitary gland. It regulates the production of sperm, as well as the growth and maintenance of the male sex organs. Testosterone also stimulates the development of the male secondary sex characteristics, including growth of facial and pubic hair. It causes the deepening of the male voice by enlarging the larynx.

The estrogens are stimulated by follicle-stimulating hormone (FSH) in the pituitary. They help regulate the menstrual cycle and the development of the mammary glands and female secondary sex characteristics. Luteinizing hormone (LH) stimulates the secretion of progestins. Progesterone prepares the uterus for the arrival of a developing embryo in case fertilization occurs. It also accelerates the movement of an embryo to the uterus. Relaxin helps enlarge and soften the cervix and birth canal at the time of delivery. It causes the ligaments of the pubic symphysis to be more flexible at the time of delivery.

What are anabolic steroids?

Properly called anabolic-androgenic steroids, anabolic steroids are hormones that work to increase synthesis reactions, particularly in muscle. They are synthetic ver-

sions of the primary male sex hormone testosterone. They promote the growth of skeletal muscle (anabolic effects) and the development of male sexual characteristics (androgenic effects).

How are anabolic steroids abused?

Anabolic steroids are often abused by teenagers, adults, and athletes, both professional and amateur, for bodybuilding and to enhance athletic performance. Although they have clinical applications for certain medical conditions, the doses prescribed legally to treat these medical conditions are ten to one hundred times lower than the doses that are abused for performance enhancement.

Steroid abuse can lead to serious side effects, ranging from oily skin to disruption of sex hormones and the cardiovascular system, liver disease, and infections. Sometimes these problems can be lethal.

OTHER SOURCES OF HORMONES

When was the hormone leptin discovered?

Leptin is a recently discovered hormone. American geneticist Jeffrey Friedman (1954–) and his colleagues published a paper in December 1994 announcing the discovery of a gene in mice and humans called obese (ob) that codes for a hormone he later named leptin, after the Greek word *leptos*, meaning "thin." Leptin is a hormone made by the body's adipose tissue that regulates food intake and energy expenditure. Individuals who lack leptin eat tremendous amounts of food and become obese.

What is the thymus gland?

The thymus gland produces several hormones, called thymosins, which stimulate the production and development of T cells, which play an important role in immunity. (For more on this subject, see the chapter, "Lymphatic System.") The thymus gland is located in the mediastinum, generally posterior to the sternum and between the lungs. It is a double-lobed lymphoid organ well supplied with blood vessels but few nerve fibers. The outer cortex of the thymus has many lymphocytes, while the inner medulla contains fewer lymphocytes.

How is the thymus divided?

The irregular-shaped thymus is divided into two halves, called lobes. Its surface is made up of many small bumps called lobules. The gland has three main layers: the medulla (innermost part of the thymus), cortex (layer surrounding the medulla), and capsule (thin outer covering of the gland).

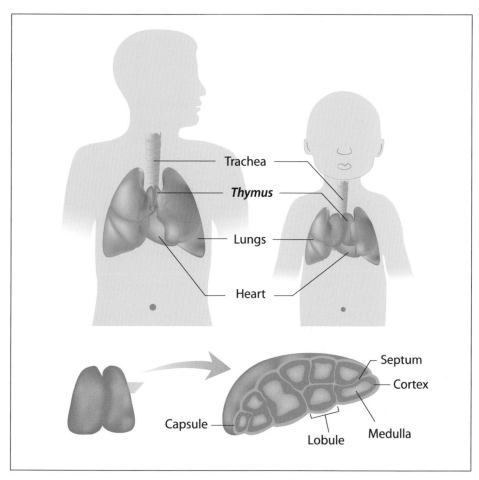

The thymus gland varies in size from infancy to adulthood.

Why is the thymus called the "shrinking gland"?

The thymus gland is largest during infancy, weighing about one ounce. It reaches its maximum effectiveness during adolescence after which time the size of the gland begins to decrease. By age fifty, the thymus gland has atrophied to a fraction of its original size and is replaced by adipose (fat) tissue.

Which other organs perform endocrine functions?

Many organs perform endocrine functions in addition to their main functions. Some of these have been discussed in detail in this chapter; others are described in the following chart of hormones produced by other organs:

Organ	Hormone(s)	Effects
Intestines	Gastrin	Promotes secretion of gastric juice and increases motility in the stomach
	Secretin	Stimulates secretion of pancreatic juice and bile

Organ	Hormone(s)	Effects
Cholecystokinin	Stimulates secretion of pancreatic juice; regulates release of bile from the gall bladder	Brings about a feeling of fullness after eating
Kidneys	Erythropoietin	Stimulates red blood cell production
	Calcitrol	Stimulates calcium and phosphate absorption; stimulates calcium ion release from the bone and inhibits PTH release
Heart	Atrial natriuretic peptide (ANP)	Increases water and salt loss at kidneys; increases thirst; suppresses secretion of ADH and aldosterone
Adipose tissue	Leptin	Suppresses appetite

COMPARING OTHER ORGANISMS

How are some birds affected by their hormones?

Many birds are affected by their natural hormones. For example, migrating birds move twice a year, with most species flying great distances to and from their ancestral breeding grounds, usually from cold to warm areas. In the warmer spring in the Northern Hemisphere, many species of birds return to the north to mate, lay eggs, and raise their young. In the fall, shortening days and colder weather trigger hormonal changes that signal to the birds to return to the warmer climates in the south, where food is more plentiful.

Who discovered the first known animal hormone?

British physiologists William Bayliss (1860–1924) and Ernest Starling (1866–1927) discovered secretin in 1902. They used the term "hormone" (from the Greek word *horman*, meaning "to set in motion") to describe this chemical substance that stimulated an animal's organ at a distance from the chemical's site of origin. Their famous experiment using anesthetized dogs demonstrated that diluted hydrochloric acid, mixed with partially digested food, activated a chemical substance in the duodenum. The activated substance secretin was released into the bloodstream and came into contact with cells of the pancreas; in turn, in the pancreas, it stimulated secretion of digestive juice into the intestine through the pancreatic duct.

What is bovine growth hormone?

One of the earliest applications of biotechnology was the genetic engineering of a growth hormone produced naturally in the bovine pituitary. The recombinant Bovine Growth Hormone (rBGH), or Bovine Somatotropin (rBST), a genetically engineered hormone manufactured by Monsanto, was reported to increase milk production in lactating cows. Using biotechnology, scientists bioengineered the gene that controls bovine growth hormone production into *E. coli* bacteria, grew the bacteria in fermentation chambers, and thus produced large quantities of hormone. The bioengineered hormone, when injected into lactating cows, resulted in an increase of up to 20 percent in national milk production. Using bovine GH, farmers were able to stabilize milk production in their herds, avoiding fluctuations in production levels. But the use of rBGH is extremely controversial, as many people believe such hormones in milk are detrimental to human health.

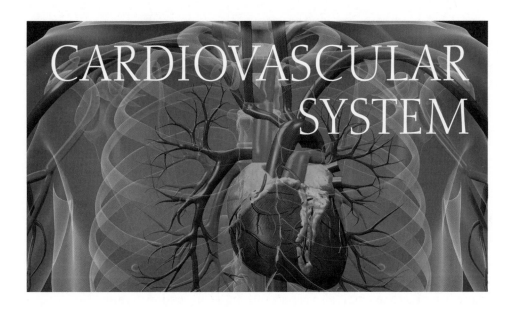

CARDIOVASCULAR SYSTEM

INTRODUCTION

What are the functions of the cardiovascular system?

The cardiovascular system provides a transport system between the heart, lungs, and tissue cells. The most important function is to supply nutrients to tissues and remove waste products.

What is the difference between the cardiovascular system and the circulatory system?

The cardiovascular system refers to the heart (cardio) and blood vessels (vascular). The circulatory system is a more general term encompassing the blood, blood vessels, heart, lymph, and lymph vessels.

What are the various types of circulation?

Several major types of circulation occur in the body. The circulation of blood through the heart is called the coronary circulation. Blood moving through the body organs is called the systemic circulation; this also often includes renal circulation (kidneys) and hepatic circulation (liver). The circulation of blood through the pulmonary artery, lungs, and pulmonary vein is called the pulmonary circulation.

Which structures and organs constitute the cardiovascular system?

Technically speaking, the structures of the cardiovascular system are the heart and blood vessels. Blood, a connective tissue, plays a major role in the cardiovascular system and is usually discussed within the context of the cardiovascular system.

189

What are some common cardiovascular diseases?

Cardiovascular disease is a generic term for diseases of the heart (cardio) and blood vessels (vascular). Some cardiovascular diseases are congenital (present at birth), while others are acquired later in life. Heart diseases affect the heart, arteries that supply blood to the heart muscle, or valves that ensure that blood in the heart is pumped in the correct direction. Examples of heart disease are coronary artery disease (diseases of the arteries, which supply the heart with blood), valvular heart disease (diseases affecting the heart valves), congenital heart disease, and heart failure. Disorders of the blood vessels include arteriosclerosis, hypertension (high blood pressure), stroke, aneurysm, venous thrombosis (formation of blood clots in a vein), and varicose veins.

How does exercise affect the cardiovascular system?

According to the American Heart Association, physical inactivity is a major risk factor for heart disease, stroke, and coronary artery disease. Regular aerobic physical activity (brisk walking, running, jogging) not only increases general fitness levels and capacity for exercise, but it also plays a role in the prevention of cardiovascular disease. Regular physical activity can also control blood lipids. Other benefits of regular aerobic physical activity include reducing high blood pressure, reducing triglyceride levels, and increasing HDL ("good") cholesterol levels.

BLOOD

What is the composition of blood?

Blood is classified as a connective tissue because it has both fluid and solid (cellular) components. The fluid is plasma, in which plasma proteins and cells (red blood cells, white blood cells, and platelets) are suspended in the watery base.

What are the functions of blood?

The functions of blood can be divided into three general categories: transportation, regulation, and protection. Transportation includes the movement of gases (oxygen and carbon dioxide), nutrients, and metabolic wastes; regulation includes stabilizing the body's temperature, blood pH, and fluid volume and pressure of the blood; and protection includes fighting infections and protecting against blood loss.

What is the normal pH of blood?

The normal pH of arterial blood is 7.4, while the pH of venous blood is about 7.35. Arterial blood has a slightly higher pH because it has less carbon dioxide. (For more about pH, see the chapter "Anatomy and Biology Basics.")

Why is blood sticky?

Blood is sticky because it is denser than water and about five times more viscous than water. Blood is viscous mainly due to the red blood cells. When the number of

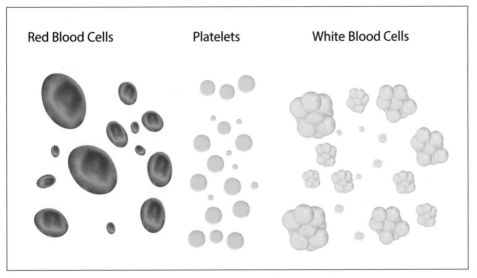

| Red Blood Cells | Platelets | White Blood Cells |

Platelets, red, and white cells comprise the blood in our veins and arteries. Red cells deliver oxygen, white blood cells fight infection, and platelets aid in healing wounds.

these cells increases, the blood becomes thicker and flows slower. Conversely, if the number of red blood cells decreases, blood thins and flows faster.

Where are red blood cells formed?

Red blood cell formation, or erythropoiesis, occurs in the red bone marrow located in the vertebrae, sternum, ribs, skull, scapulae, pelvis, and proximal limb bones. Red blood cells begin as large, immature cells (proeythroblasts), and over a seven-day period they change into a much smaller, mature, red blood cell that then enters the blood stream.

What factors can affect the rate of red blood cell formation?

The rate of red blood cell formation is stimulated by the hormone erythropoietin, which is released by the kidneys. Erythropoietin increases the rate of red blood cell division and maturation of immature red blood cells. If the blood oxygen level decreases, due to anemia, disease, or high altitude, erythropoietin is released.

How numerous are red blood cells?

Red blood cells account for one-third of all cells in the body and 99.9 percent of the cells in blood. If all the red blood cells in the human body were stacked on top of one another, they would create a tower 31,000 miles (49,890 kilometers) high.

How thick are red blood cells and how fast are they replaced in the body?

A stack of five hundred blood cells would measure only 0.04 inches (0.10 centimeters) high. Between two million and three million red blood cells enter the blood stream each second, replacing the same number that are destroyed each second.

Why do red blood cells have such a short life span?

The average red blood cell lives for 120 days. Red blood cells are subject to mechanical stress as they flow through the various blood vessels in the body, creating tremendous wear and tear. After about 120 days, the cell membrane ruptures and the red blood cell dies.

Why are red blood cells disc-shaped?

Red blood cells are perhaps the most specialized cells in the human body. They are a biconcave (donut) shape with a thin central disc. This shape is important because the disc increases the surface-area-to-volume ratio for faster exchange of gases and it allows red blood cells to stack, one on another, as they flow through very narrow vessels. Also, since some capillaries are as narrow as 0.00015748 inches (0.004 millimeters), red blood cells can literally squeeze through narrow vessels by changing shape.

How does blood transport oxygen?

Red blood cells contain hemoglobin, which is responsible for both oxygen and carbon dioxide transport in the blood. Hemoglobin is a complex protein made of four polypeptide chains, each of which has the unique ability to bind oxygen. Amazingly enough, there are about 280 million molecules of hemoglobin in each red blood cell with the potential ability to carry a billion molecules of oxygen. Each polypeptide chain has a single molecule of heme, which has an iron ion at its center. It is the iron ion that interacts with oxygen.

What are the major disorders that affect red blood cells?

The main erythrocyte disorders are anemias in which blood has a low oxygen carrying capacity. Anemia is a symptom of an underlying disease such as lowered red blood cell count, low hemoglobin levels, or abnormal hemoglobin formation. A decrease in red blood cells may be due to hemorrhage (excessive loss of blood), iron deficiency, or reduced oxygen availability. High altitudes or some cases of pneumonia may lead to reduced oxygen availability. (For more about anemia, see the chapter "Urinary System.")

Normal red blood cell
with normal hemoglobin

Mutated hemoglobin forms
strands that cause sickle shape

In sickle cell anemia, red blood cells form stiff rods that can result in clotting. Sickle cell is found more frequently in areas where malaria is prevalent, and the mutation has been found to help fend off that disease.

What is sickle cell anemia?

Sickle cell anemia is an inherited disorder in which there is a mutation in one of the 287 amino acids that make up the beta chain of hemoglobin. This slight change causes abnormal folding of the hemoglobin chains so that the hemoglobin forms stiff rods. The resultant red blood cells are crescent shaped (hence the term "sickle cells") when they unload oxygen or when the oxygen content of blood is reduced. These crescent shaped red blood cells are fragile and can rupture easily and form clots in small vessels.

What is artificial blood?

Artificial blood is a blood substitute that can be used to provide fluid volume and carry oxygen in the vessels. Two characteristics that a blood substitute should have are that it should be thinner than real blood and it should have a low affinity for oxygen so that oxygen can be delivered easily. The benefits of artificial blood are that it lessens the demand for human blood supplies and it can be given immediately without triggering a rejection in cases of massive blood loss.

Synthetic chemical compounds called perfluorocarbons are currently being studied as a substitute for red blood cells. For such a substitute to be acceptable, it needs to be: 1) able to carry oxygen and release it to tissues; 2) nontoxic; 3) storable; 4) able to function for varying periods of time in the human body; and, 5) immune-response resistant.

What is blood doping?

Blood doping refers to the use of artificial means to increase the number of red blood cells. Erythropoietin, a hormone produced by both the kidney and liver, increases the rate of maturation of erythrocytes in the red bone marrow. A genetically engineered form of erythropoietin (EPO) can be injected into athletes before an event, with a resulting increase of red blood cell count from 45 to 65 percent. With more red blood cells, there is more oxygen available for heavily exercising muscles. However, once an athlete becomes dehydrated, the blood can become too thick, causing clotting, stroke, or serious heart problems.

What are leukocytes?

Leukocytes, or white blood cells, make up less than 1 percent of total blood volume. Their main function is immunological defense. There are two major categories of leukocytes, which are based on their structural characteristics: granulocytes contain membrane-bound granules, while agranulocytes lack membrane-bound granules. The total number of white blood cells is 4,500 to 10,000 per cubic millimeter of blood. A differential white blood cell count refers to the percentage of each of the five types of white blood cells.

Where are white blood cells stored?

Most of the white blood cells in the body are stored in connective tissues or lymphatic organs. The white blood cells are released in response to areas of invasion by pathogens or injury.

How does the white blood cell count change with age?

At birth, a newborn has a higher white blood cell count (9,000 to 30,000 cells per cubic millimeter of blood), but this number falls to adult levels (4,500 to 10,000 cells per cubic millimeter of blood) within two weeks of birth. The total white blood cell count decreases slightly in the elderly.

What are the critical values for white blood cell count?

An individual with a white blood cell count of less than 500 is at high risk for infection. Leukopenia is a condition in which there are low numbers of white blood cells. A white blood cell count greater than 30,000 is an indicator of a major infection or a serious blood disorder such as leukemia. Leukocytosis is a condition in which there are very large numbers of white blood cells.

What are the functions of platelets?

Platelets are very small—1.575×10^{-3} inches (0.004 millimeters) in diameter and 3.9×10^{-5} inches (0.001 millimeter) thick. The functions of platelets include transport of enzymes and proteins critical to clotting, formation of a platelet plug to slow blood loss, and contraction of a clot after it has formed, which then reduces the size of the vessel break.

Where are platelets produced?

Platelets are produced in the bone marrow as very large cells up to 0.0063 inches (0.16 millimeters) that gradually break up cytoplasm into small packets. The small packets, four thousand of which can be released from one large cell, become the adult platelets.

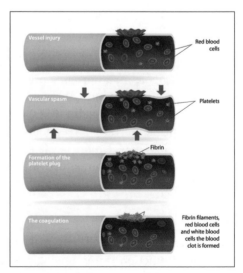

What happens if blood does not clot?

If blood clots slowly or not at all, a person is at great risk for blood loss from the smallest injury. The two most common clotting disorders are hemophilia and von Willebrand disease. Hemophilia affects mostly males and is caused by a

Hemostasis is the process of stopping the flow of blood after a cut or other injury.

deficiency in a specific clotting factor. Von Willebrand disease is due to a deficiency in a plasma protein that interacts with specific clotting proteins. In both diseases, the severity of the condition depends on how much of the specific protein is produced.

What is an anticoagulant?

An anticoagulant is any substance that prevents platelets from piling up in the inner lining (endothelium) of blood vessels. Endothelial cells naturally secrete nitric oxide and prostacyclin, which prevent platelets from sticking together. Another natural anticoagulant is heparin, which is found in basophils (a type of white blood cell) and on the surface of endothelial cells. It interferes with the process of clot formation.

What is plasma?

Plasma is the pale yellow liquid part of blood. It accounts for 46 to 63 percent of total blood volume, with an average of 55 percent. It is mostly water (95 percent) with a number of dissolved substances that add to its viscosity. The majority (92 percent) of the dissolved solutes are plasma proteins. Nonprotein components include metabolic waste products, nutrients, ions, and dissolved gases.

What are the major plasma proteins?

Plasma proteins are produced by the liver, with the exception of gamma globulins, which are produced by lymphatic tissue or tissues. There are three major plasma proteins: albumin (60 percent by weight) is important to maintain osmotic pressure in the body; globulins (36 percent by weight) include alpha and beta, which transport proteins, gamma, which are antibodies released during the immune response, and fibrinogens (4 percent by weight), which allow blood clot formation.

What is the difference between plasma and serum?

Plasma is whole blood minus cells, or the liquid in which various blood cells are suspended. It is collected by spinning fresh blood and an anticoagulant in a centrifuge in a test tube. After the cells fall to the bottom of the tube, the plasma is poured off. Serum is plasma minus the clotting proteins; it is often called "pure blood." Serum is collected by allowing a blood sample to clot. After clotting, the liquid that is left is drawn up is the serum.

How much blood does the average-sized adult human have?

An adult man has 5.3 to 6.4 quarts, or 1.5 gallons (5 to 6 liters), of blood, while an adult woman has 4.5 to 5.3 quarts, or 0.875 gallons (4 to 5 liters). Differences are due to the sex of the individual, body size, fluid and electrolyte concentrations, and amount of body fat.

What vein is usually used to collect blood for testing?

Fresh blood is usually collected from the median cubital vein (inside the elbow) in a procedure called venipuncture. If only a small amount of blood is needed—for example, to measure glucose (blood sugar) levels—the tip of a finger, an ear lobe, the big toe, or heel (in infants) are common areas for a pinprick.

How are components of blood separated?

A blood sample is centrifuged, which separates it into two components: plasma (55 percent) and formed elements (45 percent). Further centrifugation will separate plasma into proteins, water, and dissolved solutes. Additional centrifugation of the dissolved solutes will separate into platelets, leukocytes, and erythrocytes.

What are some tests that can be done on a collected sample of blood?

The table below lists common blood tests:

Blood Test	Purpose	Normal Range
Hematocrit (HCT)	Percent blood cells in whole blood	37–54%
Hemoglobin concentration	Hemoglobin concentration	12–18g/dl*
Red blood cell count	Number of red blood cells/microliter of whole blood	4.2–6.3 million/microliter
White blood cell count	Determines total number of circulating white blood cells	600–9,000

*g/dl = 5 grams/deciliter

Do males or females have a higher hematocrit?

Males have a higher hematocrit, or the volume percentage of blood cells in whole blood (also called packed cell volume, or PCV). Males have about 45 percent, whereas females have about 40 percent. This is because males have a greater capacity to carry oxygen in their blood in order to supply the greater muscle mass of their bodies.

What is the amount of carbon dioxide found in normal blood?

Carbon dioxide normally ranges from 19 to 50 millimeters per liter in arterial blood and 22 to 30 millimeters per liter in venous blood. Most of the carbon dioxide dissolved in the blood is in the form of bicarbonate (HCO_3-). If a physician wants to know about the carbon dioxide in a person's blood (also called the measure of the person's blood bicarbonate level), the serum (liquid part of the blood) is tested. The normal range is considered by most physicians to be 23 to 30 mEq/l (milliequivalents per liter).

Who discovered the ABO system of typing blood?

Austrian physician Karl Landsteiner (1868–1943) discovered the ABO system of blood types in 1909. Landsteiner had investigated why blood transfused to a patient was sometimes successful and other times resulted in death. He theorized that there must be several different blood types. If a transfusion occurs between two individuals with different blood types, the red blood cells will clump together, blocking the blood vessels. Landsteiner received the Nobel Prize in Physiology or Medicine in 1930 for his discovery of human blood groups.

What is the Rh factor?

In addition to the ABO system of blood types, blood types can also be grouped by the Rhesus factor, or Rh factor, an inherited blood characteristic. Discovered independently in 1939 by American pathologist Philip Levine (1900–1987) and American researcher Rufus E. Stetson (1886–1967) and in 1940 by Austrian physician Karl Landsteiner (1868–1943) and American physician Alexander S. Wiener (1907–1976), the Rh system classifies blood as either having the Rh factor or lacking it. Pregnant women are carefully screened for the Rh factor. If a mother is found to be Rh-negative, the father is also screened. Parents with incompatible Rh factors can have babies with potentially fatal blood problems. The condition can be treated with a series of blood transfusions.

Which of the major blood types are the most common in the United States?

The following table lists blood types and their rate of occurrence in the United States:

Blood Type	Frequency (U.S.)
O+	37.4%
O–	6.6%
A+	35.7%
A–	6.3%
B+	8.5%
B–	1.5%
AB+	2.4%
AB–	0.6%

What are the preferred and permissible blood types for transfusions?

The table below lists the blood types that are best matched with other blood types.

Blood Type of Recipient	Preferred Blood Type of Donor	Permissible Blood Type(s) of Donor in an Emergency
A	A	A, O
B	B	B, O
AB	AB	AB, A, B, O
O	O	O

How often can blood be donated?

According to the American Red Cross, as long as the potential donor is in good health, a person must wait at least 8 weeks (56 days) between donations of whole blood and 16 weeks (112 days) between double red cell donations. Platelet apheresis donors may give every 7 days up to 24 times per year.

What antigens and antibodies are associated with each blood type?

The table below explains the relationship between blood types and antigens and antibodies:

Blood Type	Antigen on Red Blood Cell Surface	Antibody in Plasma
A	A	Anti-B
B	B	Anti-A
AB	A and B	Neither anti-A nor anti-B
O	Neither A nor B	Both anti-A and anti-B

THE HEART

What is the size and location of the heart?

Heart size varies with body size. The average adult's heart is about 5.5 inches (14 centimeters) long and 3.5 inches (9 centimeters) wide, or approximately the size of one's fist. The heart is located just above the diaphragm, between the right and left lungs. One-third of the heart is located on the right size of the chest, while two-thirds is located on the left side of the chest.

How much does the heart weigh?

In an infant, the heart is about a thirtieth of total body weight. In an average adult, the heart is about one three-hundredth of total body weight; this equals about 11 ounces (310 grams) in males and 8 ounces (225 grams) in females.

How is cardiac muscle different from skeletal muscle?

Cardiac muscle, called the myocardium, is composed of a number of long, branching cells that are joined by intercalated discs. An intercalated disc is an area where cell membranes of adjacent cardiac muscle cells are joined. There are also small spaces in cardiac muscle cells that create a direct electrical connection between cells by allowing ions to move freely between cells. The interconnecting matrix joins cardiac muscle cells into a single, very large muscle cell called a syncytium (Latin for "joined cells"). Another difference between skeletal muscle

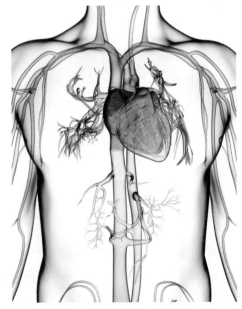

The heart is located, more of less, in the center of the chest and is approximately the size of a fist.

and cardiac muscle is that cardiac muscle has pacemaker cells, which initiate contractions rhythmically rather than through neural stimulation. Cardiac muscle contraction lasts about ten times longer than skeletal muscle contractions, and cardiac muscle cannot produce sustained contractions as skeletal muscles do.

What are the three layers of the heart wall?

The wall of the heart is composed of three distinct layers. They include an outer epicardium, a middle myocardium, and an inner endocardium.

What is pericarditis?

Pericarditis is an inflammation of the pericardium, a membrane that surrounds the heart. It is frequently due to viral or bacterial infections, which produce adhesions that attach the layers of the pericardium to each other. This is a very painful condition and interferes with heart movements. Mild cases of pericarditis may resolve themselves with little treatment other than bed rest and anti-inflammatory medications. More severe cases of percarditis may require hospitalization and/or surgical removal and drainage of the fluid. If a bacterial infection is the underlying cause of pericarditis, antibiotics will be prescribed to treat the infection.

What is angina pectoris?

Angina pectoris (from the Latin, meaning "strangling" and "chest") is severe chest pain that occurs when the heart muscle is deprived of oxygen. It is a warning that coronary arteries are not supplying enough blood and oxygen to the heart.

How is the heart protected from injury?

Since the heart is continuously moving, it is protected against friction by a large pericardial sac with an outer fibrous layer and an internal serous layer. The internal layer produces fluid, which lubricates the sac in which the heart moves. The syncytium of the myocardium wraps the cavities of the heart in a continuous muscular sheet.

How much pressure does the human heart create?

The human heart creates enough pressure when it pumps blood out of the body to squirt blood about thirty feet (ten meters).

What are the various chambers of the heart?

The heart is divided into two upper chambers called atria (singular, atrium) and two lower chambers called ventricles. The atria are receiving chambers, where blood is delivered via large vessels, and the ventricles are pumping chambers, where blood is pumped out of the heart via large arteries.

What is mitral valve prolapse?

Mitral valve prolapse (MVP) is a condition in which the mitral (bicuspid) valve extends back into the left atrium when the heart beats, causing blood to leak from the atrium to the ventricle. It may be due to genetic factors or a bacterial infection, caused by *Streptococcus* bacteria. Mitral valve prolapse affects up to 6 percent of the U.S. population. Surgery is sometimes required to repair a valve, although most people do not require any treatment.

Why is the left ventricle larger than the right ventricle?

Although the right and left ventricles contain equal amounts of blood, the left ventricle is larger because it has thicker walls. These thicker walls enable pressure to develop enough to push blood throughout the body. Since the right ventricle only pumps the blood to the neighboring lungs, the same ejection force is not required.

What are the main vessels entering and leaving the heart?

The main vessels entering the right side of the heart are the inferior and superior vena cava, which return low-oxygen blood to the right atrium. Blood leaves the right ventricle through the pulmonary artery to the lungs. High-oxygenated blood returns to the left atrium through the left and right pulmonary veins. All blood exits the left side of the heart through the aorta.

How fast and how often does the human heart beat?

The human heart beats 130 times per minute in infants and slows to 90 times per minute in a ten-year-old. By the time adulthood is reached, the heart slows to an average of 70 times per minute in men and 78 times per minute in women. The heart will beat approximately 40 million times in one year, or about 3 billion times in an average lifetime.

What are the lubb-dupp sounds that the heart makes?

Heart sounds are monitored by using a stethoscope, an instrument invented in 1816 by French physician René-Théophile-Hyacinthe Laënnec (1781–1826). Stethoscope comes from the Greek words meaning "to study the chest." The characteristic lubb-dupp sound that the heart makes is due to the closing of the two sets of valves. The "lubb" is due to closing of the atrioventricular valves, and the "dupp" is due to the closing of the semilunar valves.

How much blood is pumped by the heart?

On average, each heart contraction pumps 2.4 ounces (70 milliliters) of blood. The heart pumps 7,397 quarts (7,000 liters) of blood through the body each day. And depending on what a person is doing—whether exercising or typing at a desk on his or her computer—the heart pumps one to several gallons of blood per minute. Over the course of an average lifespan, that translates to about one million barrels of blood pumped by the heart.

How is the cardiac muscle supplied with blood?

Cardiac muscle has its own separate circulation, the coronary circulation. There are two large coronary arteries that supply the ventricles, with the most abundant blood supply going to the left ventricle, as this chamber has the most strenuous workload.

How is blood flow directed within the heart?

A system of valves prevents backflow both within chambers and in the large vessels exiting the heart. The atrioventricular valves are located between the right atrium and right ventricle (tricuspid valve) and left atrium and left ventricle (bicuspid valve). When the ventricles contract, blood moves back toward the atria, causing the flaps of these valves to close. The semilunar valves resemble a tripod and close after blood has exited the right ventricle (pulmonary semilunar valve) and left ventricle (aortic semilunar valve). When the ventricles are relaxed, the atrioventricular valves are open and the semilunar valves are closed. When the ventricles contract, the atrioventricular valves are closed and the semilunar valves are open. (For more about testing blood flow in the heart, see the chapter "Helping Human Anatomy.")

Where is the pacemaker of the heart located?

The pacemaker of the heart is located in the sinoatrial (SA) node in the right atrium. Cells of the sinoatrial node generate an impulse about seventy-five times per minute. The pacemaker coordinates heart rate through a system of nerve fibers that spread throughout the right and left atria. (For more about artificial pacemakers, see the chapter "Helping Human Anatomy.")

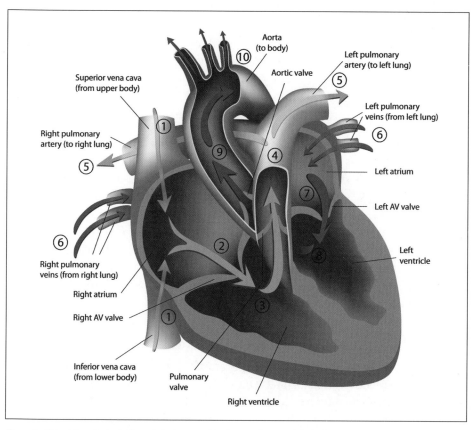

The path of blood from the body, through the heart, and back out to the rest of the body

What are the symptoms of a heart attack?

Although some heart attacks are sudden, most heart attacks start slowly, with mild pain. The following are warning signs of a heart attack:

1. Chest discomfort, usually in the center of the chest and lasting more than a few minutes

2. Discomfort in other areas of the upper body, such as one or both arms, the back, neck, jaw, or stomach

3. Shortness of breath

4. Other signs such as nausea, lightheadedness, fatigue, or breaking out in a cold sweat

There can also be some differences in heart attack symptoms between a male and a female. According to the American Heart Association, as with men, women's most common symptom of a heart attack is chest pain or discomfort. But women are more likely to experience symptoms of shortness of breath, nausea/vomiting, and back or jaw pain. (For more about heart operations, see the chapter "Helping Human Anatomy.")

How does exercise affect the heart?

Regular exercise increases the amount of blood the heart can eject with each beat, so fewer beats per minute are needed to maintain cardiac output. Exercise can increase cardiac output from 300 to 500 percent and increase heart rate up to 160 beats per minute. Individuals who exercise regularly tend to have lower resting heart rates.

BLOOD VESSELS

What are the blood vessels and their function?

Blood vessels form a closed circuit that carries blood from the heart to the organs, tissues, and cells throughout the body and then back to the heart. The blood vessels include arteries, arterioles, capillaries, venules, and veins. Arteries carry blood away from the heart under high pressure. The arteries subdivide into smaller, thinner tubes called arterioles. As the arterioles approach capillaries, the walls of the vessels become very thin. Capillaries have the smallest diameter of all the blood vessels. They connect the arterioles with the venules. Venules continue from the capillaries to form the veins.

What are the differences between arteries and veins?

Both arteries and veins have three tissue layers: the inner lining (endothelium), the middle layer (smooth muscle), and the outer layer (connective tissue). However, the walls of the arteries are much stronger and thicker to accommodate the blood under high pressure as it exits the heart. Many veins have valves to help return the blood to the heart.

What is an aneurysm?

An aneurysm is a bulge in the weakened wall of an artery, most often the aorta. It is similar to what one would see when there is a bubble in the wall of a garden hose. If an aneurysm becomes large enough, it can burst, resulting in a stroke if it is in a brain artery or massive hemorrhage if it is the wall of the aorta. If an aneurysm bursts, the massive bleeding is often fatal.

What are deep vein thrombosis and pulmonary embolism?

Deep vein thrombosis (DVT) occurs when a blood clot forms in a deep vein, usually in the lower leg, thigh, or pelvis (and less often in the arm). DVT usually occurs if there is an injury to a vein. It can happen under various conditions, including slow blood flow, for example, from bed confinement or sitting for a long time, especially with legs crossed.

The most serious complication of DVT occurs when a part of the clot breaks off and travels through the bloodstream to the lungs, causing a blockage called pulmonary embolism (PE). Because DVT and PE almost always occur together, doctors refer to the condition as venous thromboembolism, or VTE. Usually, if the clot from DVT is small, it is possible to recover from a resulting PE; if it is large, the clot can stop blood from reaching the lungs, and may be fatal.

What is the difference between arteriosclerosis and atherosclerosis?

Arteriosclerosis, also known as hardening of the arteries, occurs when the arterial walls thicken and then harden as calcium deposits form. If the coronary vessels are affected, this is known as coronary artery disease. Atherosclerosis is another type of hardening of the arteries in which lipids, particularly cholesterol, build up on the side arterial walls. Risk factors for atherosclerosis include cigarette smoking, a high fat/high cholesterol diet, and hypertension.

What is the largest artery in the human body?

The aorta is the largest artery in the human body. In adults, it is approximately the size of a garden hose. Its internal diameter is 1 inch (2.5 centimeters) and its wall is about 0.079 inches (0.2 centimeters) thick.

What is the largest vein in the human body?

The largest vein in the human body is the inferior vena cava, the vein that returns blood from the lower half of the body back to the heart.

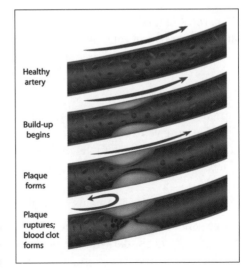

The stages of atherosclerosis

What are the functions of capillaries?

Capillaries are perhaps the most important of the blood vessels because they are the primary exchange points of the cardiovascular system. Gases, nutrients, and metabolic byproducts are exchanged between the blood in capillaries and the tissue fluid surrounding body cells. The materials exchanged move through capillary walls by diffusion, filtration, and osmosis.

How big are capillaries?

The diameter of a capillary is about 0.0003 inches (0.0076 millimeters), which is just about the same as a single red blood cell. A capillary is only about 0.04 inches long (1 millimeter). If all the capillaries in a human body were placed end to end, the collective length would be approximately 25,000 miles (46,325 kilometers), which is slightly more than the circumference of the earth at the equator: 24,900 miles (46,139 kilometers).

Where is most of the blood in the body at any one time?

Blood volume is not evenly distributed among the different types of vessels. Due to the expandable properties of veins, a vein will stretch about eight times more than an artery of corresponding size. At rest, the venous system thus contains about 65 to 70 percent of total blood volume, with the heart, arteries, and capillaries containing 30 to 35 percent of total blood volume. About one-third of the blood in the venous system is found in the liver, bone marrow, and skin.

What are varicose veins?

Varicose veins are distended veins, usually in the superficial veins of the thighs and legs. Varicose veins are caused by the valves inside the veins becoming stretched so they no longer close completely. The affected veins then become filled with blood. There are usually no serious medical problems associated with varicose veins.

Why does blood in the veins look blue?

Since venous blood is oxygen-poor blood, it is not as bright red as arterial blood. It appears as a deep, dark-red, almost purplish color. Seeing "blue blood" in veins through the skin is a combination of light passing through the skin and the oxygen-poor blood.

How does arterial bleeding differ from venous bleeding?

Each type of vessel (artery, vein, capillary) has distinct bleeding characteristics. Arterial bleeding is characterized by bright-red blood spurting each time the heart beats. Arterial bleeding is serious and difficult to control. Venous bleeding occurs in a steady flow, and the blood is dark-red, almost maroon, in color. Since capillaries are so small, capillary bleeding is a slow, oozing flow that carries with it a higher risk of infection than either arterial or venous bleeding.

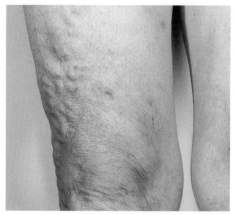

Varicose veins are caused when valves in the vein do not close properly and excess blood collects in the veins.

CIRCULATION

What is pulse?

Pulse is the alternate expansion and recoil of an artery, which can be felt in an artery close to the body's surface because of the rhythmic ejection of blood from the heart into the aorta, which causes an increase and decrease of pressure in the vessel. Pulse provides important information about the heart action, blood vessels, and circulation. A fast pulse rate may indicate the presence of an infarction or dehydration. In a medical emergency, the pressure of a pulse will help determine if a person's heart is pumping.

What are the normal values for the human pulse rate?

The following are normal values for a human pulse rate depending on age:

Age	Pulse Rate
Newborn	100–160 beats/minute
1–10 years old	70–120 beats/minute
10 to adulthood	60–100 beats/minute
Athletes	40–60 beats/minute

Where is the pulse most easily found?

Pulse can be found anywhere an artery is near the surface or over a bone. The body sites where pulse is most easily felt are:

- Wrist (radial artery)
- Temporal artery (in front of ear)

205

- Common carotid artery (along the lower margin of the jaw)
- Facial artery (lower margin of the lower jawbone)
- Brachial artery (at the bend of elbow)
- Popliteal artery (behind the knee)
- Posterior tibial artery (behind the ankle)
- Dorsalis pedis artery (on the upper surface of the foot)

How fast does blood move in vessels?

Blood flow refers to the volume of blood flowing through a vessel or group of vessels during a specific time. It is measured in milliliters per minute (mL/min). Blood flow during rest averages three to four mL/minute per 100 grams of muscle tissue but may increase to 80 mL/minute or more during exercise. The distance blood flows during a specific time period is its velocity. Blood velocity is measured as centimeters/second (cm/s). In general, blood velocity is greatest in larger vessels and decreases in vessels with a smaller diameter. The velocity of blood in the aorta is approximately 30 cm/s, in arterioles 1.5 cm/s, in capillaries 0.04 cm/s, in venules 0.5 cm/s, and in the venae cavae 8 cm/s.

What are the sounds of Korotkoff?

The sounds of Korotkoff are produced when blood pressure is taken. They are named after Nikolai Korotkoff (1874–1920), a Russian physician who first described them in 1905, when he used a stethoscope to listen to the sounds of blood flowing through an artery.

How do veins return blood to the heart?

Veins in the chest rely on the processes of inspiration and expiration of the lungs and the subsequent movement of the diaphragm to "pump" blood back to the heart. Other "pump" mechanisms are contractions of the skeletal muscles, which squeeze the veins within the muscles throughout the body. However, it is the semilunar valves found in veins that prevent blood backflow when the skeletal muscles relax.

What do the readings of blood pressure mean?

Blood pressure is monitored using a sphygmomanometer (*sphygmos*, meaning "pulse," and *manometer*, meaning a device for measuring pressure). Blood pressure is the pressure in the arterial system in the largest vessels near the heart as the heart pushes blood through vessels. It is measured in millimeters of mercury. The two numbers for blood pressure reflect two different pressures: systolic pressure and diastolic pressure. Systolic pressure (the upper number in a blood pressure reading) is the pressure of the blood against the arterial walls when the ventricles are contracting. Diastolic pressure (the lower number in a blood pressure reading) is the lowest point at which sounds can be heard as the blood flows through the smaller arteries.

What new guidelines were recently suggested concerning blood pressure readings?

According to the American Heart Association, a blood pressure reading below 120/80 mm Hg for an adult over age twenty is recommended. But these numbers are currently being debated. In 2013, research in the *Journal of the American Medical Association (JAMA)* suggested new guidelines for blood pressure readings depending on age and/or health conditions. For example, it was suggested that among adults age sixty and older with high blood pressure, aim for a target blood pressure under 150/90; among adults age thirty to fifty-nine with high blood pressure, aim for a target blood pressure under 140/90; and among adults with diabetes or chronic kidney disease, aim for a target blood pressure under 140/90. Even though the numbers are

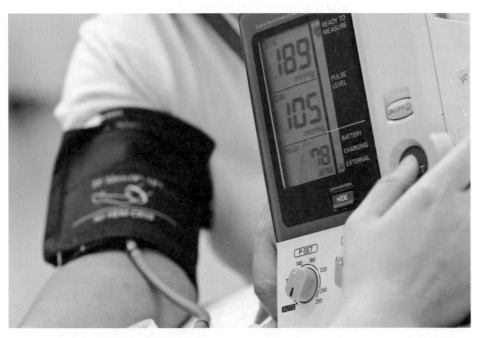

When you have your blood pressure taken, the high number is the "systolic" pressure that occurs when the ventricles contract, and the low number is the "dystolic" pressure that results when they relax.

"higher," doctors still recommend keeping pressure within the ranges mentioned above and to make lifestyle changes if necessary to keep it low.

What are hypertension and hypotension?

Hypertension is a sustained condition when the blood pressure exceeds 140/90 mm Hg. It is estimated that about 30 percent of people ages fifty and over have hypertension. Hypotension, or low blood pressure, is a systolic pressure below 100mm Hg. Most commonly, it is due to overly aggressive treatment for hypertension.

COMPARING OTHER ORGANISMS

Do all animals have blood?

Some invertebrates, such as flatworms and cnidarians, lack a circulatory system that contains blood. These animals possess a clear, watery tissue that contains some phagocytic cells, a little protein, and a mixture of salts similar to seawater. Invertebrates with an open circulatory system have a fluid that is more complex and is usually referred to as hemolymph (from the Greek term *haimo*, meaning "blood," and the Latin term *lympha*, meaning "water"); their hemoglobin is not concentrated in cells within the hemolymph but rather is found floating in the hemolymph. Invertebrates with a closed circulatory system have blood that is contained within blood vessels. Other examples of circulatory systems are found in squids, octopi, and crustaceans—animals that also have oxygen-carrying molecules in their plasma, but their bodies use the copper-based molecule hemocyanin to carry oxygen instead.

Do all organisms have red blood?

In all organisms that have blood, the blood's color is related to the compounds that transport oxygen. For example, hemoglobin containing iron is red and is found in all

---

What special adaptation do giraffes have in terms of blood flow to their brain?

Contrary to gravity, when a long-necked giraffe lowers its head to drink or eat, there is no excess blood flow to the brain. The giraffe has a special pressure regulation system in its upper neck called the *rete mirabile* (Latin for "wonderful net," a term coined by Roman physician and anatomist Galen [c. 130–c. 200 C.E]) that equalizes its blood pressure. Giraffes are not the only animals with this feature, but it is used for different reasons. For example, birds with webbed feet have a rete mirabile in their feet and legs that acts like a heat exchanger, keeping the animal's feet closer to the surrounding temperature; and penguins also have a rete mirabile in their nasal passages and flippers.

vertebrates and a few invertebrates (animals without backbones). Annelids (segmented worms) have either a green pigment, chlorocruorin, or a red pigment, hemerythrin. Some crustaceans (arthropods having divided bodies and generally having gills) have a blue pigment, hemocyanin, in their blood.

How does a human's heartbeat compare with those of other mammals?

The resting heartbeat of a human often differs from that of other mammals, mainly because certain animals have different needs in terms of their circulatory systems. For example, the average human has a resting heart rate (in beats per minute) of around 75. Other animals differ in their average resting heart rates, such as a horse (48), cow (45 to 60), dog (90 to 100), cat (110 to 140), rat (360), and mouse (498).

What are the differences between an open and a closed circulatory system?

In an open circulatory system, found in many invertebrates (for example, spiders, crayfish, and grasshoppers), the blood is not always contained within the blood vessels. Periodically, the blood leaves the blood vessels to bathe the tissues with blood and then returns to the heart; thus, no interstitial body fluid is separate from the blood. A closed circulatory system, also called a cardiovascular system, is found in all vertebrate animals—such as humans—and many invertebrates. In a closed system, the blood never leaves the blood vessels.

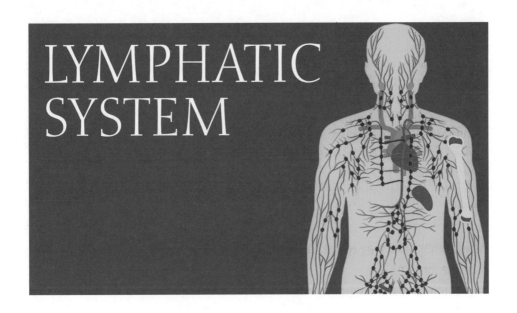

LYMPHATIC SYSTEM

INTRODUCTION

Which branch of medicine deals with the responses of the body to foreign substances?

Immunology (from the Latin *immunis*, meaning "free from services or obligations," and the Greek *ology*, meaning "the study of") is the study of cells and tissues that carry out immune responses.

What are the primary functions of the lymphatic system?

The lymphatic system is responsible for maintaining proper fluid balance in tissues and blood, in addition to its role defending the body against disease-causing agents. The primary functions of the lymphatic system are: 1) to collect the interstitial fluid that consists of excess water and proteins and return it to the blood; 2) to transport lipids and other nutrients that are unable to enter the bloodstream directly; and 3) to protect the body from foreign cells and microorganisms.

What are the three major components of the lymphatic system?

The lymphatic system consists of the lymphatic vessels, lymph, and lymphoid organs. Together these components form a network that collects and drains most of the fluid that seeps from the bloodstream and accumulates in the space between cells.

What are the primary cells of the lymphatic system?

The primary cells of the lymphatic system are lymphocytes. There are three types of lymphocytes: T cells, B cells, and NK cells. T cells account for approximately 80 percent of the circulating lymphocytes. They are thymus-dependent and are the primary cells that provide cellular immunity. B cells, which are derived from the bone marrow, account for 10 to 15 percent of the circulating lymphocytes. They are responsible for antibody-mediated immunity. NK (natural killer) cells account for the remaining 5 to 10 percent of the circulating lymphocytes. They attack foreign cells, normal cells infected with viruses, and cancer cells that appear in normal tissues.

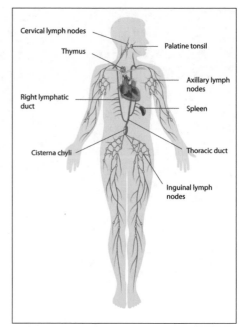

The lymphatic system

Which other cells play a role in the immune system?

White blood cells, including macrophages, neutrophils, eosinophils, basophils, and mast cells, all have active roles in the immune system. For example, macrophages, in particular, are important phagocytic cells that destroy many pathogens. Mast cells are specialized cells of connective tissue. They release heparin, histamine, leukotrienes, and prostaglandins to stimulate the inflammatory response.

How many lymphocytes are in the body?

There are approximately ten trillion lymphocytes, weighing over 2.2 pounds (1 kilogram) in the body. They account for 20 to 30 percent of the circulating white blood cell population.

How long do lymphocytes survive?

Lymphocytes have relatively long life spans. Nearly 80 percent survive for four years and some last as long as twenty years or more. They spend varying amounts of time in the blood, lymphatic vessels, and lymphatic organs. New lymphocytes are produced in the bone marrow and lymphoid tissues.

Why is phagocytosis so important to the human body?

Not only does phagocytosis allow bodies to remove potentially deadly invaders, it is also important in the maintenance of healthy tissues. Without this mechanism, non-functional materials would accumulate and interfere with the body's ability to function. A good example of the importance of phagocytosis is provided by the macrophages of the human spleen and liver, which dispose of more than ten billion aged blood cells daily.

What is the composition of lymph?

Lymph is similar in composition to blood plasma. The main chemical difference is that lymph does not contain erythrocytes. It also contains a much lower concentration of protein than plasma since most protein molecules are too large to filter through the capillary wall. Lymph contains water, some plasma proteins, electrolytes, lipids, leukocytes, coagulation factors, antibodies, enzymes, sugars, urea, and amino acids.

How much lymph is in the body?

The body contains approximately 1 to 2 quarts (1 or 2 liters) of lymph, which accounts for 1 to 3 percent of body weight. The lymphatic system returns slightly more than 3 quarts (3 liters) of fluid from the tissues to the circulatory system on a daily basis.

What is an autoimmune disease?

An autoimmune disease is one in which the body triggers an immune response against the body's own cells and tissues. The cause of most autoimmune diseases is unknown. Autoimmune diseases can affect almost every organ and body system. They may be systematic (affecting and damaging many organs) or localized (affecting only a single organ or tissue). The following lists the autoimmune diseases of various body systems:

Body System	Autoimmune Diseases
Blood and blood vessels	Autoimmune hemolytic anemia; pernicious anemia; polyarteritis nodosa; systemic lupus erythematosus; Wegener's granulomatosis
Digestive tract (including the mouth)	Autoimmune hepatitis; Behçet's disease; Crohn's disease; primary bilary cirrhosis; scleroderma; ulcerative colitis
Eyes	Sjögren's syndrome; Type 1 diabetes mellitus; uveitis
Glands	Graves' disease; thyroiditis; Type 1 diabetes mellitus
Heart	Myocarditis; rheumatic fever; scleroderma; systemic lupus erythematosus
Joints	Ankylosing spondylitis; rheumatoid arthritis; systemic lupus erythematosus
Kidneys	Glomerulonephritis; systemic lupus erythematosus; Type 1 diabetes mellitus
Lungs	Rheumatoid arthritis; sarcoidosis; scleroderma; systemic lupus erythematosus
Muscles	Dermatomyositis; myasthenia gravis; polymyositis
Nerves and brain	Guillian-Barré syndrome; multiple sclerosis; systemic lupus erythematosus
Skin	Alopecia areata; pemphigus/pemphigoid; psoriasis; scleroderma; systemic lupus erythematosus; vitiligo

How does the immune system fail in immunodeficiency diseases?

The immune system may fail in one of two ways in immunodeficiency diseases. Either the immune system fails to develop normally, as in severe combined immunodeficiency disease (SCID), or the immune response is blocked in some way, as in acquired immunodeficiency syndrome (AIDS). Treatment with immunosuppressive agents, such as radiation or specific drugs, may also result in immunodeficiency diseases.

LYMPHATIC VESSELS AND ORGANS

What are the different lymph vessels?

The smallest lymph vessels are lymphatic capillaries, which originate in the peripheral tissues. They are larger in diameter than blood capillaries but have a thinner wall. The lymphatic capillaries have a unique structure that allows interstitial fluid to flow into them, but not out. Lymph flows from the lymphatic capillaries into larger lymph vessels that lead toward the trunk of the body. The lymphatics continue to join together, finally forming two large ducts: the right lymphatic duct and the thoracic duct.

Do all tissues have lymphatic capillaries?

Lymphatic capillaries are found in almost every tissue and organ of the body. They are not found in avascular tissues (tissues that lack a blood supply), such as cartilage, the epidermis and cornea of the eye, the central nervous system, portions of the spleen, and red bone marrow.

Which vessels are the large collecting ducts?

The right lymphatic duct and thoracic duct are the large collecting ducts. The right lymphatic duct drains the right side of the head, right upper limb, right thorax and lung, right side of the heart, and the upper portion of the liver. The contents of the right lymphatic duct drain into the right subclavian vein and are returned to the blood. The thoracic duct, the largest collecting duct, receives lymph from the three-quarters of the body including the left side of the head, neck, chest, the left upper limb, and the entire body below the ribs. It drains into the left subclavian vein.

What is the route the fluid travels in the body?

Blood flows from the arteries to the capillaries, with a portion leaking into the interstitial spaces. Once the fluid leaves the interstitial spaces and enters the lymphatic capillaries, it is called lymph. It

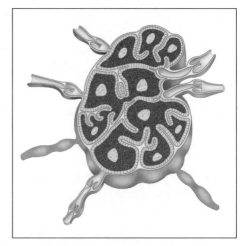

Lymphoid nodules are about 1 milimeter (about 0.04 inches) in diameter and are found in digestive, reproductive, respiratory, and urinary tracts.

flows from the lymphatic capillaries through the lymphatic vessels and lymph nodes to the lymphatic ducts, eventually entering either the right lymphatic duct or the thoracic duct. It finally drains into the subclavian veins and is returned to the blood.

Where are the most lymphoid nodules located in the body?

Lymphoid nodules are found in the connective tissues lining the digestive, urinary, reproductive, and respiratory tracts. They are small, oval-shaped, and approximately a millimeter in diameter. They are not surrounded by a fibrous capsule. The collection of lymphoid tissues lining the digestive system is called mucosa-associated lymphoid tissue (MALT) because they are found in the mucous membranes lining the digestive tract. Clusters of lymphoid tissue found in the intestine and appendix are called aggregated lymph nodules, or Peyer's patches. Tonsils are a group of lymphoid tissues found at the junction of the oral cavity, nasal cavity, and throat.

What are the cancers of the lymphatic system?

Cancers that start in the lymphoid tissue are called lymphomas (from the Latin *lympha*, meaning "water," as in the watery appearance of lymph, and from the Greek *oma*, meaning "tumor"). The two main types of lymphomas are Hodgkin's lymphoma (also called Hodgkin's disease) and non-Hodgkin's lymphoma. There are several different subtypes of both Hodgkin's lymphoma and non-Hodgkin's lymphoma.

Where are the most common sites in the body for Hodgkin's disease to begin?

Since lymphatic tissue is found throughout the body, Hodgkin's disease can start almost anywhere in the body. The most common sites for Hodgkin's disease to begin are the lymph nodes in the chest, in the neck, or under the arms. It travels through the lymph from one lymph node to another.

Who first described Hodgkin's disease?

Hodgkin's disease was first described by British physician Thomas Hodgkin (1798–1866) in 1832 in his paper "On Some Morbid Appearances of the Absorbent Glands and Spleen," published in London's *Medico-Chirurgical Transactions*. It was not until 1865 that another British physician, Samuel Wilks (1824–1911), named the medical condition Hodgkin's disease. Although Wilks described the same disease independently, he became familiar with the earlier work of Hodgkin and named the disease after him in a paper entitled "Cases of Enlargement of the Lymphatic Glands and Spleen, (or, Hodgkin's Disease) with Remarks."

What is the role of the lymphatic system in metastatic cancer?

The lymphatic vessels are often a means for transporting metastasizing cancer cells. Cancer cells often enter the lymph nodes and establish secondary cancers there. Examination and analysis of the lymph nodes provides valuable information on the spread of the cancer and helps to determine the appropriate therapy.

How many tonsils does the average person have?

Most people have five tonsils. These include a single pharyngeal tonsil, often referred to as the adenoid, located in the posterior wall of the upper part of the throat. A pair of palatine tonsils is found at the back of the mouth; a pair of lingual tonsils is located at the base of the tongue.

Which tonsils are most commonly removed during a tonsillectomy?

The palatine tonsils are the ones most commonly removed during a tonsillectomy. Normally, tonsils function to prevent infection, which is why some doctors recommend that a person keep his or her tonsils. However, when the tonsils are frequently infected, physicians may recommend removal of the infected tissues when they do not respond to noninvasive treatment, such as antibiotics.

What are "swollen glands"?

The condition commonly referred to as "swollen glands" is really enlarged lymph nodes. The lymph nodes were originally referred to as lymph glands (from the Latin *glans*, meaning "acorn") because they resembled acorns. Unlike true glands, the lymph nodes do not secrete fluids, so they are now called lymph nodes (from the Latin *node*, meaning "knob"). The term "swollen glands" has been retained to describe the condition of a slight enlargement of the lymph nodes along the lymphatic vessels draining a specific region of the body. It generally indicates an inflammation or infection of peripheral structures.

What is appendicitis?

The appendix is a small, blunt-ended tube composed of lymphatic tissue. It is considered part of the large intestine, although the tissues are different from the tissues of the large intestine. Appendicitis is an infection of the appendix that begins in the lymphoid nodules. Treatment often includes removal of the appendix, either laproscopically or with traditional surgery.

What are the major lymphoid organs?

The major lymphoid organs are the lymph nodes, thymus, and spleen. There are approximately 600 lymph nodes, ranging in diameter up to 1 inch (about 25 millimeters), scattered along the lymphatic vessels. The greatest concentrations of lymph nodes are found at the neck, armpit, thorax, abdomen, and groin.

The thymus serves both as an endocrine organ that secretes hormones and as a lymphoid organ. It consists of a large number of lymphocytes, many of which become specialized T cells. The spleen is the largest lymphoid organ. It is about 5 inches (12 centimeters) long and weighs 5.6 ounces (160 grams).

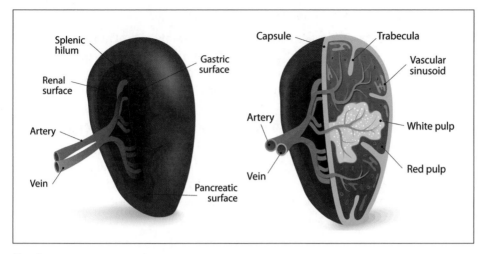

The spleen

What are the functions of the spleen?

The primary function of the spleen is the filtering of blood and removal of abnormal blood cells by phagocytosis. The spleen also stores iron from worn-out blood cells, which is then returned to the circulation and used by the bone marrow to produce new blood cells. The immune reaction begins in the spleen with the activation of immune response by B cells and T cells in response to antigens in the blood.

Can a person function without a spleen?

Individuals without a spleen are able to function without difficulty because the bone marrow and liver can perform many of the functions of the spleen. However, these individuals are at a greater risk for bacterial infections.

How serious is damage to the spleen?

Damage to the spleen, which is often the result of an injury caused by a blow to the left side of the abdomen, may be life-threatening. Since the spleen is a fragile organ, an injury can easily rupture it, resulting in serious internal bleeding, hemorrhaging, circulatory shock, and even death. Once the spleen ruptures, the only remedy is to surgically remove it in a procedure called a splenectomy.

NONSPECIFIC DEFENSES

What are nonspecific defenses?

Nonspecific defenses do not differentiate between various invaders. Barriers such as skin and the mucous membrane lining the respiratory and digestive tracts; phagocytic white blood cells; inflammation; fever; and chemicals are nonspecific defenses. The nonspecific defenses are the first to respond to a foreign substance in the body.

How does the skin act as a barrier to protect the body?

The tightly packed cells of the dermis and epidermis become a barrier that does not permit pathogens to enter the body. The acidity of the skin (pH 3 to 5) and the sebum create an environment that does not welcome microorganisms. Perspiration helps to wash away microbes from the skin. Similarly, tears wash away foreign bodies from the eyes.

How do the mucous membranes provide a defense against microorganisms?

Mucous membranes line the epithelial layer of many of the body cavities, such as the nasal passages, respiratory tract, and digestive tract. Since the mucous is viscous and slightly sticky, it traps many microorganisms, preventing them from attaching to the epithelium or entering the tissue. Some of these membranes are also ciliated, which helps to move trapped particles away from the body.

What are the four most common symptoms of inflammation?

The symptoms of inflammation, or the inflammatory response, are redness, heat, swelling, and pain. Inflammation is a response to tissue damage caused by injury, irritants, pathogens, distortion or disturbance of cells, and extreme temperatures. There are three stages to the inflammatory response: 1) vasodilation, allowing more blood to flow to the damaged tissue, and increased permeability of the blood vessels; 2) migration of phagocytes to the site of tissue damage; and 3) repair.

What is the role of fever in infection?

Normal body temperature is 98.6°F (37.2°C). Fever is defined as a higher-than-normal body temperature. Certain pathogens and bacterial toxins may stimulate the release of pyrogens (proteins that regulate body temperature) such as interleukin-1. Increases in body temperature increase the metabolic rate and may speed up body reactions that aid to resolve infections. Fever may also inhibit the growth of certain microbes. Fever appears to stimulate the liver to hoard substances that bacteria require, helping to decrease bacterial growth.

SPECIFIC DEFENSES

What is specific resistance?

Specific resistance, or immunity, is the production of specific types of cells or specific antibodies to destroy a particular antigen. An antigen is a substance that triggers the immune response, causing the body to form and produce specific antibodies. An antibody is a protein produced by B cells in response to an antigen. Antibodies are able to neutralize the antigens that provoke their production.

How many classes of antibodies have been identified?

There are five classes of antibodies, known as immunoglobulins (Igs). The following lists the known classes:

Class	Description
IgG	Accounts for 80% of all antibodies in the blood; found in blood, lymph, and the intestines; the only antibody that crosses the placenta from mother to fetus; provides resistance against many viruses, bacteria, and bacterial toxins
IgA	Accounts for 10% to 15% of all antibodies in the blood; found mostly in secretions such as sweat, tears, saliva, and mucus; attack pathogens before they enter internal tissues; levels reduce under stress-lowering resistance
IgM	Accounts for 5% to 10% of all antibodies in the blood; found in blood and lymph; first antibody secreted after exposure to an antigen; includes the anti-A and anti-B antibodies of ABO blood, which bind to A and B antigens during incompatible blood transfusions
IgD	Accounts for 0.2% of all antibodies in the blood; found in blood, lymph, and the surfaces of B cells; plays a role in the activation of B cells
IgE	Accounts for less than 0.1% of all antibodies in the blood; found on the surfaces of mast cells and basophils; stimulates cells to release histamine and other chemicals that accelerate inflammation; plays a role in allergic reactions

How do T cells differ from B cells?

T cells and B cells are both lymphocytes, which respond to the presence of specific antigens. T cells arise from cells that originate in the bone marrow and then migrate

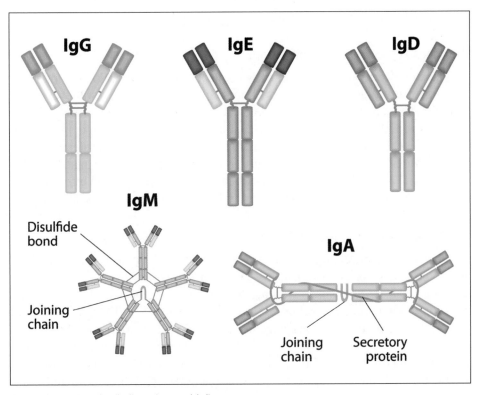

The five classifications of antibodies, or immunoglobulins

to the thymus gland, where they mature. They are responsible for cell-mediated immunity and attack specific foreign cells in the body. B cells develop in the red bone marrow. They are responsible for producing and secreting specific antibodies.

Do all B cells produce the same antibodies?

No, each B cell is programmed to produce one specific antibody. For example, one B cell produces the antibody that blocks a virus that is responsible for the common cold, while a different B cell produces the antibody that attacks the bacterium that causes pneumonia.

What are plasma cells?

Plasma cells are large, antibody-producing cells that are derived from B cells. Each plasma cell that is derived from a single B cell manufactures millions of identical antibodies to fight the antigen (virus, microbe, or other foreign tissue/substance).

What are the two forms of immunity?

Immunity may be either innate or acquired. Innate immunity is present at birth and has no basis to prior exposure to the antigen involved. Certain innate immunity is species-specific. For example, humans are not susceptible to diseases of cats and dogs. Additionally, certain individuals are not susceptible to some diseases that other individuals are susceptible to. Acquired immunity is not present at birth, but rather is acquired following exposure to a particular antigen.

How does active immunity differ from passive immunity?

Active immunity develops when the body manufactures its own antibodies following exposure to an antigen. Active immunity may occur naturally as a result of exposure to foreign antigens or artificially via vaccinations. Passive immunity is produced by transferring antibodies from one individual to another. One example of passive immunity is when maternal antibodies cross the placenta and provide protection to the fetus. Maternal antibodies are also passed from mother to infant during breastfeeding. Artificial passive immunity occurs when antibodies are injected into the body. Artificial passive immunity is used to treat rabies, tetanus, and rattlesnake bites.

When was the term "antibiotic" first used?

Antibiotics are chemical products or derivatives of certain organisms that inhibit the growth of, or destroy, other organisms. The term "antibiotic" (from the Greek *anti*, meaning "against," and *biosis*, meaning "life") refers to its purpose in destroying a life form. In 1889 French mycologist Jean Paul Vuillemin (1861–1932) used the term "antibiosis" to describe bacterial antagonism. He had isolated pyocyanin, which inhibited the growth of bacteria in test tubes but was too lethal to be used in disease therapy. In was not until the mid-1940s that Ukraninan born, Jewish-American inventor, biochemist, and microbiologist Selman Waksman (1888–1973) used the term "antibiotic" to describe a compound that had therapeutic effects against disease. Waksman received the Nobel Prize in Physiology or Medicine in 1952 for his discovery of streptomycin. Streptomycin was the first antibiotic effective against tuberculosis.

How do antibiotics destroy an infection?

Antibiotics function by weakening the cell wall, or interfering with the protein synthesis or RNA synthesis of the bacterial cell. For example, penicillin weakens the cell wall to the point that the internal pressure causes the cell to swell and eventually burst. Various antibiotics are more effective against different bacteria.

What naturally occurring substance provides protection against viral infections?

Interferons protect the adjacent cells against viral penetration. Interferons are glycoproteins produced by body cells upon exposure to a virus. In 1957, British virologist Alick Isaacs (1921–1967) and Swiss virologist and immunologist Jean Lindenmann (1924–2015) identified a group of over twenty substances that were later designated as alpha, beta, and gamma interferons.

What is HIV?

HIV—human immunodeficiency virus—is a virus that attacks CD4 cells (T cells), weakening the body's ability to fight off infection. If left uncontrolled, eventually so many T cells are destroyed that almost any type of infection can become lethal. When this point is reached, HIV is then called AIDS, or acquired immunodeficiency syndrome.

The term AIDS was first used in 1982 by U.S. public health officials to describe the occurrences of opportunistic infections such as Kaposi's sarcoma (a kind of cancer) and *Pneumocystis carinii* pneumonia in previously healthy people.

What are the symptoms and signs of HIV?

The warning signs of HIV infection include night sweats; prolonged fevers; severe weight loss; persistent diarrhea; red, brown, pink, or purplish blotches on or under the skin or inside the mouth, nose, or eyelids; persistent dry cough; swollen lymph nodes in the armpits, groin, or neck; white spots or unusual blemishes on the tongue, in the

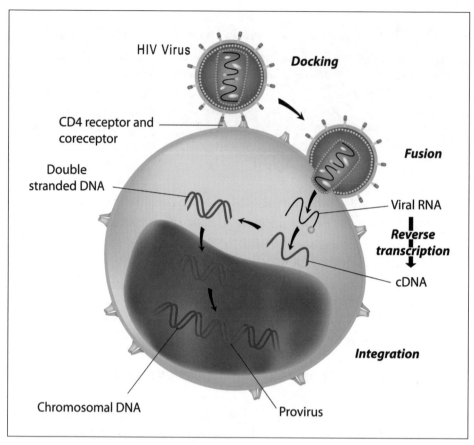

The human immunodeficiency virus (HIV), as the name indicates, causes sickness by entering and destroying T cells, which are the body's defense system against infecitons.

mouth, or in the throat; and memory loss, depression, and other neurological disorders. However, the only way to determine if you are infected is to be tested for HIV.

How is HIV diagnosed?

The only accurate way to diagnose HIV is through antibody testing. The immune system of an individual infected with HIV will begin to produce HIV antibodies to fight the infection. Although these antibodies are ineffective in destroying the HIV virus, their presence is a positive indication of the presence of HIV.

What is the difference between HIV and AIDS?

The term AIDS applies to the most advanced stages of HIV infection. The Centers for Disease Control and Prevention's (CDC) definition of AIDS includes all HIV-infected people who have fewer than 200 CD4+ T cells per cubic millimeter of blood. (Healthy adults usually have CD4+ T cell counts of 1,000 or more.) The definition also includes twenty-six clinical conditions (mostly opportunistic infections) that affect people with advanced HIV disease.

What are some statistics concerning individuals infected with HIV/AIDS?

According to the World Health Organization (WHO), in 2013 there were 35 million people worldwide living with HIV and 1.5 million people died of HIV-related illnesses. Since the beginning of the epidemic, almost 78 million people have been infected with the HIV virus and about 39 million people have died of HIV-related illnesses. An estimated 0.8 percent of people fifteen to forty-nine years old worldwide are living with HIV, and the epidemic continues to vary considerably between countries and regions. The most severely affected is Sub-Saharan Africa, with nearly one in every twenty adults living with HIV and accounting for nearly 71 percent of the people living with HIV worldwide.

What are some complications of AIDS?

The immune system of individuals with AIDS is severely compromised and weak, so the body cannot fight off certain bacteria, viruses, fungi, parasites, and other microbes. Opportunistic infections can easily establish themselves. In addition, people with AIDS are prone to developing various cancers, especially those caused by viruses such as Kaposi's sarcoma and cervical cancer, or cancers of the immune system known as lymphomas. These cancers are usually more aggressive and difficult to treat in people with AIDS.

How is the HIV virus transmitted from one person to another?

HIV is transmitted via unprotected sexual contact with an infected partner or through contact with infected blood. Rigorous screening of the blood supply and heat-treating techniques for donated blood have reduced the rate of transmission via blood transfusions to a very small percentage. However, sharing needles and/or syringes with someone who is infected is still a mode of transmission of the HIV virus. In the past, HIV was frequently passed from a mother to her baby during pregnancy and/or birth. Treatments are now available that reduce the chances of a mother passing the virus to her child to 1 percent.

When was the earliest documented case of HIV?

The earliest known case of HIV-1 in a human was from a blood sample collected in 1959 from a man in Kinshasa, Democratic Republic of Congo. The virus has existed in the United States since the mid-1970s. Between 1979 and 1981, physicians in New York City and Los Angeles reported treating a number of male patients who had engaged in sex with other men for rare types of pneumonia, cancer, and other illnesses not usually found in individuals with healthy immune systems.

Is there a vaccine to prevent HIV?

Researchers are working to develop vaccines to prevent HIV in HIV-negative individuals. They are also trying to develop therapeutic vaccines for HIV-positive individ-

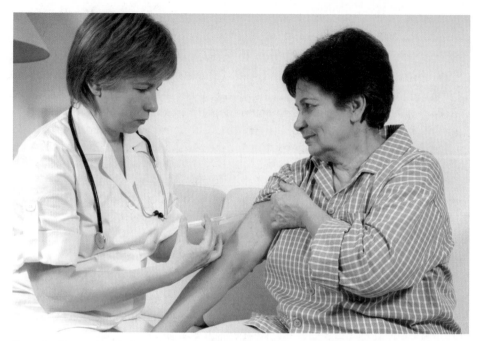
You need a different type of flu shot each year because the viruses that cause influenza are constantly changing.

uals to improve their immune system. There is no projected date when a vaccine will be approved for use, and to date, there is no true HIV vaccine.

What is the purpose of vaccination?

The purpose of vaccination, or immunization, is to artificially induce active immunity so there will be resistance to the pathogen upon natural exposure in the future. Vaccinations are prepared under laboratory conditions from either dead or severely weakened antigens.

Why is a flu vaccine required to be prepared each year?

A new flu vaccine is prepared every year because the strains of flu viruses change from year to year. Nine to ten months before the flu season begins, scientists prepare a new vaccine made from inactivated (killed) flu viruses. The vaccine preparation is based on the strains of the flu viruses that are in circulation at the time. It includes those A and B viruses expected to circulate the following winter. Another reason to get vaccinated for the flu every year is that immunity afterwards declines and may be too low to provide protection after one year.

Why are there some "bad" flu years even with a flu vaccine?

Yes, there are still some "bad" flu years even with a developed flu vaccine. One reason is that each year, when scientists prepare a new vaccine they are guessing that the next strain of flu will be similar to the more recent flu. But in some years, the "guess" of the right vaccine can be wrong, or the flu virus mutates, and many people contract the flu even with a vaccination. For example, during the 2014–2015 flu season in the United States, about half of the viruses were different than the ones that were included in that year's vaccine, thus it did not protect people against those viruses. Overall, researchers estimated that the flu vaccine that season was only about 50 to 60 percent effective.

Which vaccinations are given during childhood?

Childhood vaccinations begin within the first couple of months after birth. The recommended childhood immunization schedule for children in the United States includes: hepatitis B, diphtheria, tetanus, pertussis (DTaP), Haemophilus influenzae type b (Hib), poliovirus, measles, mumps, rubella (MMR), varicella (chickenpox), pneumococcal, and influenza.

How does a live, attenuated vaccine differ from an inactivated (killed) vaccine?

Live, attenuated vaccines contain a version of living microbes that have been weakened (attenuated) in a laboratory setting so they can no longer cause disease. Since they are very close to the actual infection, they elicit strong cellular and antibody responses. Live, attenuated vaccines often provide lifelong immunity with only one or two doses. Live, attenuated vaccines are usually more successful with viruses than bacteria.

Inactivated (killed) vaccines are generally better for use against bacteria. Scientists produce inactivated vaccines by killing the disease-causing microbe with chemicals, heat, or radiation. Most inactivated vaccines, however, stimulate a weaker immune system response than do live vaccines. Therefore, it may take several additional doses, called booster shots, to maintain a person's immunity to a particular bacterium.

What are some diseases that are preventable by vaccination?

Many diseases, such as polio and smallpox, have been eradicated due to vaccination programs. Other diseases, both bacterial and viral, that are preventable by vaccination include:

What was the earliest known vaccination?

In 1796 Edward Jenner (1749–1823) inoculated a young boy with cowpox contracted by a dairymaid. (The word "vaccination" is derived from the Latin *vacca*, meaning "cow," in recognition of this fact.) Several weeks later, the young boy was variolated with smallpox. (Variolation was a procedure in which an individual was given smallpox by scratching into their skin scab material from someone infected with the disease.) The young boy did not develop smallpox, thus Jenner was convinced that vaccination against disease was possible.

Anthrax	Bacterial meningitis
Chickenpox	Cholera
Diphtheria	Haemophilus influenzae type b (Hib)
Hepatitis A	Hepatitis B
Influenza	Measles
Mumps	Pertussis (whooping cough)
Pneumococcal pneumonia	Rabies
Rubella (German measles)	Tetanus
Yellow fever	

How do physicians determine whether an organ donor and recipient are compatible?

Researchers have developed a tissue typing technique to determine the compatibility between organ donors and organ recipients. Each individual has a unique set of cell and tissue "markers" called HLA antigens. American geneticist and immunologist George Snell (1903–1996), French immunologist Jean Dausset (1916–2009), and American immunologist Baruj Benacerraf (1920–2011) shared the Nobel Prize in Physiology or Medicine in 1980 for their research on identifying and understanding the genetic structure of cell and tissue markers. The more similar the cell and tissue markers are between donor and recipient, the less likely there will be tissue rejection or graft-versus-host disease.

ALLERGIES

What is an allergic reaction?

An allergic reaction is a reaction to a substance that is normally harmless to most other people. Allergens, the antigens that induce an allergic reaction, may be foods, medications, plants, animals, chemicals, dust, or molds.

Which antibody is responsible for most allergic reactions?

Immunoglobulin E (IgE) is responsible for most allergic reactions. Each type of IgE is specific to a particular allergen. When exposed to an allergen, IgE antibodies attach themselves to mast cells (normal body cells that produce histamines and other chemicals) or basophils. When exposed to the same allergen at a later time, the individual may experience an allergic response when the allergen binds to the antibodies attached to mast cells, causing the cells to release histamine and other inflammatory chemicals.

Are allergies an inherited medical disorder?

Hypersensitivity to specific allergens is not inherited. However, the predisposition to developing allergies may be inherited in many people. Studies have found that if neither parent suffers from allergies, the chances of a child developing allergies is only 10 to 20 percent. If one parent has allergies, the chances increase to 30 to 50 percent and if both parents have allergies, the chances increase to 40 to 75 percent. One explanation for this is in the ability to produce higher levels of IgE in response to allergens. Individuals who produce more IgE will develop a stronger allergic sensitivity.

How do immediate allergic reactions differ from delayed allergic reactions?

Immediate allergic or hypersensitive reactions are mediated by mast cells. The reactions occur within minutes following contact with an allergen. Inhaled or ingested allergens usually cause immediate allergic reactions. Delayed allergic or hypersensitive reactions are mediated by T

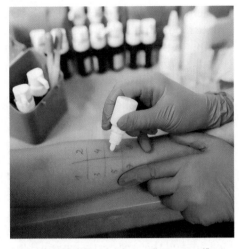

In an allergy skin prick test a doctor will expose specific parts of the skin to different potential allergens to see which ones cause a reaction.

cells. The reaction occurs hours and possibly even days after contact with the allergen. Contact on the skin with an allergen is more likely to cause a delayed allergic reaction.

What are common allergic reactions?

Allergic reactions include a variety of symptoms. Some common allergic reactions are allergic rhinitis, or "hay fever"; allergic conjunctivitis (an eye reaction); asthma; atopic dermatitis (skin reactions); urticaria, also known as hives; and severe systemic allergic reactions such as anaphylaxis.

What is anaphylactic shock?

Anaphylactic shock is a severe, potentially fatal systemic allergic reaction. It usually involves several organ systems, including the skin, respiratory tract, and gastrointestinal tract. Treatment includes an injection of adrenaline.

What allergies can a human develop?

There are a huge number of allergens that can affect humans; the most common allergens are dust and mites. People can also be allergic to various foods, pollen, chemicals in cosmetics, medications, fungal spores, insect venom, and various microorganisms. For example, there are many common food allergy triggers. They include eggs, peanuts, wheat, soy, fish, shellfish, and tree nuts. If a person is sensitive to one of these foods, allergies can develop. In people who are not sensitive, there will be no response.

How many humans have food allergies?

Although it may seem as if many people have an allergy to some type of food, in reality, only 2 to 8 percent of children (and most of them lose their sensitivity to the problem foods over time) and 1 to 2 percent of adults have a definite allergic reaction to foods. But those percentages still translate into a large number of people. According to the Harvard Medical School, to date, about fifteen million people in the United States suffer with some type of food allergy.

What's the difference between an intolerance and allergy?

An allergy is caused by a reaction to a food because of a person's immune system. For example, a person may have an allergy to ragweed that causes them to sneeze or have a runny nose or watery eyes. Intolerances are usually associated with food and the digestive tract, and they are usually caused by the inability to absorb or digest certain foods. For example, if people have lactose intolerance, they lack a certain enzyme in their digestive tract that helps digest lactose (milk sugar) in the dairy product. When they eat something such as ice cream, their gastrointestinal tract can become bloated, and they may even have diarrhea and vomiting. For the majority of people, food intolerances are usually caused by an enzyme deficiency. In other cases, they are caused by the intestines being supersensitive to certain foods, most often a reaction to the protein content of the food.

Which drug causes the most allergic reactions?

Penicillin is a common cause of drug allergy. One research study found that approximately 7 percent of normal volunteers react to penicillin allergy skin tests (IgE antibodies). Anaphylactic reactions to penicillin occur in 32 of every 100,000 exposed patients.

Why do some individuals experience allergic reactions after visits to doctors and dentists?

Some individuals are allergic to latex, a component of most rubber gloves. They may experience skin rashes, hives, eye tearing and irritation, wheezing, and itching of the skin when exposed to latex. The best therapy for latex allergy is to avoid products containing latex. Therefore, it is important to notify health care providers if you suffer from latex allergy so they may use non-latex products.

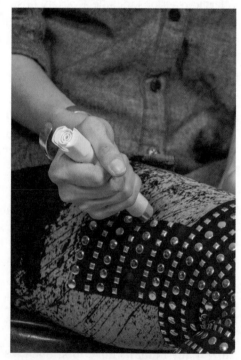

An epipen™ is a way for people with severe allergies to inject themselves with epinephrine to stave off a potentially life-threatening reaction.

What are epipens™?

Injections of epinephrine are often used to treat systemic, life-threatening allergic reactions. Epipens™ are needle and syringe medical devices that deliver a premeasured, single dose of epinephrine as a self-delivered automatic injection.

COMPARING OTHER ORGANISMS

Do other animals have immune systems?

Yes, all animals have immune systems, although they are often different than a human immune system. Most of the systems act similar to a human's immune system, recognizing foreign substances that can be harmful to the animal's body, and reacting to protect their body from the invading pathogens. And not all animals are susceptible to the same pathogens. For example, humans are not affected by the viruses that cause canine distemper or mouse pox. Conversely, horses and dogs cannot catch the human virus that causes AIDS in humans.

What are two types of immunity found in most animals?

There are two types of immunities—or systems against the invasion of pathogens, including microorganisms such as bacteria and viruses, and parasites—that most animals possess. The innate, or natural, immunity includes two divisions. The first is the humoral innate immunity, in which certain substances in the body fluids interfere with the growth of pathogens. The second is the cellular innate immunity, in which the body's cells called phagocytes ingest or degrade the pathogens that enter the body. The more sophisticated type of immunity is the adaptive immunity that can respond, recognize, and destroy specific substances called antigens. Called the immunity response, it is directed toward specific antigens; it also has an immunologic memory, and thus, the immune system will respond if there is any exposure to the same antigen.

Do other animals experience allergies?

Yes, some animals seem to have allergies. In particular, we know more about domestic than wild animal allergies (although some researchers claim that wild animals

---

Do some cats have a form of immunodeficiency virus?

Yes, some cats suffer from what is called feline immunodeficiency virus (FIV), much like the human HIV (for more about HIV in humans, see above). The virus is usually transferred from cat to cat through biting and thus it is found mostly in feral or free-roaming cats (humans are not affected by this cat virus). Although the infected cat may appear normal for many years, the infection eventually affects the immune system, making the cat unable to fight off other infections. FIV-infected cats are found worldwide, but the concentration of cats that are infected varies. For example, in the United States, to date, about 1.5 to 3 percent of healthy cats are infected with FIV.

would have fewer allergies because of their "wild," more natural diets). For example, veterinarians report that dogs, cats, and even pet birds suffer from allergies. They may be allergic to food, insect bites, dust, household chemicals, or pollen. Instead of having runny noses and watery eyes, most domestic animals experience itchy skin conditions, difficulty in breathing, or disruptions in the digestive tract.

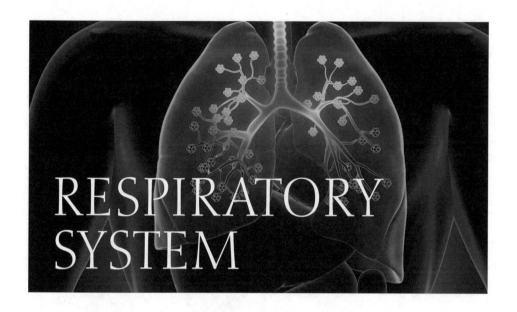

RESPIRATORY SYSTEM

INTRODUCTION

What are the major functions of the respiratory system?

The major functions of the respiratory system are as follows:

1. Gas exchange: the respiratory system allows oxygen in the air to enter the blood and carbon dioxide to leave the blood and enter the air. The cardiovascular system transports oxygen from the lungs to the cells of the body and carbon dioxide from the cells of the body to the lungs.
2. Regulation of blood pH: this can be altered by changes in blood carbon dioxide levels.
3. Voice production: as air moves past the vocal cords, it results in sound and speech.
4. Olfaction: the sense of smell occurs when airborne molecules are drawn into the nasal cavity.
5. Innate immunity: protection is provided against some microorganisms by preventing their entry into the body or by removing them from respiratory surfaces.

What are the two divisions of the respiratory system?

The respiratory system is divided into the upper respiratory system and the lower respiratory system. The upper respiratory system includes the nose, nasal cavity, and sinuses. The lower respiratory system includes the larynx, trachea, bronchi, bronchioles, and alveoli.

What is the basic difference between respiration and cellular respiration?

Respiration is the entire process of gas exchange between the atmosphere and the body cells. Cellular respiration is the process of oxygen utilization and carbon dioxide at the cellular level.

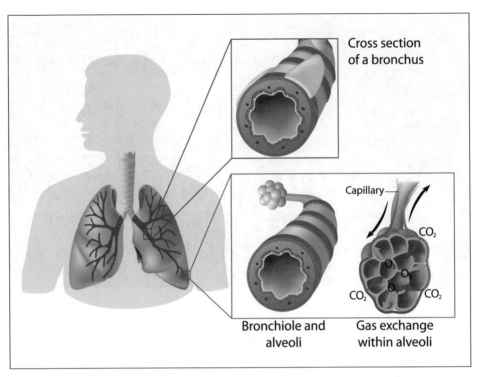

Lung anatomy and physiology

Why do we die within minutes if we are deprived of oxygen?

We die within minutes of being deprived of oxygen because each of the trillions of cells in the human body requires oxygen to survive. In particular, after around six minutes, the brain and other body cells begin to die, as the cells are deprived of oxygen.

What are non-respiratory air movements?

Air movements that occur in addition to breathing are called non-respiratory movements. They are used to clear air passages, as in coughing or sneezing, or to express emotions, as in laughing or crying. Also included in this category are hiccupping and yawning, as well as speech, where air is forced through the larynx, causing the vocal cords to vibrate. In speech, words are formed by lips, the tongue, and the soft palate.

STRUCTURE AND FUNCTION

What are the functions of the nose?

The nose is the primary entry point for air into the respiratory system. It is supported by bone and cartilage. The many hairs that line the nostrils help to filter large particles from air. From the nose, the air is then further filtered and humidified by passages through the sinuses.

How effective are the structures in the nose at removing dust, bacteria, and other particles?

As the cilia of the mucous membrane cells in the nose move, they push a thin layer of mucus and trapped particles, including dust and other small particles such as bacteria, toward the pharynx, where it is swallowed. In the stomach, gastric juice destroys the bacteria and other microorganisms trapped in the mucus. However, some bacterial spores, including those from the bacterium that causes anthrax, are very small and bypass the hairs and mucus in the nose. Spores are able to reach the lungs, where they release a toxin causing inhalation anthrax, and ultimately cause death.

What is a deviated septum?

The nasal septum, which is composed of bone and cartilage, divides the nostrils and the nasal cavity into right and left portions. It is usually straight at birth and sometimes bends as the result of a birth injury. The nasal septum is usually straight throughout childhood, but as a person ages the septum tends to bend towards one side or the other. This deviated septum may make breathing difficult by obstructing the nasal cavity.

What is the sneeze reflex?

The sneeze reflex functions to dislodge foreign substances from the nasal cavity. Sensory receptors detect the foreign substances and stimulate the trigeminal nerves to the medulla oblongata, where the reflex is triggered. During the sneeze reflex, some air from the lungs passes through the nasal passages, although a significant amount passes through the oral cavity. When the cells lining the nose are irritated, the response is a rapid rush of air to force the irritant out of the nasal passages. A sneeze can travel as fast as 100 miles (161 kilometers) per hour. It is impossible to sneeze and keep one's eyes open at the same time.

What causes a nosebleed?

The inside of the nose is heavily supplied with blood vessels. A blow to the nose, excessive nose blowing, infection, allergies, clotting disorders, or hypertension can cause a nosebleed.

Where are the sinuses located?

The sinuses are cavities located in the bones surrounding the nasal cavity. The sinuses lighten the skull, and as air flows through the cavities it is warmed and humidified. The nasal cavities are also important in voice production.

What are the functions of the paranasal sinuses?

The paranasal sinuses are membrane-lined, air-filled cavities in the bones of the skull that are connected to the nose by passageways. The four paranasal sinuses and the bones in which they are located are the frontal, ethmoid, maxillary, and sphenoid. These cavities lighten the weight of the bones and add resonance to the voice.

233

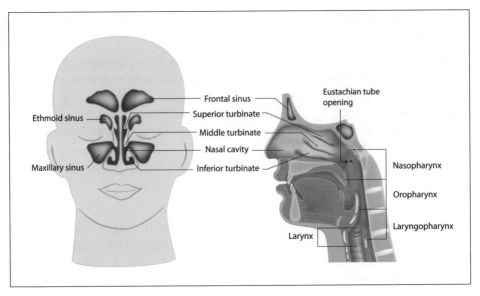

Anatomy of the nose

Why is crying often accompanied by a runny nose?

A nasolacrimal (from the Latin *nasus*, meaning "nose" and lacrima, meaning "tear") duct leads from each eye to the nasal cavity. Excessive secretion of tears in the eyes is drained into the nose, often producing a watery flow from the nose at times of crying.

Where is cold air warmed as it is breathed into the body?

Cold air is warmed in the nasal cavity, where the blood vessels in this large surface area provide the capacity for warming inhaled air. As heat leaves the blood in this extensive network of blood vessels, the air is warmed to body temperature.

Why should you not inhale super-cooled air?

Inhaled super-cooled air decreases the fluidity of the mucus that lines the respiratory passageways. Super-cooled air can even cause ice crystals to form in the nasal mucosa.

What causes a sinus headache?

A painful sinus headache can result from blocked drainage of the nasal sinuses. This is usually caused by an infection of the sinuses. It can also be caused by an allergic reaction that triggers the release of histamines, a natural inflammatory chemical in the body that responds to the allergen. The result is often blocked nasal sinuses and headaches.

What is the pharynx?

The pharynx, commonly called the throat, is connected to the nasal cavity and to the oral cavity. It serves as a passageway for food traveling from the oral cavity to the esophagus and for air traveling from the nasal cavity to the larynx. The pharynx has three divisions: 1) the nasopharynx, 2) the oropharynx, and 3) the laryngopharynx.

Do nasal strips really work to stop snoring?

Nasal strips may make a difference for the average snorer, if the snoring originates in the nose. They keep the nostrils wide apart, and by doing so they help prevent snoring and move nitric oxide from the nasal passage into the lungs, improving lung function. Snoring that originates from the base of the tongue or the soft palate is not relieved by nasal strips.

What causes a sore throat?

Most sore throats are caused by viruses, the same germs that cause colds and influenza (the flu). Some sore throats are caused by bacteria such as *Streptococcus*. Other common causes of a sore throat include allergies, dryness of indoor air, pollution and other irritants, muscle strain in the throat, acid reflux disease, HIV infection, and oral tumors.

What is the larynx?

The larynx is the passage located between the pharynx and trachea that houses the vocal cords. After air leaves the pharynx, it enters the larynx. The larynx is a cartilaginous structure that encloses the glottis. It is formed by three cartilages: the thyroid, cricoid, and the epiglottic. In addition to forming the passageway for air to travel from the pharynx to the rest of the respiratory tract, the larynx is the site of most sound production.

What is the Adam's apple?

The Adam's apple is a bulge in the throat due to the prominent surface and angle of the thyroid cartilage. It is more visible in men because the male larynx is larger and the angle of the cartilage is less. Females also have an Adam's apple. It is much less prominent because the angle is greater (thus, "flatter") and is covered by fatty tissue.

Why is it dangerous to talk while eating?

If a person talks while eating, food may be inhaled into the lungs. Normally, after food is swallowed, it passes into the pharynx and then into the esophagus. Food is prevented from entering the larynx (the passageway to the lungs) by the epiglottis, a spade-shaped cartilage flap that covers the pharynx. If food does enter the larynx, a cough reflex is usually initiated, although food may lodge in the larynx, causing a blockage of the airway.

What should you do if someone is choking on food or something else lodged in the throat?

Abdominal thrusts (formerly called the Heimlich maneuver) should be used in situations in which a person has a foreign object lodged in his or her larynx or trachea.

RESPIRATORY SYSTEM

235

According to the American Red Cross, if a person is choking, first give five back blows with the heel of your hand; if that doesn't work, make five quick abdominal thrusts to the victim's abdomen just below the diaphragm and above the navel. This usually generates enough pressure to push out the foreign object.

What is laryngitis?

Laryngitis is an inflammation of the larynx. It can be caused by smoking, exposure to irritants, or an infection. Because the larynx is the site of sound production, laryngitis usually results in hoarseness or inability to produce audible sounds.

What makes the trachea flexible?

The trachea, or windpipe, is a flexible, cylindrical tube approximately 1 inch (2.5 centimeters) in diameter and 5

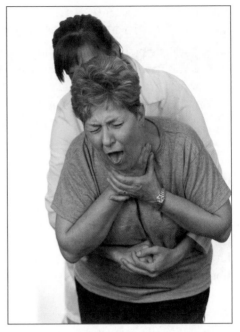

When someone is choking, quickly compressing the diaphragm can cause air to eject whatever is clogging the windpipe.

inches (12.5 centimeters) in length. Within the tracheal wall there are about twenty C-shaped pieces of cartilage, stacked one above the other. The remaining part of the tracheal ring is composed of smooth muscle and connective tissue. This soft tissue allows the nearby esophagus to expand as food moves through it into the stomach.

Are both of the lungs identical?

The lungs are cone-shaped organs in the thoracic cavity. The right lung consists of three lobes (right superior lobe, right middle lobe, and right inferior lobe) while the left lung has only two lobes (left superior lobe and left inferior lobe) and is slightly smaller than the right lung. Although relatively large, each lung weighs only 1 pound (2.2 kilograms).

What are some characteristics of the lungs?

The average human lung has several characteristics. For example, an adult lung has approximately three hundred million alveoli, or air sacs. The total surface area of the lungs is approximately the size of a tennis court. The consistency of the lungs, when inflated, is very porous, much like whipped gelatin. Each alveoli measures approximately 0.000004 inches (0.00001 centimeters) thick.

What is the essential relationship between the heart and lungs?

The teamwork of the heart and lungs ensures that the body has a constant supply of oxygen for metabolic activities and that the major waste product of metabolism, carbon dioxide, is continuously removed. This occurs through the pulmonary circulation, with the heart supplying blood that has moved through the body to the lungs.

How do the lungs connect to the heart?

The lungs connect to the heart through blood vessels. The pulmonary artery delivers deoxygenated blood to the lungs from the right ventricle and the pulmonary vein delivers oxygenated blood to the left atrium of the heart. In fact, around once every minute, the heart pumps the body's entire blood supply through the lungs.

What prevents the lobes of the lungs from rubbing against each other?

The lobes of the lungs do not normally rub against each other due to the presence of pleural fluid. Pleural fluid is secreted by the membranes that cover the lungs. It reduces the friction between the lobes of the lung and the body cavity.

What is pleurisy?

Pleura is a thin, transparent, two-layered membrane that covers the lungs and also lines the inside of the chest wall. The layer that covers the lungs is in close contact with the layer of the chest wall. Between these two thin layers is a small amount of fluid that lubricates each one as they slide over one another with each breath. Pleurisy is a painful condition that results when there is an abnormal amount of fluid between these two layers. The inflammation from this condition results in friction during inhalation and exhalation.

What is pneumothorax, or collapsed lung?

Pneumothorax refers to the presence of air in the pleural cavity. It occurs when air leaks from the lungs and gets between the lung and the chest wall. It is also known as a collapsed lung and can result from a penetrating chest injury or changes in pressure during diving, flying, or stretching.

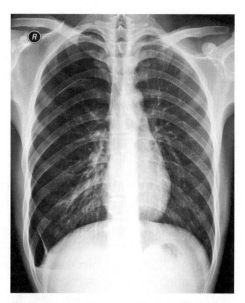

What is COPD?

COPD (chronic obstructive pulmonary disease) refers to a group of diseases, primarily bronchitis and emphysema. Bronchitis is a chronic inflammation of the bronchi caused by irritants such as cigarette smoke, air pollution, or infections. The inflammation results in swelling of the mucous membrane lining the bronchi, increased mucus pro-

This X-ray shows a pneumothorax (the white area to the right of the spine), where air has leaked from the lung and entered the chest cavity.

duction, and decreased movement of mucus by cilia. Emphysema usually follows bronchitis and involves the destruction of the alveolar walls. Ultimately, the lungs lose their elasticity and become less efficient.

What is emphysema?

Emphysema is a progressive disease that destroys alveoli, particularly the walls. As a result, groups of small air sacs combine to form larger chambers that have significantly less surface area. This reduces the volume of gases that can be exchanged through the cell membranes. Emphysema is usually associated with people who smoke, as the smoking damages the alveoli (air sacs) of the lungs making it difficult to breathe—and the body does not get the oxygen it needs.

What is a cough?

A cough is a sudden, explosive movement of air to clear material from the airways. According to the National Heart, Lung, and Blood Institute, a cough is a natural reflex that protects your lungs. It helps clear a person's airways of lung irritants, such as smoke and mucus, and helps prevent infection. A cough also can be a symptom of a medical problem, such as asthma or upper respiratory infections such as a cold.

How does smoking affect the lungs?

Tobacco smoking, especially cigarettes, is linked to a decrease in longevity and lung cancer, which is the most common cause of cancer death in most countries. In fact, it is estimated that for each pack of cigarettes smoked per day, life expectancy is decreased by seven years. A two-pack-a-day smoker will likely die at age sixty, compared to seventy-four for a nonsmoker in the United States.

Eighty percent of lung cancer cases are due to cigarette smoking. The toxins from cigarettes damage the alveoli so they leak and pop, causing emphysema. Lung cancer usually develops in the epithelium of the airways, spreads through the walls of the airways, and enters the bloodstream and lymphatic system. These changes are usually not detected until secondary growths or other symptoms occur. Only 13 percent of lung cancer patients live as long as five years after the initial diagnosis.

What is pulmonary fibrosis?

Pulmonary fibrosis occurs when chronic inflammation of the lungs results in the replacement of normal, elastic lung tissue with nonelastic scar tissue. Exposure to coal dust, asbestos, and silica are the most common causes of pulmonary fibrosis.

What is pneumonia?

Pneumonia refers to an infection (viral or bacterial) within the lungs. The resulting inflammation from the infection causes fluid accumulation and breathing difficulties. According to recent information from the Centers for Disease Control and Prevention, there were 53,667 deaths attributed to pneumonia/influenza, making this the eighth leading cause of death in the United States in 2011.

RESPIRATION AND BREATHING

What are the two phases of breathing?

Breathing, or ventilation, is the process of moving air into and out of the lungs. The two phases are: 1) inspiration, or inhalation; and 2) expiration, or exhalation. Inspiration is the movement of air into the lungs, while expiration is the movement of air out of the lungs.

What is the respiratory cycle?

The respiratory cycle consists of one inspiration followed by one expiration. The volume of air that enters or leaves during a single respiratory cycle is called the tidal volume. Tidal volume is typically five hundred milliliters, meaning that five hundred milliliters of air enters during inspiration and the same amount leaves during expiration.

What is the diaphragm?

The diaphragm is a muscular separation between the thoracic and abdominal cavities and is the principal muscle involved in inspiration. The contractions of the diaphragm enlarge the thoracic cavity, allowing inspiration or inhaling to move air into the lungs. During expiration, the diaphragm returns to its original position. (For more about the diaphragm, see the chapter "Anatomy and Biology Basics.")

Where are gases actually exchanged in the lungs?

The alveoli (air sacs) are the structures that carry out the vital process of exchanging gases (oxygen and carbon dioxide) between the air and the blood. The tiniest pas-

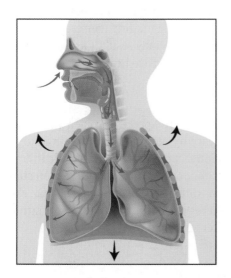 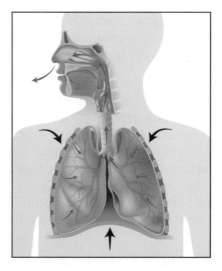

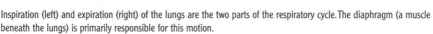
Inspiration (left) and expiration (right) of the lungs are the two parts of the respiratory cycle. The diaphragm (a muscle beneath the lungs) is primarily responsible for this motion.

sageways in the lungs are the alveoli. Each pulmonary lobule ends with alveoli, which are then interconnected with one another. There are approximately 150,000,000 alveoli in each lung.

How does the body introduce oxygen to the blood and where does this happen?

Blood entering the right side of the heart (right atrium) contains carbon dioxide, a waste product of the body. The blood travels to the right ventricle, which pushes it through the pulmonary artery to the lungs. In the lungs, the carbon dioxide is removed and oxygen is added to the blood. Then the blood travels through the pulmonary vein carrying the fresh oxygen to the left side of the heart. It first enters left atrium, where it goes through a one-way valve into the left ventricle, which must push the oxygenated blood to all portions of the body (except the lungs) through a network of arteries and capillaries. The left ventricle must contract with six times the force of the right ventricle, so its muscle wall is therefore twice as thick as that of the right ventricle.

How does external respiration differ from internal respiration?

External respiration is the exchange of gases between the alveoli and lung capillaries. Oxygen diffuses from the alveoli into the blood, while carbon dioxide moves from the blood to the alveoli. Internal respiration, in contrast, is the exchange of gases in body tissues.

Why is carbon monoxide deadly to breathe?

Carbon monoxide is a colorless, odorless gas that has the unique ability to compete with oxygen for binding sites on the hemoglobin molecule. Carbon monoxide binds to the iron in hemoglobin in red blood cells about two hundred times as readily as does oxygen, and it tends to stay bound. As a result, hemoglobin bound to carbon monoxide can no longer transport oxygen. Prolonged exposure to carbon monoxide results in carbon monoxide poisoning, with symptoms including nausea, headache, and eventually unconsciousness. If left untreated, death may result. Carbon monoxide is commonly emitted by automobiles and fuel-fired space heaters.

What is hyperbaric oxygen therapy?

Hyperbaric oxygen therapy is when a patient is given 100 percent oxygen gas at two to three atmospheres pressure to breathe for varying lengths of time. It is used to treat carbon monoxide poisoning, decompression sickness, severe traumatic injury, infections that could lead to gas gangrene, and other conditions. Normal oxygen concentration is 0.3 milliliters oxygen per 100 milliliters of blood, but breathing 100 percent oxygen at a pressure of three atmospheres raises the plasma concentration to about 6 milliliters of oxygen per 100 milliliters of blood.

What muscles help in breathing?

Breathing is caused by the actions of the muscles between the ribs, the external intercostal muscles, and the diaphragm. When air is breathed in, the intercostal muscles move the ribs upward and outward, and the diaphragm is pushed downward, thus taking air into the expanded lungs. If a person needs to take a deeper than normal breath, the diaphragm and external intercostal muscles contract more forcefully. Additional muscles, such as the pectoralis minor and sternocleidomastoid, can also pull the thoracic cage further upward and outward, enlarging the thoracic cavity and decreasing internal pressure. Breathing out is a passive process. The intercostal muscles return to their resting position, returning the thoracic cage to its original size and expelling air from the lungs as a result.

What is respiratory distress syndrome (RDS)?

In a developing embryo, pulmonary surfactant is not produced in sufficient quantities until the seventh month of gestation. As the fetus matures, the amount of pulmonary surfactant increases. In premature infants, respiratory distress syndrome (RDS), or hyaline membrane disease, is caused by too little surfactant. This is common, particularly for infants delivered before the seventh month of pregnancy.

How is air cleaned before reaching the lungs?

There are many ways by which air is cleansed before it reaches the lungs. As air is inhaled through the nose, the outer part of the nostril has visible hairs (vibrissae) that filter larger particles. In addition, the nasal cavity has a mucous lining that traps smaller particles or microorganisms. The mucus then flows toward the pharynx, where it is swallowed; any trapped microbes are usually killed when exposed to the acidic conditions in the stomach. The cells lining the respiratory system have cilia (tiny, oar-like appendages) on the surface that act as an additional filter to trap debris.

What part of the brain controls respiratory rate?

The major center of respiratory activity is the medullary respiratory center in the medulla oblongata. It consists of two dorsal respiratory groups and two ventral respiratory groups. The dorsal groups stimulate contraction of the diaphragm and the ventral groups stimulate various groups of muscles. In addition, the pontine respiratory group, located in the pons, functions in switching between inspiration and expiration.

RESPIRATORY SYSTEM

How is breathing related to age?

As we get older, we tend to breathe slower, as shown by the table below. Gender also is a factor in breathing rates. The following lists the average breathing rate depending on age in years:

Age in years	Breaths per minute
Infant	40–60
5	24–26
15	20–22
25 (male)	14–18
25 (female)	16–20

After the age of 25, breathing rates level off. In the average lifetime, an individual takes more than 600 million breaths of air.

What is a normal respiratory rate?

The average person breathes about 16 times each minute and takes in one pint of air (0.5 liters) with each breath. In an average lifetime, we breathe over 75 million gallons (2.85 trillion liters) of air.

How much air is required by the body for general activities?

The body requires 8 quarts (7.6 liters) of air per minute when lying down, 16 quarts (15.2 liters) when sitting, 24 quarts (22.8 liters) when walking, and 50 quarts (47.5 liters) when running.

What is the volume of air in the lungs?

The total volume of air the lungs of an average young adult can hold, also called total lung capacity (TLC), is 5,800 milliliters. This is the combination of the vital capacity (VC) (4,600 milliliters) and the residual volume (RV) (1,200 milliliters). The vital capacity is the maximum volume of air that can be exhaled after taking the deepest breath possible. The residual volume is the volume of air that remains in the lungs even after maximal expiration.

Do people have different lung capacities?

Yes, people have different lung capacities depending on certain factors. For example, human females usually have about 20 to 25 percent lower lung capacity than human males. Taller people have more capacity than shorter people. If a person lives at sea level all his or her life, he or she will usually have less lung capacity than a person who has lived in a high-mountain region all his or her life. This is because the mountain air is thinner, with less oxygen, making the body expand its lung capacity (take in more oxygen) to compensate.

What technique is frequently used to hold one's breath longer?

An individual can hold his or her breath longer by breathing deeply for one minute and saturating the lungs with oxygen before holding it. Professional swimmers can hold their breath three times as long as usual this way.

Why is it more difficult to breathe at high altitudes?

It is difficult to breathe at high altitudes because there is less oxygen available in the atmosphere. If the concentration of oxygen in the alveoli drops, the amount of oxygen in the blood drops. At altitudes of 9,843 feet (3,000 meters) or more, people often feel lightheaded, especially if they are exercising and placing extra demands on the cardiovascular and respiratory systems.

SOUND PRODUCTION

How do human beings create sound?

Air passing over the vocal cords causes them to vibrate from side to side generating sound waves. The frequency of vibration may fluctuate between 50 hertz in a deep bass to 1,700 hertz in a high soprano.

Why is a man's voice usually lower than a woman's voice?

The pitch of the voice—how high or low it sounds—depends on the length, tension, and thickness of the vocal cords. Because males have longer vocal cords of up to 1 inch (2.54 centimeters) in length, the male voice is deeper in pitch, while women and children with shorter cords have higher-pitched voices. Vocal cords in women aver-

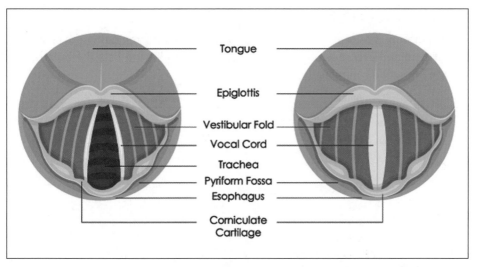

Tongue

Epiglottis

Vestibular Fold

Vocal Cord

Trachea

Pyriform Fossa

Esophagus

Corniculate
Cartilage

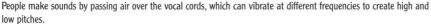
People make sounds by passing air over the vocal cords, which can vibrate at different frequencies to create high and low pitches.

age 0.167 inches (0.42 centimeters) in length. Testosterone is the hormone that is responsible for the increase of length of male vocal cords during puberty.

Why does the voice sound different when someone has a cold?

When the nasal cavity and paranasal sinuses are filled with mucus instead of air, the quality of the sound that is produced changes. Even though a person's speech is controlled by the vocal cords, the sound gets amplified by the hollow parts (sinuses) of the head. Thus, if those parts are blocked, a person's voice will sound different not only to the person, but to others.

What is phonation?

Sound production in the larynx is called phonation. It is one component of speech production that also requires articulation, or the modification of those sounds by other anatomical structures. These structures include the pharynx, oral cavity, nasal cavity, and paranasal sinuses. This combination determines the particular and distinctive sound of an individual's voice.

What determines the pitch of the voice?

The pitch of the voice depends on the diameter, length, and tension in the vocal cords. The diameter and length are directly related to the size of the larynx. The tension is controlled by the contraction of voluntary muscles associated with various cartilages of the larynx.

COMPARING OTHER ORGANISMS

Compared to humans, how fast do some animals breathe?

The respiratory rate of an animal depends on many factors, including the creature's metabolism and even how the animal is built. The following lists the breaths per minute for various animals:

Animal	Breaths Per Minute
Diamondback snake	4
Horse	10
Human	12
Dog	18
Pigeon	25–30
Cow	30
Giraffe	32
Shark	40
Trout	77
Mouse	163

How do certain animals use lungs to breathe?

Lungs are internal structures found in most terrestrial animals in which gas exchange occurs. The lungs are lined with moist epithelium (outer cells) to avoid becoming dried out. Some animals, including certain lungfish, amphibians, reptiles, birds, and mammals, have special muscles to help move air in and out of the lungs; other animals have lungs connected to the outside surface with special openings and do not require special muscles to move air in and out of their lungs.

Do insects breathe?

Insects have a system of internal tubes (called tracheae) that lead from the outside to internal regions of the body by what are called spiracles, where gases are exchanged. Some insects rely on muscles to pump the air in and out of the tracheae, while in other insects, the process is a passive exchange of gases. In addition, some insects, such as spiders, have "lungs" in addition to tracheae—hollow, leaflike structures through which the blood flows. These lungs hang in an open space that is connected to a tube, with the other side of the tube in open contact with the air.

RESPIRATORY SYSTEM

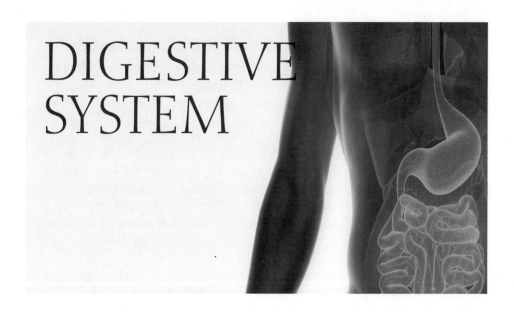

DIGESTIVE SYSTEM

INTRODUCTION

What is digestion?

The purpose of digestion is to break down large food particles into smaller molecules that can be absorbed (as nutrients) and used by the cells of the body as a source of energy for growth and reproduction. For example, fats are broken down into glycerol and fatty acids and proteins into amino acids. The digestive tract, also called the alimentary canal, is approximately thirty feet (nine meters) long from the mouth to the anus. It is lined mostly with smooth muscles (involuntary) that push the food through the body in a process called peristalsis.

What are the steps in the digestive process?

There are five major steps in the process of digestion.

1. Ingestion: the eating of any food
2. Peristalsis: the involuntary muscle contractions that move the ingested food through the digestive tract
3. Digestion: the conversion of the food molecules into nutrients that can then be used by the body
4. Absorption: the passage of the nutrients into the bloodstream and/or lymphatic system to be used by the body's cells
5. Defecation: the elimination of the undigested and unabsorbed ingested materials

What is the total average time from eating to excreting?

The total time from eating to excreting takes an average of fifty-three hours, although this number of hours is often debated. This is because food moves through the digestive system at different speeds. In addition, food is not eliminated in stool in the same order in which people ingest them. For example, vegetables are easier to digest than meat, so fibrous plants typically move through the large intestines much faster.

What are the major organs of the digestive system?

The digestive system consists of the upper gastrointestinal tract, the lower gastrointestinal tract, and the accessory organs. The organs of the upper gastrointestinal tract are the oral cavity, esophagus, and stomach. The organs of the lower gastrointestinal tract are the small intestine and large intestine (also called the colon). The accessory organs are the salivary glands, the liver, gall bladder, and pancreas.

UPPER GASTROINTESTINAL TRACT

What is the major function of the upper gastrointestinal tract?

The upper gastrointestinal tract is the site of food processing. Most mechanical digestion occurs in the upper gastrointestinal tract. It includes the mouth, teeth, and the mechanisms that help a person swallow. (For more about swallowing, see the chapter "Muscular System.")

Which structures form the oral cavity?

The oral cavity, also called the buccal (from the Latin *bucca*, meaning "cheek") cavity, is formed by the mouth, lips, and cheeks. The teeth, tongue, palate, and salivary glands are associated with the oral cavity.

How many different types of teeth are in the mouth?

The three major types of teeth are incisors, cuspids (canines), and molars. All teeth have the same basic structure, consisting of a root, a crown, and a neck. The root is embedded in the socket in the jaw. The crown is the portion that projects up from the gum. The neck, surrounded by gum, forms the connection between the root and the crown.

The different types of teeth perform different functions. The incisors, located at the front of the mouth, are blade-shaped and suited for clipping or cutting. Incisors are important to bite off pieces of food. Located next to the incisors are the cuspids or canines. Their character-istic pointed tips make them suitable for tearing, shearing, and shredding food. Both premolars (also called bicuspids) and molars have flattened crowns with prominent ridges. They are essential for crushing and grinding food.

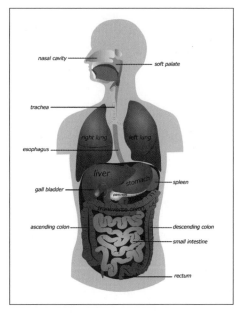

The gastrointestinal tract

What is the purpose of primary teeth?

Primary teeth, also known as baby, deciduous (from the Latin *deciduus* or *decider*, meaning to "fall down or off"), temporary, or milk (for their milk-white color) teeth serve many of the same purposes as permanent teeth. They are needed for chewing and they are necessary for speech development. They also prepare the mouth for the perma-nent teeth by maintaining space for the permanent teeth to emerge in proper align-ment. Each individual has twenty primary teeth followed by thirty-two permanent teeth.

What is the tongue?

The tongue is composed mostly of striated muscle. It is divided into two major sections: the oral, or anterior body, part and the pharyngeal, or posterior, part. The oral part is cov-ered with small projections called papillae. These papillae give the tongue its character-istic rough texture. There is also a series of taste buds on the tongue. The tongue aids mechanical digestion and is important for sensory input and speech production.

How much force does a human bite generate?

All the jaw muscles working together can close the teeth with a force as great as 55 pounds (25 kilograms) on the incisors or 200 pounds (90.7 kilograms) on the molars. A force as great as 268 pounds (122 kilograms) for molars has been reported.

How are the teeth and tongue involved in chewing?

The first stage of mechanical digestion is mastication, or chewing. Initially, the teeth tear and shred large pieces of food into smaller units. The muscles of the tongue, cheeks, and lips help keep the food on the surfaces of the teeth. The tongue then

249

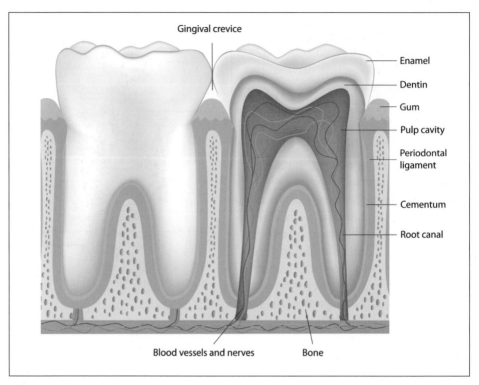

Anatomy of a tooth

compacts the food into a small round mass of material called the bolus. The salivary glands help lubricate the food with secretions so it is moist.

What is the composition of saliva?

Nearly 99.5 percent of the total composition of saliva is water. The remaining 0.5 percent consists of ions, such as potassium, chloride, sodium, and phosphates, which serve as buffers and activate the enzymatic activity. An important enzyme in saliva is salivary amylase. It breaks down complex carbohydrates, such as starches, into smaller molecules that can be absorbed by the digestive tract.

How much saliva does a person produce in a day?

Saliva is a mixture of mucus, water, salts, and the enzymes that break down carbohydrates. Awake individuals secrete saliva at a rate of approximately 0.5 milliliters per minute, or an average of 480 milliliters of saliva in a sixteen-hour waking day. Various activities such as exercise, eating, drinking, and speaking increase salivary volume.

When a person swallows solid or liquid food, what prevents it from going down the windpipe?

Once food is chewed, voluntary muscles move it to the throat. In the pharynx (throat), automatic, involuntary reflexes take over. The epiglottis closes over the larynx (voice box), which leads to the windpipe. A sphincter at the top of the esophagus relaxes, allowing the food to enter the digestive tract. On the average, a person swallows (called deglutition) 2,400 times a day.

What is peristalsis?

Peristalsis is the movement that propels food particles through the digestive tract. Rhythmic waves of smooth muscle contractions perform this action.

What is the role of the esophagus in digestion?

The esophagus is a muscular tube approximately 10 inches (25 centimeters) long and 0.75 inches (2 centimeters) in diameter that allows solid food and liquids to pass from the pharynx to the stomach. (On the average, food stays in the esophagus around five to nine seconds.) It is lined with cells that secrete mucus to lubricate the tube and allow for smoother passage of food through the esophagus. Peristaltic action moves the food through the esophagus.

What ensures that food moves only one way through the esophagus?

A sphincter muscle is found at each end of the esophagus to ensure one-way movement of food. The sphincter at the upper end of the esophagus is the pharyngoesophageal sphincter. The gastroesophageal sphincter, also called the cardiac sphincter, is at the lower end of the esophagus. It is important because it prevents acids from the stomach from being forced into the esophagus and irritating the lining of the esophagus.

Which medical condition may occur when the lower esophageal sphincter does not close properly?

Gastroesophageal reflux disease (GERD) occurs when the lower esophageal sphincter does not close properly and stomach contents leak back, or reflux, into the esophagus. When refluxed stomach acid touches the lining of the esophagus, it causes a burning sensation in the chest or throat called heartburn. The fluid may even be tasted in the back of the mouth, and this is called acid indigestion.

How is gastroesophageal reflux disease (GERD) treated?

Lifestyle changes help prevent GERD. Among these are quitting smoking, weight loss, not eating within two to three hours before bedtime, eating more frequent but smaller meals, and avoiding certain foods that may aggravate heartburn, including greasy or spicy foods, coffee, alcohol, chocolate, and tomato products. If lifestyle changes alone do not alleviate the symptoms of GERD, medications, including antacids, acid blockers, or proton pump inhibitors, may be beneficial.

What functions are performed by the stomach?

The stomach serves several important functions in the digestion of food. It stores ingested food; it is the site of mechanical and enzymatic digestion of ingested food;

and it produces and secretes intrinsic factor. Intrinsic factor is a compound secreted by cells of the stomach that facilitates the absorption of vitamin B12 in the small intestine.

What are the four regions of the stomach?

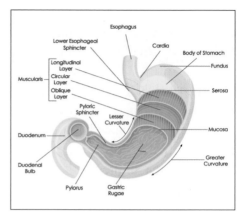

Anatomy of the stomach

The stomach, which is located directly under the dome of the diaphragm and protected by the rib cage, is a J-shaped organ that has a maximum length of 10 inches (25 centimeters) and a maximum width of 6 inches (15 centimeters). It is divided into four regions: 1) the cardia, 2) the fundus, 3) the body, and 4) the pylorus. Each region is slightly different anatomically. The cardia is located near the gastroesophageal junction. The fundus is the small, rounded part of the stomach located above the gastroesophageal sphincter. The body is the main region of the stomach. It is the area between the fundus and the "J" shape of the stomach. Most food storage and mixing occur in the body. The pylorus is the bottom curve of the "J" shape. It is located at the junction between the stomach and the small intestine.

How does the volume of the stomach change from when it is empty to when it is full?

The inner mucous membrane of the stomach contain branching wrinkles called rugae (from the Latin, meaning "folds"). As the stomach fills, the rugae flatten until they almost disappear when the stomach is full. An empty stomach has a volume of only 0.05 quarts (50 milliliters). A full stomach expands to contain 1 to 1.5 quarts (a little less than 1 to 1.5 liters) of food in the process of being digested.

What are gastric juice and chyme?

Gastric juice is a clear, colorless fluid secreted by specialized cells in the fundus of the stomach. It contains hydrochloric acid, pepsinogen (an inactive proenzyme that is converted to pepsin), mucus, and intrinsic factor. An average of 1.5 quarts (a little less than 1.5 liters) of gastric juice is secreted daily. The major function of gastric juice is to digest protein. The acidity of gastric juice denatures proteins and inactivates most of the enzymes in food. The acidity of gastric juice creates an environment that is unfriendly to many microorganisms ingested with food that may be harmful.

Chyme is the soupy, semi-fluid mixture of partially digested food that forms in the stomach. It is highly acidic and contains the nutrients that will be absorbed in the small intestines. It is expelled by the stomach into the duodenum, then through the small intestines by peristalsis during digestion.

Why is the stomach lining not harmed by gastric juice?

The stomach is covered by a protective, alkaline coat of mucus, so it is not harmed by gastric juice. In addition, since pepsin is secreted as pepsinogen, which is inactive, it cannot digest the cells that produce it.

How frequently are the cells of the stomach lining renewed?

The lining of the stomach sheds about 500,000 cells per minute. It is completely renewed every three days.

What causes a stomach ulcer?

Historically, doctors thought that genetics, anxiety, or even spicy foods caused stomach ulcers. While these may worsen the pain of an ulcer, scientists now believe that the gastric ulcer is caused by a bacterium called *Helicobacter pylori*. Australian physician and researcher Barry J. Marshall (1951–) observed that many ulcer patients had these bacteria present in their systems. In 1984 Marshall designed an experiment to determine whether there was a link between *Helicobacter pylori* and stomach ulcers. He consumed a large amount of the bacteria and waited. After ten days, he developed ulcers. Marshall shared the 2005 Nobel Prize in Physiology or Medicine with Australian pathologist J. Robin Warren (1937–) for their discovery of the bacterium *Helicobacter pylori* and its role in gastritis and peptic ulcer disease.

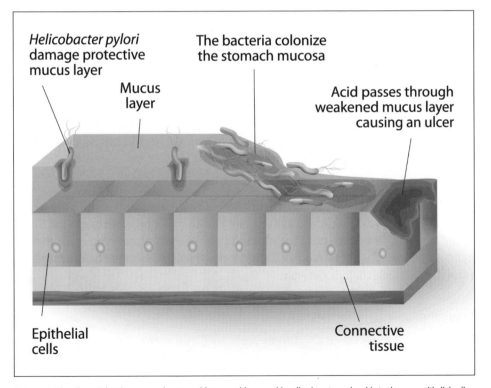

Helicobacter pylori damage protective mucus layer

The bacteria colonize the stomach mucosa

Mucus layer

Acid passes through weakened mucus layer causing an ulcer

Epithelial cells

Connective tissue

In a peptic ulcer, bacterial action causes the stomach's mucosal layer to thin, allowing stomach acids to damage epithelial cells.

Does the stomach have a memory?

Yes, the stomach has a kind of memory, but it is much different than the brain's memory. Often referred to as the gut brain, it is made up of about 500 million nerve cells. These cells control the stomach's muscular contractions, moving the ingested food and drink to the small intestines. Researchers in the relatively new field of neurogastroenterology now know that the nerves also communicate with the brain signaling if a person is full or hungry. It is thought that evolutionarily, humans developed the gut behavior independent of the brain so digestion could happen somewhat autonomously and knock out the "middleman" of the spinal cord. This may be true, since about 90 percent of the nerves carry information from the gut to the brain, not vice versa.

How long does food remain in the stomach—and how does food leave and move to the intestines?

Food remains in the stomach for one to three hours as it becomes partially digested and forms chyme. The pyloric sphincter, located between the stomach and duodenum, or small intestine, is never completely closed. Water and other fluids pass continually from the stomach to the duodenum. A small amount of chyme moves through the pyloric sphincter into the duodenum with each peristaltic contraction. The remainder of the chyme is forced back into the pyloric region of the stomach for further mixing.

Does absorption of nutrients occur in the stomach?

Absorption of nutrients does not occur in the stomach. Most carbohydrates, lipids, and proteins are only partially broken down by the time they leave the stomach and cannot be absorbed. In addition, the lining of the stomach is covered with alkaline cells so the acidic chyme is not in direct contact with the stomach lining. It is relatively impermeable to water. Absorption of nutrients occurs after the chyme has left the stomach mainly in the small intestines.

Are drugs absorbed through the stomach?

Some drugs, such as aspirin and alcohol, are able to be absorbed through the lining of the stomach. They penetrate the mucous layer of the stomach and enter the circulatory system. Therefore, alcohol in the stomach will be absorbed before nutrients from a meal reach the bloodstream.

Why do people burp (or belch)?

Burping, technically called eructation (from the Latin *ructare*, meaning "belch") is a normal occurrence that results from an abundance of air in the stomach. Nearly a half a quart of air is typically swallowed during a meal. Much of this air is relieved as a burp or belch. Belches can also occur when a person is hungry, tense, when they swallow their saliva, or when the intestines produce gases from fermentation (due to the digestive tract's natural, internal bacteria). This is why a person sometimes burps without eating, as the upper digestive tract tries to eliminate the gas swallowed during the day or the gas from the intestines.

LOWER GASTROINTESTINAL TRACT

What are the major functions of the lower gastrointestinal tract?

The lower gastrointestinal tract, consisting of the small intestine and large intestine, is the main location where nutrient processing and absorption occurs. In general, a human's small intestine is about 22 feet (7 meters) long. The large intestine is about 5 feet (1.5 meters) long.

How is the small intestine divided into segments?

The small intestine is divided into three segments: 1) the duodenum, 2) the jejunum, and 3) the ileum. The duodenum is approximately 10 inches long (25 centimeters) and begins at the junction between the stomach and small intestine at the pyloric sphincter. The jejunum is about 8 feet (2.5 meters) long. The longest segment of the small intestine is the last segment, the ileum. It is 12 feet (3.5 meters) long. Each segment has the same basic anatomical structure and appearance.

Does each segment of the small intestine have the same function?

The final digestive processing of chyme takes place in the duodenum. Once the chyme passes from the stomach to the duodenum, it is mixed with digestive secretions from the liver and pancreas in the final steps in digestion. Most of the absorption of nutrients into the bloodstream and lymphatic system takes place in the jejunum and ileum.

How does chyme move through the small intestine?

It takes an average of one to six hours for chyme to move the length of the small intestine. It moves through the small intestine by two different types of contractions: peristalsis and segmentation. Peristalsis is the rhythmic contractions that move chyme through the gastrointestinal tract. Segmentation involves localized contractions of small segments of the small intestine. These contractions mix the chyme with the secretions of the small intestine, gall bladder, and pan-

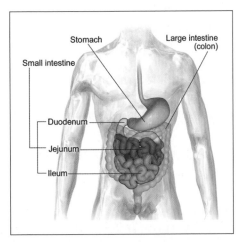

The small intestine is divided into the duodenum, jejunum, and ileum.

creas. The nutrients are brought into contact with the microvilli in the small intestine. The chyme is slowly propelled toward the ileocecal valve. Unlike peristaltic contractions, which are directional, the contractions of segmentation are not directional. Therefore, in order to keep the chyme moving downward, the duodenum contracts more frequently than the jejunum or ileum.

What is the purpose of villi?

The mucosa layer of the small intestine has many fingerlike projections called villi. Villi (from the Latin, meaning "shaggy hairs"), and the smaller microvilli, increase the surface area of the small intestine dramatically, allowing for greater absorption of nutrients. If the small intestine were a smooth tube without villi, it would have a total absorptive area of 3.6 square feet (3,344 square centimeters). The existence of villi effectively increases the absorptive area of the small intestine by a factor of nearly 600 to more than 2,200 square feet (2,043,800 square centimeters).

Why is the small intestine longer than the large intestine?

The descriptive terms "small" and "long" refer to the diameter of the intestine rather than the length. The diameter of the small intestine is only approximately 1 inch (2.5 centimeters), while the diameter of the large intestine is 2.5 to 3.0 inches (6.5 to 7.5 centimeters).

What is the role of the small intestine in nutrient processing?

The small intestine is the site of most nutrient processing in the body. The first step is to break down the large complex structures of all nutrients, including carbohydrates, lipids, proteins, and nucleic acids, into smaller units. Most absorption of these nutrients also takes place in the small intestine.

What is "gluten intolerance"?

Some people have a slight to moderate intolerance to grain products (especially wheat), a condition called gluten intolerance or gluten sensitivity. It usually becomes apparent when a person eats foods containing gluten, mainly cereal grains, such as wheat, rye, and barley. The major culprit is one of the proteins, gliadin, collectively called gluten. It combines with antibodies in the digestive tract. Once this happens, sometimes gradually, the gliadin causes damage to the intestinal walls, which, in turn, interferes with the absorption of many necessary nutrients.

What type of diet is recommended for individuals with celiac disease?

A gluten-free diet is the only treatment for individuals with celiac disease. Celiac disease is an autoimmune digestive disease that damages the small intestine and interferes with absorption of nutrients. The villi in the small intestine are damaged or destroyed whenever sufferers of celiac disease eat products that contain gluten. Gluten is found in wheat, rye, and barley. Once the villi are damaged, they cannot take in nutrients to be absorbed by the bloodstream, leading to malnutrition.

What valve is located where the small and large intestines meet?

The ileocecal valve is a sphincter muscle valve that serves as the boundary between the small intestine and the large intestine. The valve is closed most of the time but opens when material passes from the small to the large intestines. The valve's main purpose is to prevent any backflow of material from the large intestines into the small intestines.

How many distinct regions are in the large intestine?

Overall, the large intestine frames the small intestine on three sides. The large intestine consists of three distinct regions: 1) the cecum, 2) the colon, and 3) the rectum. The cecum is the first section of the large intestine below the ileocecal valve. The appendix is attached to the cecum. Since the colon (ascending, transverse, descending, and sigmoid colon) is the largest region of the large intestine, the term "colon" is often applied to the entire large intestine. The rectum (rectum, anal canal, and anus) is the final region of the large intestine and the end of the digestive tract.

What are the functions of the large intestine?

The large intestine is mostly a storage site for undigested materials until they are eliminated from the body via defecation. Although digestion is complete by the time the chyme enters the large intestine and most absorption has occurred in the small intestine, water and electrolytes are still absorbed through the large intestine.

What is the difference between diverticula, diverticulosis, and diverticulitis?

Diverticula are bulging, sac-like pouches in the wall of the colon that protrude outward from the wall of the colon. Diverticula appear most often in individuals over forty whose diet is low in fiber. In diverticulosis the pouches are present but the individuals do not

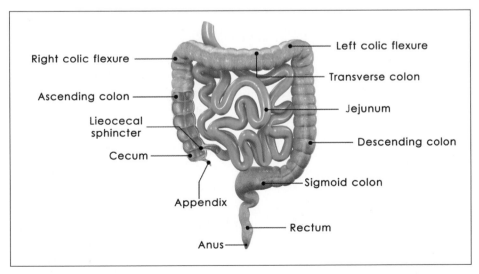

The large intestine is shorter and thicker than the small intestine that it surrounds. It is the last part of the digestive tract through which waste passes before exiting the anus.

have any symptoms or discomfort. In diverticulitis, the diverticula are inflamed and often infected when undigested food and bacteria are caught in the diverticula. Patients experience pain, either constipation or increased frequency of defecation, nausea, vomiting, and low-grade fever. In severe cases, surgical removal of the infected area of the colon is necessary. Changing to a high-fiber diet usually relieves the symptoms.

How much water enters the large intestine daily?

Nearly 95 percent of the water that enters the digestive tract is absorbed by the small intestine. Only 0.5 quarts (0.47 liters) of water enters the large intestine daily. Absorption of water in the large intestine helps to avoid dehydration. Unabsorbed water is excreted with feces.

How long do the undigested remains stay in the large intestine?

The undigested remains of the digestion process stay in the large intestine (colon) an average of 40 hours, though this varies significantly with gender. For women, it takes approximately 47 hours, while for men it takes roughly 33 hours for food to pass through the colon.

How much flatulence (gas) is produced daily?

Most individuals produce nearly 1 pint (473 milliliters) of flatus (gas) daily, a gas that consists of mostly of nitrogen, carbon dioxide, and hydrogen, with a small amount of oxygen, methane, and hydrogen sulfide. Passing gas 10 to 20 times per day is considered normal. Diets that consist of an abundance of carbohydrate-rich foods (such as beans) produce more gas because a greater amount of undigested carbohydrates enter the large intestine. In addition, some gases result from the breakdown of foods people ingest by certain natural bacteria in the large intestines. The mixture of different gases causes the characteristic odor associated with flatulence.

What is the role of bacteria in the large intestine?

Many bacteria, such as *Escherichia coli* and *Clostridium*, are normally found in the colon. Some bacteria are able to break down undigested fiber into glucose. Bacteria are also able to synthesize certain vitamins, especially vitamins K and B12. The presence of large colonies of beneficial bacteria inhibits the growth of pathogenic bacteria. In addition, the bacteria ferment some of the indigestible carbohydrates, including cellulose, contributing to the production of intestinal gas, or flatus (from the Latin, meaning "blowing").

What is feces?

Feces (from the Latin *faex*, meaning "dregs") is the remaining portion of undigested food. Approximately 5.3 ounces (150 grams) of feces are produced daily. Feces normally consists of 3.5 ounces (100 grams) of water and 1.7 ounces (50 grams) of solid material. The solid material is composed of fat, nitrogen, bile pigments, undigested food such as cellulose, and other waste products from the blood of intestinal wall.

Why do feces vary in color?

Feces color is usually influenced by what a person eats as well as by the presence of two bile pigments, bilirubin and biliverdin. Bilirubin is formed by the breakdown of red blood cells (hemoglobin) in the liver and the oxidized form of bilirubin is biliverdin. These are secreted into the bile, which, in turn, enters the intestines. The bilirubin is orange or yellow; the biliverdin is green. In a healthy person, when mixed with the contents of the intestines, the bilirubin and biliverdin give a person's feces its normal light to dark brown color. If the contents pass faster through the bowel the feces may appear green. (Note: The Mayo Clinic recommends that if a person's feces is bright red or black to seek medical attention, as it may indicate the presence of blood, but not always. In some cases, if a person ingests iron supplements, bismuth [such as in Pepto Bismol], or black licorice, it may also turn the feces black.)

What causes the characteristic odor associated with feces?

The characteristic odor associated with feces is caused by indole and skatole. These two substances result from the decomposition of undigested food residue, unabsorbed amino acids, dead bacteria, and cell debris. High-protein diets produce larger quantities of indole and skatole, accompanied by an increase in the odor of feces.

What is the process of defecation?

Defecation is the final step in the digestive process that removes all undigested materials from the body. Defecation involves both voluntary and involuntary actions. When feces fills the rectum, it triggers the defecation reflex. The urge to defecate can be controlled in most individuals except young children and others who have suffered spinal cord injuries.

What is constipation?

Constipation may be defined as the passage of small amounts of hard, dry bowel movements. Constipation is the result of too much water being absorbed into the colon. In addition, when the fecal material moves through the colon at a slow rate, more water is absorbed, resulting in hard, dry feces. Diets high in fiber help prevent constipation.

What are some common causes of diarrhea?

Diarrhea, which is frequent, loose, watery bowel movements, may be caused by infections or other intestinal disorders. Diarrhea may be associated with both bacterial and viral infections. Common bacteria, such as *Campylobacter*, *Salmonella*, *Shigella*, and *Escherichia coli*, consumed in contaminated food and/or water will cause diarrhea. Many viruses cause diarrhea, including rotavirus, Norwalk virus, cytomegalovirus, herpes simplex virus, and viral hepatitis. In addition to bacterial and viral infections, parasites, including *Giardia lamblia*, *Entamoeba histolytica*, and *Cryptosporidium*, may enter the body through food and water and cause diarrhea. Several disorders, including irritable bowel syndrome, inflammatory bowel disease, celiac disease, and side effects of medication, may also cause diarrhea. Most cases of diarrhea often resolve themselves without medical intervention. It is important to prevent dehydration by replacing fluids and electrolytes. It is usually recommended to avoid milk products, greasy foods, very sweet foods, and foods that are high in fiber until the diarrhea has subsided. Bland foods may then be slowly reintroduced to the diet.

Is the large intestine essential for life?

Since the role of the large intestine is mainly as a storage site for fecal material and the elimination of it from the body, it is not essential for life. Individuals who suffer from colon cancer or other diseases will often have their colon removed. The end of the ileum is brought to the abdominal wall. Food residues are eliminated directly from the ileum into a sac attached to the abdominal wall on the outside of the body. Alternatively, the ileum may be connected directly to the anal canal.

Which two diseases are considered inflammatory bowel diseases?

Inflammatory bowel disease (IBD) is the general term for diseases that cause inflammation in the intestines. Crohn's disease and ulcerative colitis belong to the group of illnesses known as IBDs. Crohn's disease may affect any part of the digestive tract, but it most often affects the ileum. Ulcerative colitis occurs only in the inner lining of the colon (the large intestine) and the rectum. Abdominal pain and diarrhea are the most common symptoms of both ulcerative colitis and Crohn's disease. Both diseases are chronic, ongoing diseases, although periods of remission are not uncommon.

How does irritable bowel syndrome differ from inflammatory bowel disease?

Irritable bowel syndrome (IBS) is not a disease but a disorder that disrupts the functions of the colon. There is no inflammation of the digestive tract. It is characterized by a group of symptoms, including crampy abdominal pain, bloating, constipation, and diarrhea.

ACCESSORY ORGANS

Which organs are considered accessory organs to the digestive system?

The pancreas, liver, and gall bladder are accessory organs in digestion. None of these organs are a part of the digestive tract that begins at the mouth and ends at the anus, but they contribute important chemicals, enzymes, and lubricants necessary for the functioning of the digestive system.

Which cells of the pancreas secrete enzymes?

The pancreas, which also functions as an endocrine gland as described in the chapter on the endocrine system, consists of both endocrine and exocrine cells. The acinar cells (also called *acini*, from the Latin meaning "grapes" because their structure resembles clusters of grapes) are responsible for secreting digestive enzymes.

How do the pancreatic enzymes reach the small intestine?

The pancreatic enzymes reach the small intestine via the heptatopancreatic duct. This duct is formed by linking the bile duct with the pancreatic duct. The pancreatic secretions are highly alkaline in order to neutralize the acidic chyme (for more about chyme, see this chapter).

What is the importance of pancreatic digestive juices?

The pancreatic digestive juices are an alkaline solution (pH 8) composed of many enzymes. These enzymes are able to break down all categories of food. In fact, nearly 1.6 quarts (1.5 liters) of pancreatic digestive juices are secreted by the cells of the pancreas daily.

How large is the liver?

The liver is the second largest organ in the body (the skin is the largest). It weighs 3 pounds (1.4 kilograms) in adults, representing about 2.5 percent of the total body weight. In children, the liver accounts for 4 percent of total body weight. The characteristic pudgy abdomen of infants is a result of the size of the liver.

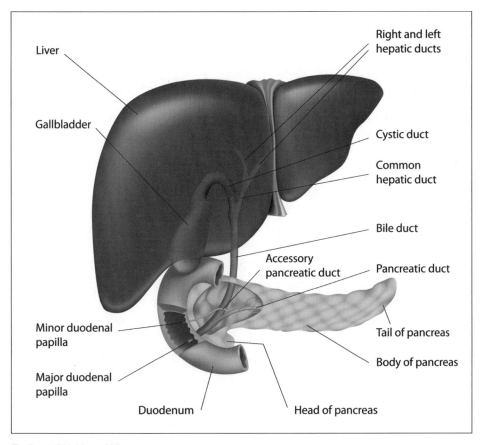

The liver, gall bladder, and bile passage

What is a unique feature of the liver?

The liver is the only organ that can regenerate itself. As much as 75 percent of the liver may be removed, and it will still grow back to the same shape and form within a few weeks.

What are the digestive functions of the liver?

The liver has more than five hundred vital functions. Its major function as a digestive organ is to produce and secrete bile. Other functions of the liver include separating and filtering waste products from nutrients, storing glucose, and producing many chemical substances, such as cholesterol and albumin.

What is bile?

Bile is an alkaline liquid composed mostly of water, bile salts, bile pigments (bilirubin), fats, and cholesterol. It is essential for digestion of fats because it breaks down fats into fatty acids, which can then be absorbed by the digestive tract. Bile gets its color from bilirubin. Bilirubin is a waste product from the breakdown of worn-out red blood cells.

What are some causes of cirrhosis of the liver?

The most common causes of cirrhosis in the United States are alcoholism and hepatitis C. Worldwide, hepatitis B is probably the most common cause of cirrhosis. In cirrhosis of the liver, scar tissue replaces normal, healthy tissue, blocking the flow of blood through the organ and preventing the liver from working as it should. Although liver damage from cirrhosis cannot be reversed, treatment can stop or delay further progression and reduce complications.

What is hepatitis?

Hepatitis is inflammation of the liver and commonly results from a viral infection. Hepatitis may either be acute or chronic. Acute hepatitis is short-lived, while chronic hepatitis is an inflammation of the liver that lasts for at least six months. Symptoms of acute hepatitis usually begin suddenly and include loss of appetite, nausea, vomiting, abdominal pain, low-grade fever, fatigue, and jaundice. Jaundice (from the French *jaune*, meaning "yellow") is a yellow discoloration of the skin and the whites of the eyes. Many individuals with chronic hepatitis have mild symptoms. Depending on the type of viral hepatitis, it may resolve itself on its own without medical intervention.

How many different types of viral hepatitis have been identified?

There are five main types of viral hepatitis: hepatitis A, hepatitis B, hepatitis C, hepatitis D, and hepatitis E. The following lists the type of hepatitis (A through E), transmission, and treatment and prevention:

Type of Hepatitis	Transmission	Treatment	Prevention
Hepatitis A	Food or water contaminated by feces from an infected person due to poor hygiene	Usually resolves itself without medical intervention within a few weeks	Practice good hygiene and sanitation; hepatitis A vaccine; avoid contaminated water (even tap water) in areas where the disease is widespread
Hepatitis B	Contact with infected blood; sexual relations (both heterosexual and homosexual) with an infected person; from mother to child during birth	Acute hepatitis B usually resolves itself without medical intervention; chronic hepatitis B may be treated with medications	Hepatitis B vaccine; avoid high-risk behaviors such as sharing needles and multiple sexual partners
Hepatitis C	Contact with infected blood (e.g. sharing needles); rarely through sexual contact and childbirth	Acute hepatitis C may resolve itself within a few months; chronic hepatitis C is treated with medications	Avoid high-risk behaviors such as sharing needles; avoid sharing personal items, such as razors and toothbrushes, with infected individuals; there is no vaccine
Hepatitis D	Contact with infected blood; only occurs in people who are already infected with hepatitis B	Chronic hepatitis D is treated with the drug alpha interferon	Avoid exposure to infected blood, contaminated needles, and an infected person's personal items (toothbrush, razor, nail clippers); immunization against hepatitis B
Hepatitis E	Food or water contaminated by feces from an infected person; uncommon in the United States	Usually resolves itself within several weeks or months	Reduce exposure to the virus by avoiding tap water when traveling internationally and practicing good hygiene and sanitation; there is no vaccine

What are the main causes of nonviral hepatitis?

The two main types of nonviral hepatitis are alcoholic hepatitis and toxin/drug-induced hepatitis. Alcoholic hepatitis is the result of excessive drinking and often leads

263

to cirrhosis. The inhalation or ingestion of certain toxins, such as carbon tetrachloride, vinyl chloride, and poisonous mushrooms, are causes of hepatitis. Certain drugs, including large dosages of the pain reliever acetaminophen, may cause hepatitis.

What is the major purpose of the gall bladder?

The gall bladder (from the Latin *galbinus*, meaning "greenish yellow"), a pear-shaped, small sac, is mainly a storage vessel. It is connected by ducts to both the liver and small intestine. It stores bile until it is needed in the duodenum. Its name is derived from its usual color of green from the accumulation of bile.

Is the gall bladder a necessary organ?

The gall bladder is a nonessential organ. It may be removed surgically—a procedure called a cholecystectomy—if it is diseased or injured. Once the gall bladder is removed, bile flows directly to the small intestine. Excess bile is then stored in the bile duct. Individuals who have had their gall bladder removed lead a normal life and enjoy their regular diet.

What are the two types of gallstones?

Gallstones are hardened masses (stones) of bile. Gallstones form when bile contains too much cholesterol, bile salts, or bilirubin. The two types of gallstones are choles-

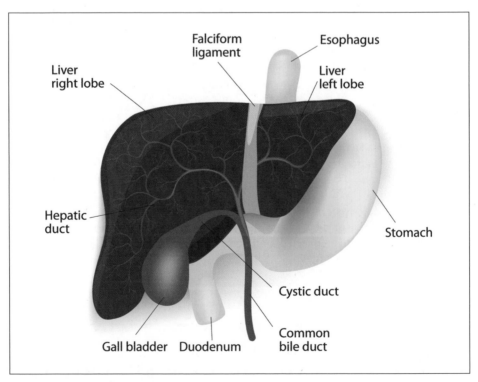

The gall bladder is located under the liver; its main purpose is to store bile.

terol stones and pigment stones. Cholesterol stones are more common, accounting for nearly 80 percent of all instances of gallstones. They are usually yellow-green in color and are made primarily of hardened cholesterol. An insufficient amount of water may also contribute to the development of cholesterol gallstones. Pigment stones are small, dark stones made of bilirubin.

How serious are gallstones?

Many individuals who have gallstones are asymptomatic and treatment is not necessary. However, if the stones block a duct, bile may be prevented from entering the small intestine. Surgery is then often recommended to remove the gall bladder. Women between twenty and sixty years of age are twice as likely to develop gallstones as men.

METABOLISM AND NUTRITION

What is metabolism?

Metabolism (from the Greek *metabole*, meaning "change") refers to the physical and chemical processes involved in the activities of the body. It includes the conversion of nutrients into usable energy contained in ATP, the production and replication of nucleic acids, the synthesis of proteins, the physical construction of cells and cell parts, the elimination of cellular wastes, and the production of heat, which helps regulate the temperature of the body.

How do catabolic reactions differ from anabolic reactions?

Catabolic and anabolic reactions are metabolic processes. A catabolic reaction is one that breaks down large molecules to produce energy, for example, in digestion. An anabolic reaction is one that involves creating large molecules out of smaller molecules, for example, when the body makes fat out of extra nutrients a person eats.

What are the essential nutrients?

There are six essential nutrients: carbohydrates, fats (lipids), proteins, water, vitamins, and minerals. Each has a specific function and relationship to the body and not one of these nutrients can act independently of the others. In addition, scientists now know that each nutrient is equally valuable for human health. Thus, although factors can alter the amounts a person ingests—based on such things as our age, body volume, gender, and lifestyle—deficiencies in any of these nutrients can lead to an imbalance. (For more about nutrients, see the chapter "Anatomy and Biology Basics.")

Which nutrients are energy nutrients?

Energy nutrients are those that provide the body with the majority of the energy needed for daily metabolic reactions. Carbohydrates, fats, and proteins are energy nutrients. The table below shows comparative values of energy sources.

265

Energy Source	Energy Yield (kcal/g)
Carbohydrate	4
Fat	9
Protein	4

What is a calorie?

A calorie is the amount of energy (heat) required to raise 1 gram (1 milliliter) of water by 1°C. A kilocalorie (kcal) is the amount of energy required to raise 1 kilogram (1 liter) of water by 1°C. The kilocalorie is the unit used to describe the energy value in food, since the calorie is a relatively small unit of measurement. For example, if a chocolate chip cookie were completely incinerated, the amount of heat energy released would be enough to raise the temperature of one liter of water by approximately 300°C.

What are vitamins and minerals?

A vitamin is an organic, nonprotein substance that is required by an organism for normal metabolic function but that cannot be synthesized by that organism. In other words, vitamins are crucial molecules that must be acquired from outside sources. While most vitamins are present in food, vitamin D, for example, is produced as a precursor in our skin and converted to the active form by sunlight. Minerals, such as calcium and iron, are inorganic substances that also enhance cell metabolism. Vitamins are often classified as fat-soluble or water-soluble based on their ability to be absorbed or stored in the body. For example, vitamins A, D, E, and K are soluble only in fats, thus are called fat soluble; the C and B vitamins are soluble in water, thus called water soluble.

What vitamins are essential for human health?

The following lists the fat- and water-soluble vitamins that are essential for human health:

Vitamin	Essential for Health
Fat Soluble Vitamins	
Vitamin A	Beta carotene, retinols: Needed for growth and cell development; prevents night blindness; helps fight some cancers; helps the cardiovascular system; needed to maintain healthy gums, glands, bones, teeth, nails, skin, and hair; beta carotene is also considered to be an antioxidant
Vitamin D	Calciferol: Needed for calcium absorption, and helps build strong bones and teeth; helps the brain, pancreas, and reproductive organs; also targets the kidneys and intestines
Vitamin E	Tocopherols: Helps maintain muscles and red blood cells; it's also a major antioxidant
Vitamin K	Necessary for efficient blood clotting

Vitamin	Essential for Health
Water-Soluble Vitamins	
Biotin	Needed for energy and metabolism
Folate	Folic acid, folacin (some call this vitamin F; it is more often thought of as B_9): necessary to make DNA, RNA, red blood cells, and to synthesize certain amino acids
Niacin	Vitamin B_3, nicotinic acid, nicotinamide: necessary to metabolize energy and to promote normal growth; for some people, larger doses often help lower cholesterol
Pantothenic acid	Vitamin B_5: helps to metabolize energy; normalizes blood sugar levels, and helps the body to synthesize antibodies, some hormones, cholesterol, and hemoglobin (in the blood)
Riboflavin	Vitamin B_2: necessary to metabolize energy and helps the body's adrenal function
Thiamine (thiamin)	Vitamin B_1; necessary to metabolize energy; needed for proper nerve function, normal digestion, and appetite
Vitamin B_6	Pyridoxine, pyridoxamine, pyridoxal: helps to metabolize proteins and carbohydrates (for energy) in the body; good for proper nerve function and helps synthesize red blood cells
Vitamin B_{12}	Cyano-cobalamin: necessary to make DNA, RAN, red blood cells, and myelin (for the body's nerve fibers)
Vitamin C	Ascorbic acid; helps to build blood vessel walls and promote wound healing; necessary for iron absorption; and is said to help prevent arthrosclerosis; it is also considered to be an antioxidant

What are the best sources for some essential vitamins?

There are many sources for the essential vitamins. The following lists some of the more common sources of the fat- and water-soluble vitamins:

Vitamin	Best Food Sources
Fat Soluble Vitamins	
Vitamin A	Beta carotene: orange and yellow fruits and vegetables, such as carrots and squash; green leafy vegetables
Retinols	Found in liver, salmon, many cold water fish; egg yolks; and enriched and/or fortified soy, cow, and other dairy products
Vitamin D	Calciferol: enriched and/or fortified dairy (cow, goat, sheep), and soy products; egg yolks; fish liver oils
Vitamin E	Tocopherols: eggs, mayonnaise, nuts, seeds, and certain vegetable oils; fortified and/or enriched cereals
Vitamin K	Green leafy vegetables, such as spinach and cabbage; pork and liver; and green tea
Water-Soluble Vitamins	
Biotin	Egg yolks, soybeans, certain cereals, soy milk, and yeast

Vitamin	Best Food Sources
Folate	Folic acid, folacin, B_9: liver; yeast; cruciferous vegetables, such as broccoli and cabbage, along with many raw vegetables and avocados
Niacin	Vitamin B_3, nicotinic acid, nicotinamide: lean meats, poultry and game meats, and certain seafood; milk, eggs, and fortified and/or enriched cereals, breads, and flours; certain legumes, such as black beans
Pantothenic acid	Vitamin B_5: almost all food
Riboflavin	Vitamin B_2: fortified and/or enriched cereals, grains, and flours; lean meats; milk and other dairy products; certain mushrooms
Thiamine	Vitamin B_1: lean pork; nuts and seeds; legumes; and fortified and/or enriched cereals and grains
Vitamin B_6	Pyridoxine, pyridoxamine, pyridoxal: lean meats, fish, and poultry; grains, cereals, and flours; green leafy vegetables, potatoes, and soybeans (and some soy products)
Vitamin B_{12}	Cobalamins: all animal products
Vitamin C	Ascorbic acid: citrus fruits, juices, and dried fruit; melons, berries, peppers, potatoes, broccoli, cabbage, and many other fruits and vegetables

Which diseases are caused by vitamin deficiencies?

Certain diseases are linked to dietary deficiencies of certain vitamins. Insufficient quantities of vitamin C in the diet can lead to scurvy. Pellagra results from a lack of niacin (vitamin B_3) in the diet. Rickets in children and the related disease of osteomalacia in adults is caused by a lack of vitamin D. Even a chronic dietary deficiency in calcium can lower the amount of calcium in a person's bones, which can affect bone strength.

Is it possible to have an overabundance of vitamins?

Excessive intake of vitamins may cause health complications as serious as vitamin deficiencies. The clinical term for excessive intake of vitamins is hypervitaminosis. It occurs when the dietary intake of a vitamin exceeds the body's ability to store, utilize, or excrete the vitamin. Hypervitaminosis is most common among the fat-soluble vitamins because the excessive quantities of the vitamins may be stored in lipid tissues. Water-soluble vitamins do not accumulate in the body since they are excreted in urine.

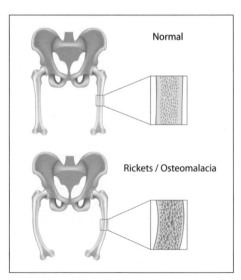

Lack of vitamin D in one's diet can cause rickets in children, a weakening of the bones that causes them to become bent.

Are vitamin supplements necessary?

Vitamin supplements may be a useful addition to the diet of individuals who do not receive all of the nutrients they need from their diet. These individuals cannot or do not eat enough, or do not eat enough of a variety of healthy foods. But for most people who eat a balanced, healthy diet, vitamin supplements are usually not necessary.

How do saturated fats differ from unsaturated fats?

In general, saturated fats are solid at room temperatures (except palm, palm kernel, and some coconut oils); most animal fats (meat, poultry, eggs, and dairy, such as beef, butter, and cheeses) are also saturated. The main problem with saturated fats is thought to be the raising of blood cholesterol. This occurs because these fats interfere with the removal of cholesterol from the blood.

Unsaturated fats—or the polyunsaturated and monounsaturated fats—have been known to either lower or have no effect on the blood cholesterol. (When polyunsaturated fats are hydrogenated, which makes them firmer, they become more like saturated fats. They can then affect the body's blood cholesterol in the same way as saturated fats.) Monounsaturated fats are liquid at room temperature and are found in olive, canola, and peanut oils, along with in some nuts, seeds, and avocados. These fats are also considered helpful to human health, as they also lower the bad LDL cholesterol levels in the blood.

What are trans fatty acids?

Trans fatty acids, or trans fats, are made when manufacturers add hydrogen to liquid vegetable oil—a process called hydrogenation—creating solid fats like shortening and hard margarine. Hydrogenation increases the shelf life and flavor stability of foods containing these fats. Diets high in trans fat raise the LDL (low density lipoprotein) or "bad" cholesterol, increasing the risk for coronary heart disease. Cakes, crackers, cookies, snack foods, and other foods made with or fried in partially hydrogenated oils are the largest source (40 percent) of trans fats in the American diet. Animal products and margarine are also major sources of trans fats. Since January 2006, the U.S. government has directed that the amount of trans fat in a product must be included in the Nutrition Facts panel on food labels.

What are eating disorders?

Eating disorders are medical illnesses in which patients become obsessed with food and their body weight. Research indicates that more than 90 percent of those who have eating disorders are women between the ages of twelve and twenty-five. The main types of eating disorders are anorexia nervosa and bulimia nervosa. A third disorder, binge eating disorder, is still being investigated by researchers.

What is anorexia?

Anorexia simply means a loss of appetite. Anorexia nervosa is a psychological disturbance that is characterized by an intense fear of being fat. This persistent "fat image," however untrue in reality, leads the patient to self-imposed starvation and emaciation to the point where one-third of the body weight is lost. Clinical diagnosis of anorexia is determined when patients weigh at least 15 percent less than the normal healthy weight for their height. Many patients do not maintain a normal weight because they refuse to eat enough or avoid eating, exercise obsessively, induce vomiting, and/or use laxatives or diuretics to lose weight. For women, this causes the menstrual period to stop.

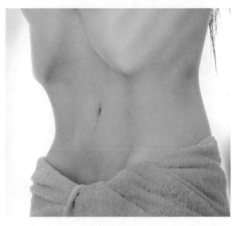

Anorexia nervosa is a psychological illness that causes a person to become obsessed with body weight and to starve herself.

What is bulimia?

Bulimia is an eating disorder in which individuals binge eat frequently, often several times a week or even several times per day. Sufferers of this illness may eat an enormous amount of food in a short time, consuming thousands of calories. Then they will purge their bodies by vomiting or using laxatives and/or diuretics.

How does binge eating disorder differ from bulimia?

Individuals with binge eating disorder also consume large quantities of food in a short period of time. However, unlike individuals with bulimia, there is no purging.

What are some causes of nausea and vomiting?

Nausea, the sensation of having the urge to vomit, and vomiting, the emptying of the contents of the stomach through the mouth, may be caused by a variety of different reasons. Some common causes of nausea and vomiting are:

- Gastroenteritis, commonly known as "stomach flu," from bacterial or viral infections
- Food poisoning
- Overeating
- Migraine headaches
- Brain injury or concussions
- Inner ear and balance disorders
- Motion sickness
- Hormonal imbalances, especially during the first trimester of pregnancy
- Certain toxins, such as alcohol
- Drugs, such as chemotherapy agents, for treating cancer

How does the definition of "overweight" differ from "obesity"?

Both "overweight" and "obesity" describe ranges of weight that are greater than what is generally considered healthy for a given height. An objective measure of these terms is based on body mass index (BMI). An adult with a body mass index of 25 to 29.9 is considered overweight. An adult with a body mass index greater than 30 is considered obese.

What are the causes of obesity?

Obesity is a result of an individual consuming more calories than she or he burns. Genetic, environmental, psychological, and underlying medical problems may all be factors that lead to obesity. Underlying medical conditions may include hypothyroidism, Cushing's syndrome, depression, and certain neurological problems that can lead to overeating. Drugs, such as steroids, used to treat certain medical conditions may cause weight gain, too. Scientific study has indicated that obesity is linked to heredity. However, since family groups also share the same basic diet, it is difficult to separate genetics from environment. The average American diet tends to include many foods that are high in fat. The standard portion size for many foods has increased over the past several years. "Super-sized" portions available at fast-food restaurants have more calories. This diet coupled with a lack of physical activity leads to obesity. In addition, many people overeat in response to negative emotions such as sadness, anger, and boredom.

What is body mass index, or BMI?

Body mass index (BMI), as the name implies, is a statistical way of measuring a person's body mass based on his or her body weight and height. Most doctors use BMI

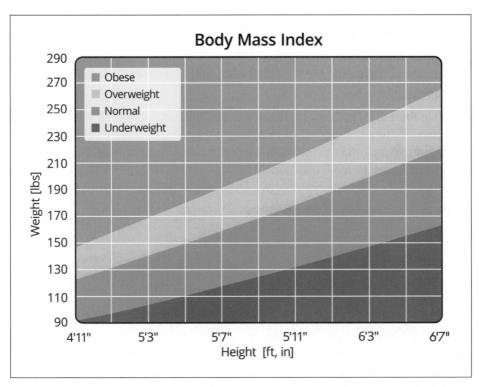

The body mass index chart is designed as a guide to maintaining a proper weight for one's height and gender. However, in some cases, such as people with a high muscle mass, the guidelines don't hold absolutely firm.

to determine a patient's healthy weight for his or her age, gender, and size. According to the Centers for Disease Control and Prevention, in general, if an adult female has a BMI less than 18.5, she is considered underweight; 18.5 to 24.9 means normal weight; 25.0 to 29.9 means overweight; and over 30.0 is considered obese. Males have slightly higher BMI numbers.

How does a person determine his or her BMI?

To determine BMI, a person takes his or her weight in pounds and height in total inches, then multiplies his or her weight by 703, and divides that number by his or her height squared (or height times height). For example, if a person is 5 feet 4 inches tall (or 64 inches tall) and weighs 133 pounds, the calculation would be as follows: $133 \times 703 = 93,449$; 64 inches squared = 4,096; divide 93,449/4,096 = 22.83—thus, that person's BMI is in the normal range.

What health risks are associated with obesity?

People who are obese are more likely to develop a variety of health problems. This includes the risk of developing hypertension, dyslipidemia (for example, high total cholesterol or high levels of triglycerides), Type 2 diabetes, coronary heart disease, stroke, gall bladder disease, osteoarthritis, sleep apnea and respiratory problems, and even some cancers (endometrial, breast, and colon).

COMPARING OTHER ORGANISMS

What is the relative gut volume when comparing chimpanzees, gorillas, and humans?

Although there are similarities between humans and other primates, there are major differences between all primates' relative gut volumes. The biggest difference between humans and other primates is the human small intestines and colon. This is mainly because of the foods consumed by humans versus other primates. The following lists some of these species and the percent of the species' total gut volumes:

Species	Stomach	Small Intestines	Cecum	Colon
Human	17	67	—	17
Chimpanzee	20	23	5	52
Gorilla	25	14	7	53

What other organisms on Earth rely on nutrients?

Besides humans, all other organisms on Earth—from animals in the Amazon and plants in the Sahara Desert to bacteria on our skin and fungi on a tree—rely on nutrients. The biggest differences are the amounts and types of nutrients each organism needs. For example, plants need nutrients mostly in the form of nitrogen, potassium, and phosphorous in order to grow strong and healthy; whereas, a human needs even more nutrients in the form of vitamins and minerals—and many non-nutrients—in order to survive.

Do cows really have four stomachs?

Cows actually have a four-chambered stomach. (Other animals also have different numbers of stomach chambers.) Cows are called ruminants, and each chamber of the cow's stomach has a certain task. Simply put, the first chamber is where the grass the cow eats (it chews the grass just enough to swallow it) is mixed with saliva; the second chamber separates the liquids from the solids. When the cow is full, it rests. From there, the cow eventually regurgitates the solids (called cud), chews the cud even more, swallows it again, and passes it to the third chamber. The inorganics and water are absorbed, and the remainder is passed into the fourth chamber where what is left is digested. The milk that comes from the cow's udder is actually some of the digested food that enters the bloodstream and travels to the udder—while the rest is for the cow's nourishment.

How do the teeth differ between some animals?

Animal teeth vary because of their use for obtaining and/or chewing food and offering protection. For example, depending on the animal's main diet, it has specialized teeth to eat. Animals called herbivores (plant-eaters, such as beavers, squirrels, and other rodents) have sharp incisors to bite off blades of grass and other plant matter, along with flat premolars and molars for grinding and crushing plants. Carnivores (meat-eaters, such as tigers, sharks, cats, wolves, and dogs) have pointed incisors and enlarged canine teeth to tear off pieces of meat; their premolars and molars are jagged to aid in chewing flesh. Omnivores (both plant- and meat-eating, such as bears and raccoons) have nonspecialized teeth to accommodate a diet of both plant material and animals.

What is unusual about shark's teeth?

Sharks were among the first vertebrates to develop teeth. The teeth are not set into the jaw, but rather sit on top of it—thus they are not firmly anchored and are easily lost. The teeth are arranged in six to twenty rows, with the ones in front doing the biting and cutting. Behind these teeth, others grow. When a tooth breaks or is worn down, a replacement moves forward. In fact, one shark may eventually develop and use more than 20,000 teeth in its lifetime.

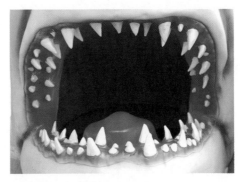

Why do some animals swallow food whole?

Some animals swallow food whole out of necessity. For example, snakes swallow their prey whole because they don't have

Sharks have rows of teeth that continue to replenish themselves throughout a shark's life.

teeth designed for chewing, but many have teeth that curl backward in order to push food toward the throat. Birds don't have teeth in the same sense as human teeth, but have beaks that can pound, rip, or shred food that they swallow. Frogs also have no teeth, so they swallow insects and small animals whole, catching their prey with a sticky tongue. And crocodiles, alligators, and most sharks don't chew, but rip and tear, then swallow the big chunks whole.

What's the connection between animals and global climate change?

One of the major concerns about global climate change—and the increase in average global temperatures—is the release of greenhouse gases. One greenhouse gas, methane, is produced by certain animals based on their digestion: ruminants. It is thought that in the United States, about 20 percent of the methane is due to cows and other ruminants. And it's not always the flatulence of the animals that is the problem, but the cows' burping because of the way they digest their food.

The true amount of methane that enters the atmosphere because of animals is still highly debated—especially when compared to other greenhouse gas emissions. For those who want to curtail the possible gas production by ruminants, it is suggested that people cut down on the amount of meat consumed; still another suggestion is to feed cattle different foods that would result in less gas production.

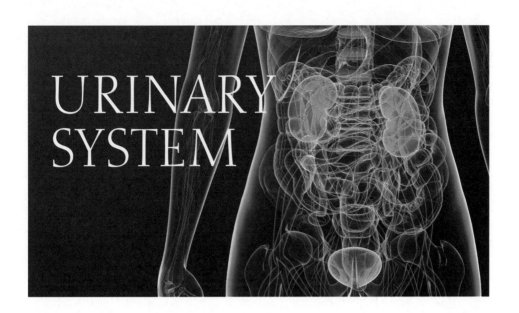

INTRODUCTION

What are the functions and major parts of the urinary system?

The functions of the urinary system include regulation of body fluids, removal of metabolic waste products, regulation of volume and chemical make-up of blood plasma, and excretion of toxins. The major parts of the urinary system are the kidneys, the urinary bladder, two ureters, and the urethra. Each component of the urinary system has a unique function. Urine is manufactured in the kidneys. The urinary bladder serves as a temporary storage reservoir for urine. The ureters transport urine from the kidney to the bladder, while the urethra transports urine from the bladder to the outside of the body.

What are the sources of water gain and loss per day?

The main sources of water gain are drinking and ingesting fluids, such as water contained in food, and water produced as a byproduct of metabolic processes. The main sources of water loss are urine formation, evaporation from the lungs (breathing), evaporation from the skin (sweating), and through the feces. The following lists the sources of water gain (intake) and loss (output) per day in milliliters:

Fluid Intake Source	mL	Fluid Output Source	mL
Ingested liquid	1,200–1,500	Urine	1,200–1,700
Ingested food	700–1,000	Feces	100–250
Metabolic oxidation	200–400	Sweat	100–150
Insensible losses			
Skin	350–400		
Lungs	350–400		
Total	2,100–2,900	*Total*	2,100–2,900

What percent of a person's intake of water comes from drinking water?

Only about 47 percent of a person's daily water intake comes from drinking. Nearly 39 percent of water intake comes from eating solid food, since water is a major component of many foods. For example, fruits and vegetables may contain more than 90 percent water.

What is cosmetic dehydration?

Cosmetic dehydration is the practice of taking large doses of diuretics to cause temporary weight loss. It has been used by fashion models and body builders, but it is a dangerous practice because it can cause electrolyte imbalance and cardiac arrest.

KIDNEYS

How big is a human kidney?

The human kidney is about the size of a human fist and weighs about 5 ounces (150 grams). Its average dimensions are 5 inches (12 centimeters) long, 3 inches (6 centimeters) wide, and 1 inch (2.5 centimeters) thick.

Where are the kidneys located?

The kidneys are located on either side of the spinal column in the lumbar region, just underneath the ribcage. Their exact anatomic position is retroperitoneal (between) the dorsal body wall and the parietal peritoneum.

What are the parts of the kidney?

The kidney has two layers: the outer layer, called the cortex, which is reddish brown and granular, and the inner zone, the medulla, which is darker and reddish brown in color. The medulla is subdivided into six to eighteen cone-shaped sections called the pyramids. The pyramids are inverted so that each base faces the cortex and the tops project toward the center of the kidney. Separating the pyramids are bands of tissue called renal columns. A renal lobe consists of a renal pyramid and its surrounding tissue.

How are the kidneys protected in the abdominal cavity?

The kidneys receive some protection from the lower part of the rib cage. In addition, three layers of connective tissue enclose, protect, and stabilize the kidneys: 1) the renal capsule that covers the outer surface of the entire organ; 2) the adipose (fatty) tissue that surrounds the capsule; and 3) the renal fascia, which is a layer of dense connective tissue that anchors the kidney to the surrounding structure.

Why is it important to protect the kidneys?

The kidneys are protected from day-to-day jolts and shocks by the adipose tissue and renal fascia. If the fibers of the dense connective tissue break, the kidneys will be

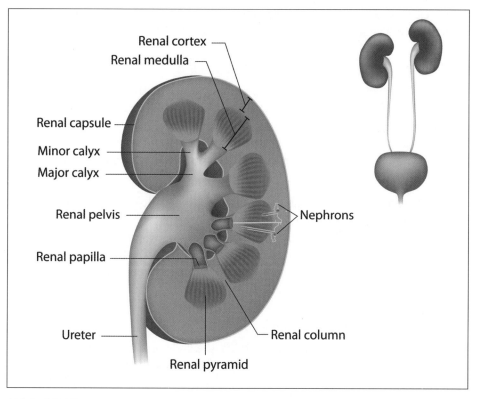

Anatomy of the kidney

able to move to the abdominal area. Movement of the kidneys, called floating kidney or nephroptosis, is dangerous because the ureters of renal blood vessels may become twisted.

What gland is found on top of the kidneys?

The adrenal glands are closely associated with the kidneys, with one adrenal gland sitting on top of each kidney and embedded in fatty tissue that encloses the kidney.

How do kidneys help vitamin D be available for bone growth?

The kidneys turn vitamin D into an active hormone called calcitrol, which helps bones absorb the right amount of calcium from blood. If the kidneys are impaired, bones do not get enough calcium, either because the kidneys fail to turn vitamin D into calcitrol or because they allow too much phosphorus to build up in the blood. The excess phosphorus draws calcium into the blood and blocks it from getting to the bones.

What is a nephron and how many are found in the average kidney?

A nephron is the functional working unit of the kidney. Blood is filtered in the nephrons and toxic wastes are removed, while water and necessary nutrients are re-absorbed into the system. Each nephron produces a minute amount of urine, which then trickles into the renal pelvis. From there it goes into the ureter and eventually collects in the bladder. There are approximately one million nephrons in each kidney.

What are the major parts of the nephron?

The major parts of the nephron are the renal corpuscle, the proximal convoluted tubule, the loop of Henle, and the distal convoluted tubule. Each part has a significant, distinct function in urine production. Inside the renal corpuscle is a tangled network of about fifty capillaries, the glomerulus, where filtration occurs. The blood pressure within the capillaries forces water and dissolved substances out of the capillaries and into the renal tubule.

The proximal convoluted tubule starts at the glomerulus and continues into the loop of Henle. About 65 percent of the original filtrate is reabsorbed in the proximal convoluted tubule. The loop of Henle is a critical region where the filtrate is further adjusted for water and solute balance. The distal convoluted tubule removes additional water from the filtrate, until the final concentration of the urine is approximately equal to that of the body fluids.

What is urea and where is it produced?

During the process of metabolizing proteins, the body produces ammonia. Ammonia combines with carbon dioxide to form urea. The human body can tolerate 100,000 times more urea than ammonia. It is the most abundant organic waste produced in the body and it is eliminated by the kidneys. Humans generate about 0.75 ounces (21 grams) of urea each day.

What are the major vessels that enter and leave the kidneys?

Each kidney receives blood from a renal artery. The kidneys receive 20 to 25 percent of the total cardiac output, or approximately 2.5 pints (1,200 milliliters) of blood per

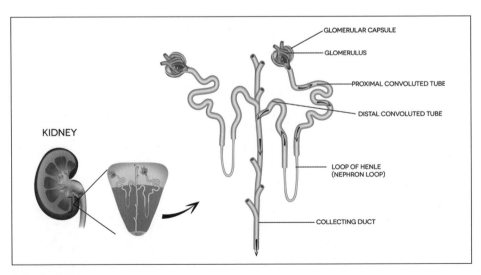

GLOMERULAR CAPSULE
GLOMERULUS
PROXIMAL CONVOLUTED TUBE
DISTAL CONVOLUTED TUBE
KIDNEY
LOOP OF HENLE (NEPHRON LOOP)
COLLECTING DUCT

Parts of the nephron

minute. After circulating through the kidney and nephrons, the blood is collected in the renal vein. The renal nerves innervate the kidneys.

How much blood is filtered daily by the kidneys and how much is filtered in an average lifespan?

The kidneys filter about 120 to 150 quarts (114 to 142 liters) of blood daily and produce about 4 ounces of filtrate per minute. Each day, about 1.5 to 2 quarts (1.4 to 1.9 liters) of urine is eliminated by the kidneys and is eventually excreted. The entire blood supply is filtered through the kidneys sixty times per day. Thus kidneys in a person living seventy-three years have filtered almost 1.3 million gallons of blood.

How does kidney function change with age?

Of all the major internal organs, the kidneys show significant deterioration with age. The capacity of the kidneys to filter waste decreases by 50 percent by age eighty. Humans, however, have four times the necessary reserve of kidney tissue, so the kidneys can still adequately perform their job and maintain normal function even at age eighty.

What is the connection between kidneys and red blood cell production?

The kidneys produce erythropoietin (EPO), a hormone that affects the number of circulating red blood cells. This is also why people with chronic kidney disease (CKD, or about 20 to 50 percent of normal kidney function) are often anemic, a condition in which the body has fewer blood cells than normal. (With anemia, red blood cells carry less oxygen to tissues and organs—particularly the heart and brain—and those tissues and organs may not function as well as they should.) Thus when the kidneys are diseased or damaged as in CKD, they do not take in enough erythropoietin. As a result, the bone marrow makes fewer red blood cells, causing anemia. (For more about anemia, see the chapter "Cardiovascular System.")

What is a kidney stone?

Kidney stones, or renal calculi, are the precipitates of substances such as uric acid, calcium oxalate, calcium phosphate, and magnesium phosphate that usually form in the renal pelvis. A stone passing into a ureter can cause very severe pain. Approxi-

mately 50 percent of kidney stones pass from the body on their own. Stones were once removed surgically, but most are now shattered with sound waves in a procedure called lithotripsy. Stones may form in the ureter or bladder, in addition to the kidneys. The process of stone formation in the kidney tract is called urolithiasis, renal lithiasis, or nephrolithiasis.

What are the symptoms of kidney failure?

Symptoms of kidney failure include excess fluid buildup due to inability of the kidneys to produce enough urine. The resulting excess fluid increases blood pressure, leading to hypertension. An indirect symptom of kidney failure is anemia due to the decreased production of erythropoietin by the kidneys. Erythropoietin controls the rate of maturation of red blood cells. Without adequate red blood cells, a person will become tired and short of breath.

What conditions can cause chronic renal disease?

Diabetes mellitus is the leading cause of chronic renal disease. It accounts for about 44 percent of new cases each year. The second leading cause is hypertension (high blood pressure).

What is kidney dialysis?

Kidney dialysis, also known as hemodialysis, is used when the kidneys are not functioning or are improperly functioning. An artificial membrane is used to regulate blood composition, particularly the removal of toxic substances. The patient's blood flows through artificial membrane tubing, which is immersed in a solution that differs in concentration from the normal concentration of blood plasma. Critical to this process is the composition of the dialysis fluid, which permits retention of needed substances while wastes are removed. Dialysis is usually carried out three times per week. The procedure usually takes about four hours, but the exact time needed for dialysis is dependent on several factors, including body weight, the amount of fluid retained, and how well the kidneys are functioning. The dialysis does not cure kidney disease, but it does the work of the person's kidneys (although it cannot replace them permanently).

Can dialysis be carried out at home?

An alternative to a dialysis machine is continuous ambulatory peritoneal dialysis (CAPD) in which the patient's own peritoneal (abdominal cavity) is used as a dialysis membrane. Dialysis fluid is introduced into the peritoneal cavity through a permanently implanted catheter. The fluid then stays in the body for four to six hours, after which it is removed and discarded. While the fluid is in the peritoneum, wastes and ions are exchanged within the capillaries.

ACCESSORY ORGANS

How long is each ureter?

Each ureter is 10 to 12 inches (25 to 30 centimeters) long. The ureters extend from the kidney into the bladder. They begin as thin, hollow, narrow tubes and widen to 0.5 inches (1.7 centimeters) as they enter the bladder. Urine is transported to the urinary bladder via the two ureters.

Where is the urinary bladder located?

The urinary bladder is located in the abdominal cavity. In males, it is anterior to the rectum and above the prostate gland. In females, it is located much lower, anterior to the uterus and upper vagina.

How much urine can the urinary bladder hold?

The urinary bladder is highly distensible and can vary in its capacity. As urine fills the bladder, it can expand to about 5 inches (12 centimeters) long and hold 1 pint (473 milliliters) of urine at moderate capacity. The bladder can expand to twice that capacity if necessary. It usually accumulates 300 to 400 milliliters of urine before emptying, but it can expand to hold 600 to 800 milliliters.

Why do pregnant women have an increased need and urge to urinate?

Early in a pregnancy, hormonal changes in the mother make the urge to urinate more frequent and more urgent. Towards the end of a pregnancy, the increased size and weight of the uterus pressing on the bladder reduces its capacity to hold urine, causing an increased need to urinate.

How does the urethra differ in males and females?

Urine is transported to the outside through the urethra, which is a thin, muscular tube that extends from the urinary bladder to the exterior of the body. The length and structure of the urethra differs between males and females. In males, the urethra is about 8 inches (20 centimeters) long and extends from the urinary bladder to the exterior. It has the dual function of transporting semen as well as urine out of the body. The female urethra is only about 1.5 inches (3 to 4 centimeters) long and extends from the bladder to the exterior opening.

Why are women more prone to urinary tract infections?

Women are more likely to suffer from urinary tract infections because the ure-

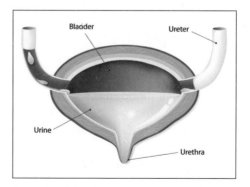

The bladder collects urine until it is ready to be expelled out the urethra. The average bladder can hold about a pint of fluid.

thra in women is much shorter. It is easier for bacteria to reach the bladder, causing an infection.

URINE AND ITS FORMATION

What are micturition and incontinence?

Micturition, or urination, is the process that expels urine from the urinary bladder. It is a complex interaction between the bladder and the nervous system, and it has two main phases, called the filling and emptying stages. It is also medically known as voiding, uresis, or, rarely, emiction. In most healthy people (and other animals) it is under voluntary control.

The involuntary response of urination is referred to as urinary incontinence (UI) or enuresis. Contributing factors to incontinence include emotional problems, pregnancy, and nervous system problems. It is common for many infants to be incontinent (they must learn bladder control, which is why children go through what is commonly termed "potty training"). In addition, age-related changes in the lower urinary tract function predispose the elderly to incontinence. It is estimated that among adults more than sixty years of age, women are two times more likely to have a UI than men.

What is the composition of urine?

Urine is mostly water and contains organic waste products such as urea, uric acid, and creatinine. It also contains excess ions, such as sodium (Na^+), potassium (K^+), chloride (Cl^-), bicarbonate (HCO_3^-), and hydrogen (H^+).

Is urine always yellow in color?

Normally, dilute urine is nearly colorless. Concentrated urine is a deep yellow; colors other than yellow are not normal. Food pigments can make the urine red, and drugs can produce colors such as brown, black, blue, green, or red. Urine may also be brown, black, or red due to disorders or diseases such as severe muscle injury or melanoma. Cloudy urine suggests the presence of pus, due to a urinary tract infection, or salt crystals from uric acid or phosphoric acid.

What is a urinalysis?

Urinalysis is the chemical and physical analysis of a urine sample. It involves the color and appearance of urine, plus a detailed list of specific compounds and their concentration found in the sample. Substances that should not be found in urine are proteins, glucose, acetone, blood, and pus. The presence of any of these substances may indicate a disease.

Why is urine used to test for drug use?

Urine drug testing is commonly used for opioids and illicit drugs. The liver is where drug detoxification occurs, but the byproducts of this process are excreted by the kidneys. There are two types of tests: a screening test and a confirmatory test. Most urine drug tests screen for marijuana, cocaine, opiates, PCP, and amphetamines.

What are the main parts of urine production?

The main parts of urine production are glomerular filtration, tubular reabsorption, and tubular secretion. In filtration, blood pressure in the glomerular capillaries forces water and other solutes across the glomerular membrane. The process is similar to what happens in a coffee maker, where water passes through a filter and it carries with it some dissolved compounds (coffee). Usually, the coffee grinds never reach the pot, unless there is a hole in the filter. Reabsorption involves the return of water and major nutrients to the blood. Secretion is the removal of harmful or excess substances from the blood and their transport into the urine.

What is a diuretic?

A diuretic is a chemical that increases urine output. Examples of diuretics include alcohol and any beverages that contain caffeine (coffee, tea, colas). Diuretics are usually prescribed for patients with high blood pressure or congestive heart failure.

What is the antidiuretic hormone (ADH)?

The antidiuretic hormone (ADH) is one of the major hormones controlling fluid balance by inhibiting diuresis or urine output. ADH is secreted by the posterior pituitary gland. For example, when a person is dehydrated, the level of ADH secretion increases so that more water is reabsorbed from the urine and returned to the blood. Conversely, a diuretic such as alcohol inhibits the release of ADH and thus increases urinary output.

What muscles control urination?

The walls of the urinary bladder have three layers: 1) a mucosa layer made of transitional epithelium; 2) connective tissue; and 3) detrusor muscle, composed of smooth muscle fibers arranged in both longitudinal and circular layers. It is the contraction of the detrusor muscle that compresses the bladder when voiding urine.

COMPARING OTHER ORGANISMS

Do other animals urinate at the same speed?

In an interesting study done in 2013, scientists found something they call the "Law of Urination" in animals. By measuring the flow rate of animals at a zoo in Atlanta, the scientists found that animals empty their bladders over nearly a constant duration of about twenty-one seconds, no matter what the volume of the bladder. That means that mice to elephants urinate at around the same speeds. The researchers believe that this is possible by the increasing length of the urethra in large animals, which translates to a high flow rate thanks to gravitational forces. In other words, the

URINARY SYSTEM

Why do kidneys fail so often in cats?

For reasons that are not clearly understood, kidney failure is one of the leading causes of death in cats. Chronic Kidney Disease (or Disorder, or CKD) is diagnosed in part by measuring the creatinine levels in the animal's blood, and it is usually uncovered during routine blood work. By the time the creatinine level is detected to be "higher," the cat may have lost up to 75 percent of its kidney function. In 2015, researchers discovered that a new kidney function test—symmetric dimethylarginine (SDMA)—can allow veterinarians to diagnose CKD months to years earlier than the traditional method, allowing for early treatment to manage the disease.

urethra in all animals seems to have evolved as a flow-enhancing device that allows the urinary system to work in the same speed and is not dependent on the size of the animal's bladder.

Why is urine important to some animals for reasons other than elimination?

Because smell is so important to communication in the animal world, most organisms respond to odors given off by their own or other species—especially the smell of urine. This is often used to indicate where a certain animal territory exists, and although there are other ways to mark territories, most animals use urine to mark their space. For example, when a dog sniffs a tree and smells the urine from another dog, it is understood that this tree is the urinating-dog's territory. In addition, some animals use urine to pass along messages, such as if there are predators nearby.

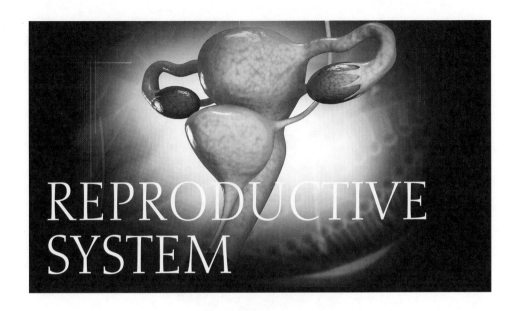

REPRODUCTIVE SYSTEM

INTRODUCTION

What are the organs of the reproductive system and their general function?

The overall function of the reproductive system is to produce new offspring. Thus, the continuity of the human species is guaranteed through reproduction. The major organs of the reproductive system are the gonads, various ducts, accessory sex glands, and supporting structures. The gonads produce gametes and secrete hormones. The various ducts store and transport the gametes. The accessory sex glands produce substances that protect the gametes. The supporting structures assist the delivery and joining of gametes.

Are the male and female reproductive systems identical?

Unlike every other organ system in the human body, the male and female reproductive systems are not identical. The specialized organs of the male and female reproductive systems are different. However, it is essential for both the male and female reproductive systems to be involved in order for reproduction to occur.

Which branches of medicine specialize in the female and male reproductive systems?

Gynecology (from the Greek *gune*, meaning "woman") is the branch of medicine that specializes in the diagnosis and treatment of conditions of the female reproductive system. Urology is the branch of medicine that specializes in the treatment of medical conditions of the male reproductive system.

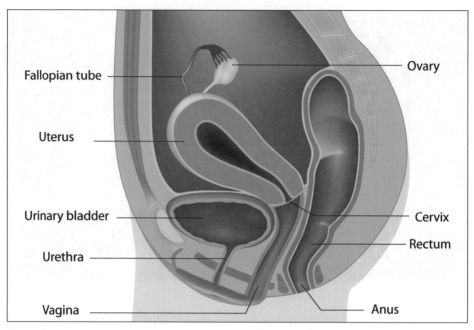

The female reproductive system

MALE REPRODUCTIVE SYSTEM

What are the male reproductive organs and structures?

The male reproductive organs and structures are the testes, a duct system that includes the epididymis and the vas deferens; the accessory glands, including the seminal vesicles and prostate gland; and the penis.

What are the testes and where are they located?

The testes are the male gonads that produce the male reproductive cells called sperm. The testes hang in a pouch called the scrotum (from the Latin *scrautum*, meaning a "leather pouch for arrows"). The scrotum is a fleshy pouch consisting of loose skin, loose connective tissue, and smooth muscle. It is divided internally by a septum into two chambers each of which contains one testis.

Why is it advantageous for the scrotum to hang outside the body?

Since the scrotum hangs outside the body, the temperature in the scrotum is lower than normal body temperature. The normal temperature in the scrotum is 2°F to 3°F (1.1°C to 1.6°C) below normal body temperature. This lower temperature is necessary for production and survival of sperm.

How does the scrotum react to changes in temperature?

Changes in the external temperature stimulate a muscle called the cremaster muscle to contract or relax, thereby moving the testes closer to or further from the body

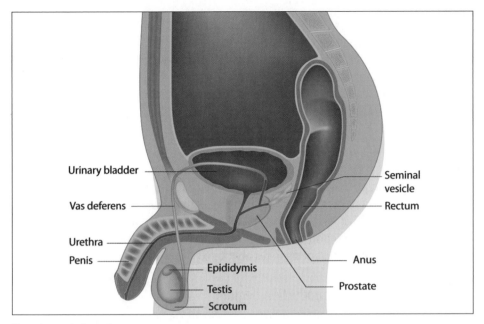

The male reproductive system

cavity. When exposed to cold temperatures, such as swimming in cold water, the cremaster muscle contracts, bringing the testes closer to the body to keep them warm. When temperatures are warm, the cremaster muscle relaxes, moving the testes farther away from the body so excess heat is lost through the increased surface area of the scrotum.

When do the testes descend into the scrotum during fetal development?

The testes begin their development in the fetus's abdomen. They descend from the abdomen into the groin (the inguinal canal) during the third month of fetal development. The descent into the scrotum usually begins during the seventh month of fetal development and is completed shortly before or after birth.

Which medical condition occurs when one or both testes do not descend into the scrotum?

Failure of the testes to descend from the abdomen to the scrotum is called cryptorchidism (from the Greek *crypto*, meaning "hidden," and *orchis*, meaning "testis"). This condition occurs in 3 to 5 percent of male babies born at full-term. In premature babies, the number of cases of cryptorchidism increases to 30 percent. In most cases, the undescended testicle(s) move into the scrotum without medical intervention. If necessary, the testicle(s) may be moved into the scrotum surgically. Testes are generally not left in the abdomen since this may interfere with their ability to produce sperm. Furthermore, testicular cancer is more common in men with undescended testicles.

Where does sperm production occur in the testes?

Sperm production, called spermatogenesis, occurs in the seminiferous tubules. There are approximately 800 seminiferous tubules in each testis. Each tubule is slender, tightly coiled, and approximately 31.5 inches (80 centimeters) long. Each testis contains nearly half a mile of seminiferous tubules. Spermatogenic (sperm-forming) cells line the seminiferous tubules.

What are the steps of spermatogenesis?

Spermatogenesis is the process whereby the seminiferous tubules produce sperm. There are three steps in spermatogenesis: 1) meiosis, during which the number of chromosomes in the cell is reduced to half or twenty-three chromosomes each; 2) meiosis II, during which each haploid cell forms spermatids; and 3) spermiogenesis, during which each spermatid develops into a sperm cell with a head and tail. The entire process of spermatogenesis takes about sixty-four days.

At what age does sperm production begin?

Sperm production begins with the onset of puberty, usually between ages eleven to fourteen in boys. It continues throughout the life of an adult male, although the volume and quality of sperm goes down. Some research also indicates that as a man ages, the sperm can undergo random mutations, increasing the chances of fathering children with genetic abnormalities.

How many sperm are produced?

It is estimated that during his lifetime a normal male will produce 10^{12} sperm. This is equivalent to about 300 million sperm per day.

What are the characteristics of sperm?

Each sperm cell has three distinct regions: the head, the middle piece, and the tail. The head consists of a nucleus with the genetic material (DNA) and an acrosome at the tip. The acrosome contains enzymes to help the sperm penetrate the oocyte (egg). The middle piece contains mostly mitochondria, which provide energy for movement. The tail moves the sperm cell from one place to another. It is also the only flagellum (a cell with a whip-like structure that allows it to move) in the human body.

How large are sperm cells?

Sperm are some of the smallest cells in the human body. A single sperm cell is about 0.002 inches (0.05 millimeters) long from the head to the tip of the tail.

What are the three ducts through which sperm travel from the testes to the urethra?

When sperm leave the testes they have the physical characteristics of mature sperm, but they are not functionally mature. Final maturation occurs in the ducts, known as the accessory ducts. The three ducts are the epididymis, the ductus deferens, also known as the vas deferens, and the ejaculatory duct.

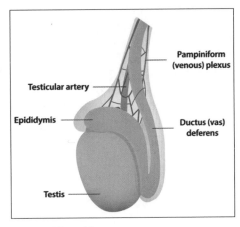

Anatomy of the testicle

What are the functions of the epididymis?

The epididymis (from the Greek, meaning "upon the twin," with twin being the testis) is a long, twisted and coiled tubule. It occupies a space of about 1.5 inches (4 centimeters), but if uncoiled it would measure 20 to 23 feet (6 to 7 meters) long. While in the epididymis, sperm mature and gain motility. The epididymis stores sperm until they are ready to be ejaculated. The smooth muscle of the epididymis helps propel mature sperm into the ductus deferens via peristaltic (relaxing and contracting of smooth muscles) movement.

How long does it take sperm to mature in the epididymis?

Sperm spend ten to fourteen days in the epididymis prior to reaching final maturation. They may then be stored in the epididymis for up to a month. If they are not ejaculated during that time, they degenerate and are reabsorbed by the body.

What is the function of the ductus deferens?

The ductus deferens (from the Latin *deferre*, meaning "to carry away"), also called the vas deferens, carries sperm from the epididymis to the ejaculatory duct. It is only 16 to 18 inches (40 to 45 centimeters) long, much shorter than the epididymis, and it has a larger diameter.

Which is the shortest duct of the male reproductive system?

The ejaculatory duct is only about 1 inch (2.5 centimeters) long, much shorter than either the epididymis or the vas deferens. It is formed by the duct from the seminal vesicles and the vas deferens. The ejaculatory duct ejects sperm into the urethra.

What is the reproductive function of the urethra?

The urethra is the last section of the reproductive duct system. It provides a passageway to transport sperm outside of the body during ejaculation.

What are the accessory glands of the male reproductive system?

The accessory glands are the seminal vesicles, the prostate gland, and the bulbourethral glands. Each of these glands adds secretions to the sperm. These secretions constitute the liquid portion of semen.

Which accessory gland contributes the greatest volume of secretions for seminal fluid?

The seminal vesicles contribute nearly 60 percent of the volume of seminal fluid. These secretions are an alkaline fluid containing water, fructose, prostaglandins, and clotting proteins. The fructose provides an energy source for the sperm. Prostaglandins contribute to sperm motility and viability. The clotting proteins help semen coagulate after ejaculation. The alkalinity of the fluid helps to neutralize the acidic environments of the male urethra and female reproductive tract.

How large is the prostate gland?

The prostate gland in healthy adult males is about the size of a walnut. It is located in front of the rectum and under the bladder. The prostate surrounds the urethra. When it becomes enlarged, it squeezes the urethra, restricting the flow of urine.

What is the purpose of prostate gland secretions?

Prostate gland secretions account for nearly 30 percent of the volume of seminal fluid. The slightly acidic secretions help semen clot following ejaculation and then break down the clot.

What is prostatitis?

Prostatitis is an inflammation of the prostate gland. The two types of prostatitis are acute prostatitis and chronic prostatitis. It may be caused by bacterial pathogens or may develop without evidence of an infecting organism. Symptoms include pain in the lower back, difficulty with urination often accompanied by a burning sensation and/or pain, frequent urination, especially at night, and sometimes fever.

How common is prostate cancer in the United States?

Prostate cancer is the second most common cancer (the first is lung cancer) in American men. It is the leading cause of cancer death among American men. The American Cancer Society estimated that in 2015, 220,800 new cases of prostate cancer were diagnosed in the United States and that 27,540 men in the United States would die of prostate cancer. Approximately one man in every six will be diagnosed with

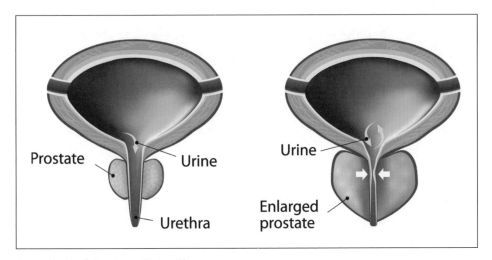

A normal bladder (left) and one with prostatitis

prostate cancer during his lifetime, but only one man in thirty-eight will die of this disease. About six in ten men age sixty-five or older are diagnosed with prostate cancer, and it is rare before age forty (the average age at the time of diagnosis is about sixty-six).

Are all tumors found in the prostate malignant?

No, benign prostatic hyperplasia is a condition involving an abnormal growth of benign (noncancerous, nonmalignant) cells in the prostate. Its most characteristic symptom is a reduced flow of urine. Most changes in the prostate are not cancer. (For more about prostate tests, see the chapter "Helping Human Anatomy.")

What is semen?

Semen (from the Latin, meaning "seed") is the combination of the secretions of the accessory glands and sperm. The composition of the seminal fluid is as follows: sixty percent of the secretions come from the seminal vesicles, 30 percent from the prostate, 5 percent from the epididymis, less than 5 percent from the bulbourethral glands, and only about 1 percent from the sperm.

How much semen is produced in a typical ejaculation?

An average, typical ejaculation contains 2 to 5 milliliters (approximately 0.5 to 1 teaspoon) of semen. There are 50 to 150 million sperm per milliliter in a typical ejaculation, or 300 to 400 million sperm in each ejaculation.

When is a man considered sterile?

Males with a sperm count below twenty million sperm per milliliter are considered functionally sterile. This is because too few spermatozoa will survive to reach and fertilize an oocyte (female egg).

Are there medical benefits to circumcision?

Circumcision is the removal of the foreskin of the penis. The foreskin is the skin that covers the end of the penis. Medical research has not found significant evidence of medical benefits from circumcision. One benefit is fewer urinary tract infections in infants. Some studies have shown that circumcision reduces the risk of sexually transmitted diseases. However, it is negligible compared to other behaviors that reduce the risk of sexually transmitted diseases.

What is the function of the penis?

The penis is an external organ of the male reproductive system. It consists of three parts: 1) the root where the penis attaches to the wall of the abdomen; 2) the shaft or body; and 3) the glans, which is the tip of the penis. The body of the penis has a tubular, cylindrical shape and surrounds the urethra. Spongy tissue that can expand and contract fills the body of the penis. There is a slit opening at the tip of the glans through which urine is excreted and semen is ejaculated through the urethra.

What is erectile tissue?

The penis contains three cylindrical columns of erectile tissue; two are the corpus carvernosa and one is the corpus spongiosum. Upon stimulation, blood flow to the erectile tissue is increased resulting in an erection.

What are some causes of erectile dysfunction?

Erectile dysfunction (ED), the inability to achieve or sustain an erection, is often the result of disease, injury, or a side effect of certain drugs, including blood pressure drugs, antihistamines, antidepressants, tranquilizers, appetite suppressants, and cimetidine (an ulcer drug). Damage to nerves, arteries, smooth muscles, and fibrous tissues of the penis are the most common causes of erectile dysfunction. Diabetes, kidney disease, chronic alcoholism, multiple sclerosis, atherosclerosis, vascular disease, and neurologic disease account for about 70 percent of ED cases. Treatment may include lifestyle changes, adjusting medications to alleviate side effects, medications to induce erection, and surgery.

FEMALE REPRODUCTIVE SYSTEM

What are the female reproductive organs?

The organs of the female reproductive system include the ovaries, the uterine tubes, the uterus, the vagina, the external organs called the vulva, and the mammary glands. The paired ovaries are the female gonads. Each ovary is approximately 1 to 2 inches (2.5 to 5.0 centimeter) long, 0.6 to 1.2 inches (1.5 to 3.0 centimeters) wide, and 0.24 to 0.6 inches (0.6 to 1.5 centimeters) thick, similar in size and shape to an unshelled almond. They produce the female gametes, called ova, and secrete the female sex hormones.

What are the steps of oogenesis?

Oogenesis is the production of ova (eggs) in the ovaries. The process begins during fetal development. Meiosis I occurs prior to birth, and then development of the primary oocytes is suspended until puberty. The secondary oocyte is released at ovulation in response to hormonal secretions. If fertilization occurs, meiosis II resumes and an ovum is formed.

How many oocytes does a girl have at birth?

There are approximately two million oocytes in the ovaries at birth. Only about 300,000 to 400,000 of the oocytes survive until puberty, and only about 400 of the oocytes that survive until puberty will be released at ovulation during a woman's lifetime. The remainder of the oocytes degenerate over time.

Are the uterine tubes connected to the ovaries?

No, the superior, funnel-shaped end of the uterine tubes opens in the abdominal cavity very close to the ovaries, but is not actually connected to the ovaries. The inferior end of the uterine tubes opens into the uterus.

What event occurs in the uterine tubes?

The uterine tubes are also called the oviducts or Fallopian tubes, named for Italian anatomist Gabriel Fallopius (1523–1562). Fertilization of the secondary oocyte by sperm usually occurs in the uterine tubes. It takes four to seven days following fertilization for a fertilized oocyte (zygote) to reach the uterus. The unfertilized oocytes degenerate after fertilization of the oocyte.

What are the functions of the uterus?

The uterus (from the Latin, meaning "womb") is an inverted, pear-shaped structure located between the urinary bladder and the rectum. Implantation of a fertilized ovum occurs in the uterus. Following implantation, the uterus houses, nourishes, and protects the developing fetus during pregnancy.

How large is the uterus?

The uterus is normally 3 inches (7.5 centimeters) long and 2 inches (5 centimeters) wide. It weighs 1 to 1.4 ounces (30 to 40 grams). During pregnancy it can increase three to six times in size. At the end of gestation, the uterus is usually 12 inches (30 centimeters) long and weighs 2.4 pounds (1,100 grams). The total weight of the uterus and its contents (fetus and fluid) at the end of pregnancy is approximately 22 pounds (10 kilograms).

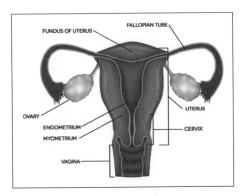

The uterus

What are the regions of the uterus?

The uterus consists of the body and the cervix. The body is the largest area of the uterus. The superior part of the body is called the fundus. The cervix connects the uterus to the vagina.

Which medical condition is caused by the abnormal location of endometrial tissue?

Endometriosis is the abnormal location of uterine endometrial tissue. Under normal conditions, endometrial tissue only lines the uterus, but in some instances the tissue migrates to remote locations of the body such as the ovaries, pelvic peritoneum, the vagina, the bladder, or even the small intestine and lining of the chest cavity. Endometriosis is associated with dysmenorrhea (painful menstruation), pelvic pain, and infertility.

What is pelvic inflammatory disease?

Pelvic inflammatory disease (PID) is an infection of the uterus, uterine tubes, or other female reproductive organs. Many women do not experience any symptoms of PID while the disease is damaging their reproductive organs. When symptoms are present, pain in the lower abdominal area is most common. Other symptoms may include fever, unusual vaginal discharge that may have a foul odor, painful intercourse, painful urination, irregular menstrual bleeding, and pain in the right upper abdomen (rare).

What are the complications of PID?

Permanent damage to the reproductive organs is a complication of PID. Scar tissue often develops in the uterine tubes. Infertility is the result of the uterine tubes becoming blocked by scar tissue. As many as one in eight women become infertile due to PID. Ectopic pregnancies are more common in women who suffer from PID since a fertilized egg may remain in the uterine tubes and begin to grow. Finally, many women with PID suffer from chronic pelvic pain.

What is the female reproductive cycle?

The female reproductive cycle is a general term to describe both the ovarian cycle and the uterine cycle, as well as the hormonal cycles that regulate them. The ovarian cycle is the monthly series of events that occur in the ovaries related to the maturation of an oocyte. The menstrual cycle is the monthly series of changes that occur in the uterus as it awaits a fertilized ovum.

What are the phases of the reproductive cycle?

The reproductive cycle consists of three phases: 1) the menstrual phase; 2) the preovulatory phase; and 3) the postovulatory phase. During each phase there are changes in the ovaries and in the uterus in response to hormonal secretions from the pituitary (follicle-stimulating hormone [FSH] and luteinizing hormone [LH]) and gonads (estrogens and progesterone). The following lists the events of the ovarian and uterine cycles during the reproductive cycle:

	Ovarian Cycle	Uterine Cycle
Menstrual phase	Twenty or more secondary follicles begin to enlarge; follicular fluid accumulates in the follicle	Menstrual flow passes from uterus to exterior via vagina
Preovulatory phase	Follicular phase because follicles are growing and developing; secondary follicles continue to grow; one follicle outgrows the others and becomes the dominant follicle; estrogen secretion increases; secretion of FSH decreases	Proliferative phase because the endometrium is growing
Ovulation	Rupture of the mature follicle; release of the secondary oocyte into the pelvic cavity	
Postovulatory phase	Luteal phase; mature follicle collapses; if fertilization occurs, the corpus luteum continues past its normal two-week lifespan; if fertilization does not occur, the corpus luteum lasts for only two weeks and then degenerates; follicular growth then resumes and a new cycle begins	Preparation of endometrium for fertilized ovum; if fertilization does not occur, the menstrual phase begins again

How long the does female reproductive cycle last?

The reproductive cycle averages twenty-eight days, but it may last from twenty-four to thirty-five days. A cycle of twenty to forty-five days is still considered within the

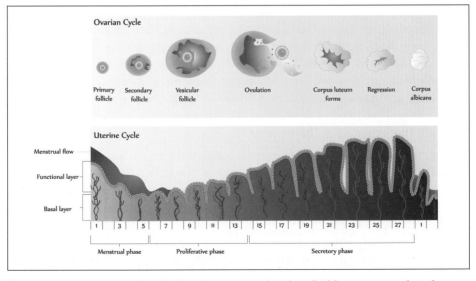

Ovarian and uterine cycles are coordinated so that when a woman ovulates the walls of the uterus are receptive to the implantation of the egg.

295

range of normalcy. The menstrual phase lasts five to seven days. Ovulation, the release of a secondary oocyte into the pelvic cavity, separates the preovulatory phase from the postovulatory phase. The preovulatory phase lasts eight to eleven days, and the postovulatory phase lasts an average of fourteen days. Ovulation, the release of a secondary oocyte into the pelvic cavity, separates the preovulatory phase from the postovulatory phase.

When does menarche occur?

Menarche (from the Greek, meaning "beginning the monthly") is the first menses, or menstrual cycle, of a girl's life. It usually begins between the ages of eleven and twelve.

How does primary amenorrhea differ from secondary amenorrhea?

Amenorrhea is the absence of a menstrual period not due to pregnancy, breastfeeding, or menopause. Primary amenorrhea is diagnosed when girls have not had their first menstrual period by age sixteen. It may be caused by an endocrine, genetic, or developmental disorder. Secondary amenorrhea is diagnosed in previously menstruating women who have not had a menstrual period for six months or more. It may be caused by physical or emotional stress, including excessive weight loss, anorexia nervosa, depression, or grief. Many female athletes have secondary amenorrhea due to decreased levels of body fat.

How much blood is lost during menstruation?

Approximately 1.2 to 1.7 ounces (35 to 50 milliliters) of blood is lost during menstruation. In addition to blood, degenerating tissue cells from the endometrium of the uterus are also lost. They are replaced during the next uterine cycle.

What is menopause?

Menopause is the cessation of ovulation and menstrual periods. The supply of follicles in the ovaries is depleted, increasing the amount of follicle-stimulating hormone (FSH), while decreasing the amount of estrogen and progesterone. The process may take one to two years. The few years following the final menstrual period are known as perimenopause. Menopause usually occurs between ages forty-five and fifty-five; the average age in the United States is fifty-one to fifty-two years.

What are the functions of the vagina?

The vagina is a muscular tube that extends between the uterus and the external genitalia. Its main functions are: 1) to serve as the passageway for the elimination of menstrual fluids; 2) to receive the penis during sexual intercourse and hold sperm prior to its passage into the uterus; and 3) to serve as the birth canal for the fetus during childbirth and delivery.

How does the environment of the vagina protect it against infections?

The acidic environment of the vagina—its normal pH is 3.5 to 4.5—restricts the growth of many pathogenic organisms. If the acidity of the vagina is reduced, bacterial, fungal, or parasitic organisms may increase, thus increasing the risk of infections known as vaginitis. Yeast infections are common infections of the vagina.

Which organs comprise the external female genitalia?

The external female genitalia, collectively known as the vulva or pudendum, consist of the mons pubia, labia majora, labia minora, vestibular glands, clitoris, and vestibule of the vagina. The *mons pubis* (from the Latin, meaning "mountain" and "pubic") is a bulge of adipose tissue covered by pubic hair. The labia majora, two longitudinal folds of skin below the mons pubia, form the outer borders of the vulva. The labia minora lie between the labia majora. The labia majora and minora surround and protect the vaginal and urethral openings. The labia merge at the anterior point to form the clitoris. The clitoris becomes enlarged during sexual stimulation. The area between the labia minora is the vestibule. The vestibule contains glands and the openings of the vagina and urethra.

What are the mammary glands?

The mammary glands, located in the breasts, are modified sweat glands. They produce and secrete milk following childbirth in response to hormonal stimulation. Each breast has a pigmented projection, the nipple, surrounded by a darker pigmented area called the areola.

Where do most breast cancers begin?

Most breast cancers begin in the cells that line the ducts. The ducts are the passages that connect the lobules (the milk-producing glands) to the nipples. Some cancers begin in the cells of the lobules.

What routine screening tests are recommended to detect breast cancer?

Three screening tests are often recommended to detect breast cancer: 1) a screening mammogram, 2) clinical breast exam, and 3) monthly breast self-exam. The National Cancer Institute recommends that women in their forties and over have a screening mammogram every one to two years, unless they have certain risk factors that would require screening to begin at a younger age or more frequently.

Are all lumps found in the breasts cancerous?

Most lumps found in breast tissue are benign and not cancerous. Many of these lumps are fibrous, scarlike tissue and cysts (fluid-filled sacs).

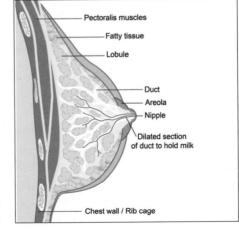

Anatomy of the breast

CONCEPTION

What is the human sexual response?

The act of sexual intercourse, also called coitus (from the Latin *coire*, meaning "to come together"), involves a sequence of physiological and emotional changes in males and females. These changes are known as the human sexual response. Human sexual response (both male and female) may be divided into four stages: 1) excitement, 2) plateau, 3) orgasm, and 4) resolution.

Which reflexes are activated during the excitement stage of the human sexual response?

During excitement, the initial stage of the human sexual response that is also called arousal, parasympathetic reflexes of the autonomic nervous system are activated. The parasympathetic impulses increase blood flow to the genitalia and secretion of lubricating fluids. In males, the penis becomes erect. In females, the clitoris and nipples become erect. These responses continue through the plateau stage, when the penis is inserted into the vagina.

What are the responses of males and females during orgasm?

Orgasm is marked by rhythmic contractions of the genital organs in both males and females. In males, ejaculation occurs during the orgasm stage of sexual response. During ejaculation, sperm are released into the vagina and begin to swim to the uterus. Although in females there is no counterpart to ejaculation, rhythmic contractions of the vagina, uterus, and perineal muscles do occur.

How many sperm are required to ensure fertilization?

Although millions of sperm may enter the vaginal canal during sexual intercourse, only one sperm enters and fertilizes an ovum. The additional sperm are necessary to increase the chance that one sperm will survive the acidic conditions of the vagina and will successfully complete the journey to the uterine tubes for fertilization. Once the sperm reaches the ovum it releases an enzyme that dissolves the outer wall of the ovum, allowing the sperm to enter. Following fertilization the outer membrane of the ovum thickens to prevent other sperm from entering the now fertilized egg.

What are some causes of infertility?

Infertility is defined as the inability to become pregnant after one year of trying. The cause of infertility may be with either the man or the woman or a combination of both partners. Female causes of infertility are often associated with problems with ovulation. The underlying reason may be premature ovarian failure (when the ovaries stop functioning prior to the onset of natural menopause); polycystic ovary syndrome (PCOS), when the ovaries do not regularly release an egg or do not release a viable egg; blocked fallopian tubes; physical problems with the uterine wall; or uterine fi-

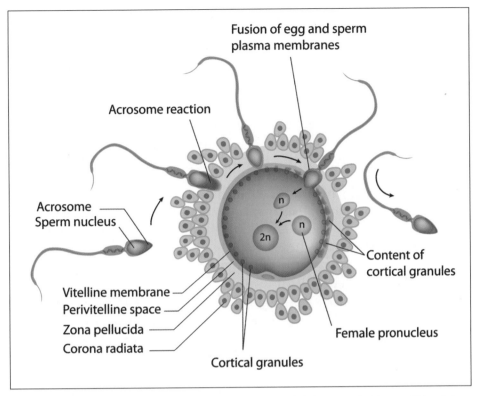

Fusion of egg and sperm plasma membranes

Acrosome reaction

Acrosome
Sperm nucleus

n

2n

n

Content of cortical granules

Vitelline membrane
Perivitelline space
Zona pellucida
Corona radiata

Cortical granules

Female pronucleus

Usually only one sperm out of millions manages to penetrate the egg, after which the outer wall of the ovum thickens to keep additional sperm out.

broids. Male causes of infertility include the inability to have or sustain an erection, not having enough sperm or enough semen to carry the sperm to the egg, or having sperm that do not have the proper shape to move in the right way. (For more about fertility treatments, see the chapter "Helping Human Anatomy.")

How is pregnancy confirmed?

Early external signs of pregnancy include a missed menstrual period, bleeding or spotting, fatigue, and tender, swollen breasts. A pregnancy test is needed to confirm pregnancy. Pregnancy tests look for the hormone human chorionic gonadotropin (hCG), also called the pregnancy hormone, in either urine or blood. This hormone is produced when the fertilized egg implants in the uterus. It is only present in pregnant women.

Are home pregnancy tests more reliable than laboratory tests?

Home pregnancy tests check the urine for hCG. The amount of hCG increases daily in pregnant women, so most tests are fairly accurate about two weeks after ovulation. Laboratories may use one of two blood tests to confirm pregnancy. A qualitative hCG blood test checks to see whether or not the hormone is present. Its reliability is comparable to a urine test. A quantitative blood test (the beta hCG test) measures the exact amount of hCG in the blood. It can detect even tiny amounts of hCG, making it extremely accurate. Blood tests may confirm a pregnancy as early as six to eight days following ovulation.

What is the purpose of contraceptive devices and methods?

Contraceptive ("against conception") devices and methods aim to prevent pregnancy. The various contraceptive devices and methods use one of two techniques to avoid pregnancy: preventing sperm and ova from meeting or making the environment unsuitable for fertilization. Hormonal methods may either prevent the ovary from releasing an egg into the uterine tubes or cause changes in the cervix or uterus, making it difficult for sperm to enter the uterus or implant a fertilized egg in the uterus. Barrier methods, withdrawal, natural family planning, and sterilization prevent sperm and ova from meeting. Intrauterine devices interfere and prevent the implantation of a fertilized ovum in the uterus.

What are the main types of birth control?

There are numerous types of birth control. Spermicides, condoms (male and female), the diaphragm, the cervical cap, and Lea's Shield are barrier methods. Oral contraceptives (birth control pills), injections of hormones, the vaginal ring, and skin patch are various types of hormonal methods of birth control.

Other methods are used in natural family planning. For example, one formerly called the "rhythm method" is also now called periodic abstinence or fertility awareness. It involves a variety of different methods, including basal body temperature method, ovulation/cervical mucus method, symptothermal method, calendar method, and lactational amenorrhea to monitor the fertile and non-fertile times during an individual woman's cycle.

Withdrawal requires the man to remove his penis from the woman's vagina prior to ejaculation thereby not allowing sperm to enter the woman's vagina. However, sperm may be present in the fluid prior to ejaculation, so this method is not very re-

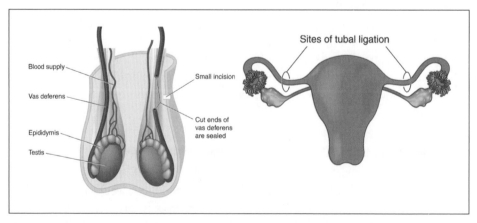

The illustration at left shows where the vas deferens is cut in the male during a vasectomy, and on the right the fallopian tubes are cut to prevent a woman's eggs from entering the uterus.

liable. Sterilization, called a vasectomy in men and tubal sterilization in women, permanently blocks the pathways for sperm and ova. In men, the vas deferens are cut so sperm cannot mix with semen. The tubes that carry sperm to the penis are clamped, cut, or sealed so that the ends do not heal and rejoin. In women, the fallopian tubes are closed by tying, banding, clipping, blocking, or cutting them, or by sealing them with electric current.

Is surgical sterilization reversible?

Sterilization, both male and female, should be considered a permanent, irreversible procedure. Tubal sterilization may be reversible, but it requires microsurgery to reconnect the uterine tubes after they have been blocked. Similarly, the reversal of a vasectomy in males is a lengthy, complicated operation requiring general anesthesia.

SEXUALLY TRANSMITTED DISEASES

How are sexually transmitted diseases (STDs) transmitted?

Sexually transmitted diseases (STDs) are usually transmitted via sexual contact between individuals. Any form of sexual activity, including anal, oral, and vaginal sex, may spread an STD. In addition, sexually transmitted diseases may be spread from a mother to her baby during childbirth.

What are the some common STDs?

STDs may be caused by bacteria or viruses. Common STDs are chlamydia, genital warts (HPV), genital herpes (HSV), syphilis, gonorrhea, and chancroid. Viral hepatitis, especially hepatitis B, and AIDS are also considered STDs since they are frequently transmitted from one individual to another via sexual contact.

What is the incidence of STDs in the United States?

According to the Centers for Disease Control and Prevention, it is estimated that nearly twenty million new sexually transmitted infections occur every year in this country, half among young people ages fifteen to twenty-four. The following lists some of the STDs in 2013 in the United States (the latest data from the CDC):

Chlamydia Syphilis (primary and secondary)—Cases reported in 2013: 1,401,906; rate per 100,000 people: 446.6; decrease of 1.5%

Gonorrhea—Cases reported in 2013: 333,004; rate per 100,000 people: 106.1; overall stable

Syphilis (primary and secondary)—Cases reported in 2013: 17,375; rate per 100,000 people: 5.5; 10.0% increase since 2012; this national rate increase was only among men, particularly gay and bisexual men

Syphilis (congenital)—Cases reported in 2013: 348; rate per 100,000 live births: 8.7; 4% increase since 2012

What complications may arise from STDs?

Complications associated with STDs include pelvic inflammatory disease in women and epididymitis (inflammation of the epididymis) in men. Complications from STDs may cause infertility and increase the risk of some cancers. Blindness, bone deformities, and mental retardation may also be caused by STDs.

COMPARING OTHER ORGANISMS

Do all animals give birth to their young?

External fertilization is common among aquatic animals including fishes, amphibians, and aquatic invertebrates. Following an elaborate ritual of mating behavior to synchronize the release of eggs and sperm, both males and females deposit their gametes in the water at approximately the same time in close proximity to each other. The water protects the sperm and eggs from drying out. Fertilization occurs when the sperm reach the eggs. Internal fertilization requires that sperm be deposited in or close to the female reproductive tract. It is most common among terrestrial animals that either lay a shelled egg, such as reptiles and birds, or when the embryo develops for a period of time within the female body. For example, certain sharks, skates, and rays have internal fertilization. The pelvic fins are specialized to pass sperm to the female. In most of these species, the embryos develop internally and are born alive.

Do all young animals resemble the adult animals?

No, not all young animals resemble the adult animals. A tadpole does not look like its parent, the frog; other examples include a chick and rooster and a caterpillar and butterfly. But most animals do resemble their parents—they are just smaller in size.

HUMAN GROWTH AND DEVELOPMENT

INTRODUCTION

What is growth and development?

Growth is an increase in size. Human growth begins with a single fertilized ovum. As it grows, the number of cells increases as a result of mitosis, while at the same time the newly formed cells enlarge and grow in size. Development is the ongoing process of change in the anatomical structures and physiological functions from fertilization through each phase of life to death.

What are the two divisions of human development?

Human development is divided into the prenatal development phase and postnatal development phase. Prenatal development begins at the time of conception and continues until birth. Postnatal development begins at birth and continues to maturity, when the aging process begins and ultimately ends in death.

What is the gestation period of human development?

The gestation period is the time spent in prenatal development. Human gestation is conveniently divided into three trimesters of three months each. The first trimester is the period of embryological and early fetal development. During the second trimester, the organs and organ systems develop. The third trimester is characterized by a period of rapid growth prior to birth.

How do the terms zygote, embryo, and fetus differ?

All three terms refer to the individual developing within the uterus of a woman following conception. Fertilization between sperm and ovum produces a zygote, a sin-

303

gle cell consisting of forty-six chromosomes. It is called a zygote during the first week of development. At the end of the first week of development, the zygote becomes an embryo. It is called an embryo from the second week of development through the eighth week of development. Beginning at the ninth week of development, it is called a fetus.

How large is a zygote at the time of conception?

The single cell that is formed at conception is the zygote. It is approximately 0.005 inches (0.135 millimeters) in diameter and weighs approximately 0.005 ounces (150 milligrams).

PRENATAL DEVELOPMENT–EMBRYONIC PERIOD

How many distinct stages of development are part of the prenatal period?

The prenatal period of development consists of two distinct stages of development: embryological development and fetal development. Embryological development begins with fertilization and continues until the end of the eighth week of development. Fetal development begins at the ninth week of development and continues until birth.

What four events follow fertilization?

The four events that occur immediately following fertilization are cleavage, implantation, placentation, and embryogenesis. Immediately after fertilization, the single cell divides into two cells. During cleavage, these cells continue to divide. Each division brings two new cells called blastomeres. Each blastomere is approximately half the size of the parent cell. Cleavage occurs as the cells move from the uterine tube to the uterine cavity.

How long does the journey take from the uterine tube to the uterine cavity?

It takes about three days for the zygote to travel from the uterine tube to the uterine cavity. At the end of the journey, the solid mass of cells is called a morula (from the Latin *morum*, meaning "mulberry") because it resembles a mulberry.

When does the blastocyst form?

The morula hollows out into a fluid-filled sphere on day four or five following fertilization. It is now called a blastocyst. By the end of the first week, the blastocyst begins to implant in the uterus.

When does implantation occur?

Implantation, the attaching of the blastocyst to the endometrium of the uterus, begins about six to seven days following fertilization. By day nine, the blastocyst is completely enclosed by endometrial cells and is implanted in the uterus. Only after the blastocyst is implanted in the uterus can a pregnancy continue.

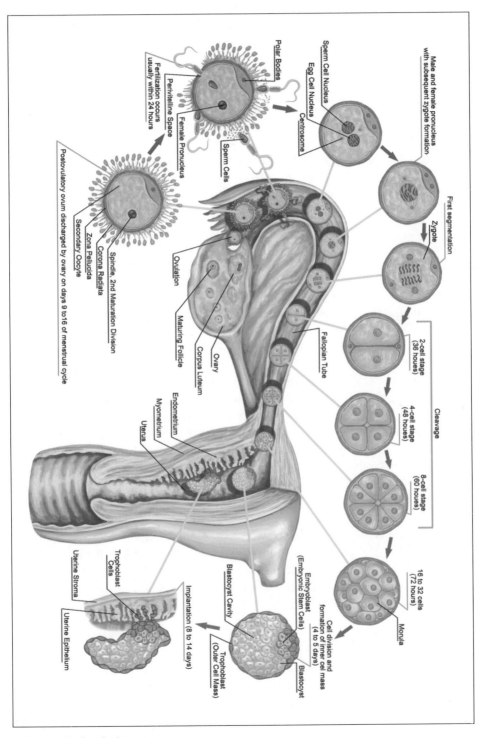

Early stages of embryo development

What is an ectopic pregnancy?

An ectopic pregnancy occurs when the fertilized ovum is implanted in an area other than the uterus. Most frequently, the fertilized ovum is implanted in the uterine tubes. A fetus in an ectopic pregnancy cannot survive because it does not receive nourishment from the uterus. An ectopic pregnancy must be terminated because it jeopardizes the health of the mother.

What are the functions of the placenta?

The placenta is a vascular structure that forms at the site of implantation between the cells that surround the embryo and the endometrium of the uterus. Its main functions are the exchange of gases, nutrients, and wastes between the maternal and fetal bloodstreams. In addition, the placenta secretes hormones. There is no actual blood flow between mother and fetus.

Does the placenta continue to grow throughout a pregnancy?

The placenta grows rapidly until the fifth month of a pregnancy, at which time it is nearly fully developed. At the end of a pregnancy, the placenta is about 1 inch (2.5 centimeters) thick and 8 inches (20.5 centimeters) in diameter. It weighs approximately 1 pound (0.45 kilograms).

How does the umbilical cord form?

The umbilical cord is formed from the extraembryonic or fetal membranes of the embryo during the fifth week of development. It contains two arteries and one vein. The arteries carry carbon dioxide and nitrogen wastes from the embryo to the placenta, and the vein carries oxygen and nutrients from the placenta to the embryo. The umbilical cord is usually 0.4 to 0.8 inches in diameter (1 to 2 centimeters) and 19 to 22 inches long (50 to 55 centimeters).

What is embryogenesis?

Embryogenesis is the process during which the embryo begins to separate from the embryonic disc. The body of the embryo and the internal organs begin to form at this point. The embryo becomes a distinct entity separate from the embryonic disc and the extraembryonic membranes. The left and right sides, as well as the dorsal and ventral surfaces, are now distinct.

How many embryonic layers are in the embryonic disc?

The embryonic disc contains three distinct embryonic layers: 1) the outer layer, called the ectoderm, which is exposed to the amniotic cavity; 2) the inner layer called the endoderm; and 3) the mesoderm. The mesoderm forms between the ectoderm and the endoderm. The following lists the organ and organ system formation from the three germ layers:

Organ System	Ectodermal Layer	Mesodermal Layer	Endodermal Layer
Integumentary system	follicles and hairs, nails, sweat glands, mammary glands, and sebaceous glands	Dermis	
Skeletal system	Pharyngeal cartilages, portion of thesphenoid bone, the auditory ossicles, the styloid processes of the temporal bone, neural crest (formation of the skull)	All components except some pharyngeal derivatives	
Muscular system		All components	
Nervous system	All neural tissue, including brain and spinal cord		
Endocrine system	Pituitary gland and adrenal medullae	Adrenal cortex, endocrine tissues of heart, kidneys, and gonads	Thymus, thyroid gland, and pancreas
Cardiovascular system		All components	
Respiratory system	Mucous epithelium of nasal passageways		Respiratory epithelium and associated mucous glands
Lymphatic system		All components	
Digestive system	Mucous epithelium of mouth and anus, salivary glands		Mucous epithelium (except mouth and anus), exocrine glands (except salivary glands), liver, and pancreas
Urinary system		Kidneys, including the nephrons and the initial portions of the collecting system	Urinary bladder and distal portions of the duct system
Reproductive system		Gonads and the adjacent portion of the duct systems	Distal portions of the duct system, stem cells that produce gametes
Miscellaneous		Lining of the body cavities (pleural, pericardial, and peritoneal) and the connective tissues that support all organ systems	

What is the amniotic cavity?

The amniotic cavity is a fluid-filled chamber. It contains amniotic fluid that surrounds and cushions the developing embryo. At week 10 there is approximately 1 ounce (30 milliliters) of fluid in the cavity. Towards the end of a pregnancy, between weeks 34 and 36, there is about 1 quart (1 liter) of fluid. The average temperature of the amniotic fluid is 99.7°F (37.6°C), slightly higher than the mother's body temperature.

When is amniocentesis usually performed and why?

Amniocentesis is usually done after the fifteenth week of pregnancy. It is a prenatal test used to screen for and identify genetic disorders or test for lung maturity. During the procedure, a thin needle is inserted into the amniotic cavity to remove some of the amniotic fluid, which contains fetal cells and various chemicals produced by the fetus.

When do significant changes occur during the early stages of prenatal development?

Several significant changes occur during the first two weeks of prenatal development, as shown in the accompanying table:

Time Period	Developmental Stage
12–24 hours following ovulation	Fertilized ovum
30 hours to third day	Cleavage
Third to fourth day	Morula (solid ball of cells is formed)
Fifth day through second week	Blastocyst
End of second week	Gastrula (germ layers form)

What are some major developmental events during the embryonic period?

At the end of the embryonic period (eighth week of development), all of the major external features (ears, eyes, mouth, upper and lower limbs, fingers, and toes) are formed and the major organ systems are nearing completion. The following lists the major developmental events during the embryonic period:

Time Period	Major Developments
Week 3	Neural tube, primitive body cavities and cardiovascular system form
Week 4	Heart is beating (it begins to beat by day twenty-five); upper limb buds and primitive ears visible; lower limb buds and primitive eye lenses appear shortly after ears
Week 5	Brain develops rapidly; head grows disproportionately; hand plates develop
Week 6	Limb buds differentiate noticeably; retinal pigment accentuates eyes
Week 7	Limbs differentiate rapidly
Week 8	Embryo appears human; external ears are visible; fingers, toes lengthen; external genitalia are visible, but are not distinctly male or female

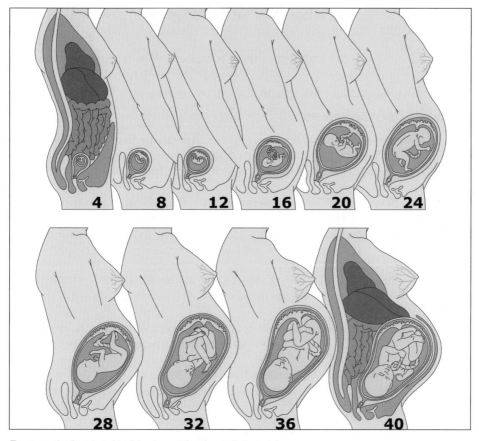

The stages of embryonic and fetal development (numbers indicate weeks)

How large is the embryo at the end of the embryonic period?

The embryo is about 0.75 inches (19 millimeters) in length at the end of the embryonic period (end of the eighth week).

How large is the fetus at the end of the first trimester?

At the end of the first three months of pregnancy, the fetus is nearly 3 inches (7.6 centimeters) long and weighs about 0.8 ounces (23 grams).

PRENATAL
DEVELOPMENT–FETAL STAGE

What is the purpose of the fetal stage of development?

During the fetal stage, beginning after the eighth week of development, the fetus increases in size. All organs formed during the embryonic stage mature to the point where they can function at birth.

What are some major developmental events during the second trimester of pregnancy?

The second trimester of a pregnancy lasts from weeks thirteen through twenty-seven. Each week brings changes and new developments in the fetus. The following lists the major developmental events during the second trimester:

Time Period	Major Developments
Week 13	Baby begins to move, although the movements are too weak to be felt by the mother; ossification of bones begins
Week 14	Prostate gland develops in boys; ovaries move from the abdomen to the pelvis in girls
Week 15	Skin and hair (including eyebrows and eyelashes) begin to form; bone and marrow continue to develop; eyes and ears are nearly in their final location
Week 16	Facial muscles are developing allowing for facial expressions; hands can form a fist; eggs are forming in the ovaries in girls
Week 17	Brown fat tissue begins to develop under the skin
Week 18	Fetus is able to hear such things as the mother's heartbeat
Week 19	Lanugo and vernix cover the skin; fetal movement is usually felt by the mother
Week 20	Skin is thickening and developing layers; fetus has eyebrows, hair on the scalp, and well-developed limbs; fetus often assumes the fetal position of head bent and curved spine
Week 21	Bone marrow begins making blood cells
Week 22	Taste buds begin to form; brain and nerve endings can process the sensation of touch; testes begin to descend from the abdomen in boys; uterus and ovaries (with the lifetime supply of eggs) are in place in girls
Week 23	Skin becomes less transparent; fat production increases; lungs begin to produce surfactant, which will allow air sacs to inflate; may begin to practice breathing
Week 24	Footprints and fingerprints begin to form; inner ear is developed, controlling balance
Week 25	Hands are developed, although the nerve connections are not yet fully developed
Week 26	Eyes are developed; eyebrows and eyelashes are well-formed; hair on head becomes fuller and longer
Week 27	Lungs, liver, and immune system are developing

What is the purpose of vernix and lanugo?

Vernix is a white, pasty, cheese-like coating on the skin consisting of fatty secretions from the sebaceous glands and dead epidermal cells. It protects the skin of the developing fetus. Lanugo is a very fine, silk-like or down-like hair that covers the skin. It may help to hold the vernix on the skin.

How does blood circulate in the fetus?

Fetal circulation differs from circulation after birth because the lungs of the fetus are nonfunctional. Therefore, blood circulation essentially bypasses the lungs in the

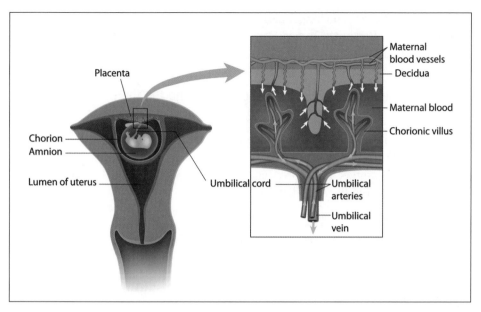

The umbilical cord connects the fetus to the placenta and is the lifeline for the growing baby, transporting nutrients and removing wastes.

fetus. The umbilical vein carries oxygenated blood from the placenta to the fetus. About half of the blood from the umbilical vein enters the liver, while the rest of the blood bypasses the liver and enters the ductus venosus. The ductus venosus joins the inferior vena cava. Blood enters the right atrium of the heart and then flows through the foramen ovale to the left atrium. Blood then passes into the left ventricle (lower portion of the heart) and then to the aorta. From the aorta, blood is sent to the head and upper extremities. It returns to the right atrium of the heart through the superior vena cava. Some blood stays in the pulmonary trunk to reach the developing lung tissues.

How does fetal blood differ from adult blood?

Fetal blood has a greater oxygen-carrying capacity than adult blood. Fetal hemoglobin can carry 20 to 30 percent more oxygen than adult hemoglobin.

What is a normal fetal heart rate?

The fetal heart rate is much faster than an adult's (or even a child's) heart rate. The average resting heart rate is 60 to 80 beats per minute. The normal fetal heart rate is 110 to 160 beats per minute.

How does the fetal circulatory system change at birth?

Immediately after birth, an infant no longer relies on maternal blood to supply oxygen and nutrients. As soon as the baby begins to breathe air, blood is sent to the lungs to be oxygenated. The ductus arteriosus, the special fetal vessel connecting the aorta and pulmonary valve, is no longer needed and closes. A separate left pulmonary artery and aorta form after birth. In addition, the foramen ovale, the special opening between the left and right atria in the heart, closes and normal circulation begins.

What are some major developmental events during the third trimester of pregnancy?

During the third trimester of a pregnancy the fetus continues to grow, while the organ systems continue to develop to the point of being fully functional. Fetal movements become stronger and more frequent. The following lists the major developmental events during the third trimester:

Time Period	Major Developments
Week 28	Eyes begin to open and close; fetus has wake and sleep cycles
Week 29	Bones are fully developed, but still pliable; fetus begins to store iron, calcium, and phosphorus
Week 30	Rate of weight gain increases to 0.5 pounds (227 grams) per week; fetus practices breathing; hiccups are not uncommon
Week 31	Testes begin to descend into scrotum in boys; lungs continue to mature
Week 32	Lanugo begins to fall off
Week 33	Pupils in the eyes constrict, dilate, and detect light; lungs are nearly completely developed
Week 34	Vernix becomes thicker; lanugo has almost disappeared
Week 35	Fetus stores fat all over the body; weight gain continues
Week 36	Sucking muscles are developed
Week 37	Fat continues to accumulate; the baby is considered to be at "full term"
Week 38	Brain and nervous system are ready for birth
Week 39	Placenta continues to supply nutrients and antibodies to fight infection
Week 40	Fetus is fully developed and ready for birth

How much does the fetus grow during each month of pregnancy?

During the early weeks of development there are great changes from the embryonic to the fetal stages, but the overall size of the embryo is very small. As the pregnancy continues, weight gain and overall size becomes much more significant. Until the twentieth week of pregnancy, length measurements are from the crown (or top) of the head to the rump. After the twentieth week, the fetus is less curled up, and measurements are from the head to the toes. The following lists the average size of a fetus during pregnancy:

Gestational Age (weeks)	Size	Weight
8	0.63 in (1.6 cm)	0.04 oz (1 g)
12	2.13 in (5.4 cm)	0.49 oz (14 g)
16	4.57 in (11.6 cm)	3.53 oz (100 g)
20	6.46 in (16.4 cm)	10.58 oz (300 g)

Gestational Age (weeks)	Size	Weight
24	11.81 in (30 cm)	1.32 lb (600 g)
28	14.8 in (37.6 cm)	2.22 lb (1 kg)
32	16.69 in (42.4 cm)	3.75 lb (1.7 kg)
36	18.66 in (47.5 cm)	5.78 lb (2.62 kg)
40	20.16 in (51.2 cm)	7.63 lb (3.46 kg)

What are the complications of premature birth?

Babies born before the thirty-seventh week of gestation (preterm babies) are very small and fragile. Birth weights are often less than two pounds. Many of the organ systems are not fully developed, which leads to complications as these infants struggle to survive. Complications include:

- Inability to breathe or breathe regularly on their own due to underdeveloped lungs
- Difficulty in body temperature regulation; the baby cannot maintain his or her own body heat
- Feeding and growth problems because of an immature digestive system
- Jaundice due to a buildup of bilirubin
- Anemia due to not enough red blood cells to carry oxygen to tissues
- Bleeding into the brain

Although after a year or two most preterm babies are developmentally the same as full-term babies, some may still experience breathing difficulties, hearing or vision problems, and learning disabilities.

How common are premature births?

According to the Centers for Disease Control and Prevention, the preterm birth rate for 2012 in the United States was 450,000, or about one in nine infants born. Preterm refers to infants delivered at less than thirty-seven completed weeks of gestation. The number of preterm babies has risen 20 percent since 1990, when it was only 10.6 percent.

BIRTH AND LACTATION

What are the maternal changes during pregnancy?

There are several physiological changes in a mother during pregnancy, in addition to changes in the size of the uterus and changes in the mammary glands. The mother must eat, breathe, and eliminate wastes for both herself and her developing fetus, which is totally dependent upon the mother. The mother's respiratory rate goes up so her lungs can deliver the extra oxygen and remove the excess carbon dioxide generated by the fetus. The maternal blood volume increases by nearly 50 percent by the end of a pregnancy, since blood flowing into the placenta reduces the volume of blood throughout the rest of the cardiovascular system. Because the mother must also nourish the fetus, she may feel hungry more often, and her nutritional requirements increase 10 to 30 percent. The maternal glomerular filtration rate increases by approximately 50 percent to excrete the fetus's waste. Consequently, the combination of increased weight and pressure on the mother's urinary bladder and the elimination of additional waste products lead to more frequent urination.

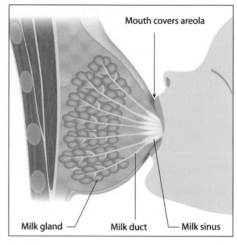

Mouth covers areola

Milk gland   Milk duct   Milk sinus

The infant receives milk from the milk ducts in the breast through the areolas.

How do the mammary glands develop during pregnancy?

The mammary glands increase in size in response to placental hormones and maternal endocrine hormones. The areola darken in color. Clear secretions, such as colostrum, are stored in the duct system of the mammary glands. The secretions may be expressed from the nipple. (For more about breast feeding, see below.)

What are the stages of labor?

The goal of labor is the birth of a new baby. Labor is divided into three stages: 1) dilation, 2) expulsion, and 3) placental. Delivery of the fetus occurs during expulsion.

How is the onset of labor identified?

Different women will experience different symptoms at the onset of labor. Some women may experience lower back pain or cramping similar to menstrual cramps. In some women, the amniotic sac ruptures early in labor with a sensation of fluid leaking either as a trickle or a large gush of fluid. Some women will lose the mucous plug that blocks the cervix with a brownish or red-tinged mucous discharge. Ultimately, as labor progresses, uterine contractions become more powerful and more frequent.

What events occur during dilation?

The purpose of dilation is to dilate (open) and thin (efface) the cervix to permit the fetus to move from the uterus into the vagina. Dilation is divided into three phases: 1) the early labor phase, 2) the active labor phase, and 3) transition. During the early phase, contractions last 30 to 60 seconds and occur every 5 to 20 minutes at regular intervals. As labor progresses, the frequency of the contractions will increase. The cervix dilates from 0 to 3 centimeters during the early phase of labor. During active labor, contractions become stronger, last longer (45 to 60 seconds or longer), and occur at more frequent intervals (as frequent as every 2 to 4 minutes). The cervix dilates from

3 to 7 centimeters. The final phase of dilation is transition. During transition, the cervix dilates from 7 to 10 centimeters. It is now fully dilated. Contractions during transition last 60 to 90 seconds with sometimes not even a minute between contractions.

What is the placental stage of labor?

During the placental stage of labor, uterine contractions separate the connections between the endometrium and the placenta. The placenta, along with the fetal membranes and any remaining uterine fluid, are ejected through the birth canal.

How does false labor differ from true labor?

Uterine contractions that are neither regular nor persistent are false labor. Oftentimes in false labor the contractions may stop when the mother walks or even shifts positions. The contractions in true labor become stronger, more frequent, and do not cease. Once true labor, begins it will continue until the fetus is delivered.

How long does labor last?

The length of labor differs with every woman. Even the same woman will experience labor differently with each pregnancy. In general, dilation is the longest stage of labor. It can last for several hours to several days, especially for the first-time mother. The early phase of labor is the longest. Active labor may last from three to eight hours, although it can be shorter or longer. Transition is the shortest part of dilation. It may last for only fifteen minutes. Expulsion (delivery) may take only a few minutes to several hours. Delivery of the placenta usually takes only five to ten minutes and is usually no longer than thirty minutes.

Why does the blood loss during labor not cause a problem for the mother?

In a normal delivery, there may be as much as 1 pint (0.473 liters) of blood loss during labor, most during the placental stage. Since the maternal blood volume increased during pregnancy, this blood loss is tolerated without difficulty.

What is the difference between fraternal and identical twins?

Fraternal, or dizygotic twins, develop when a woman ovulates two separate oocytes that are fertilized by two different sperm. Fraternal twins do not resemble each other any more than other brothers and sisters of the same parents

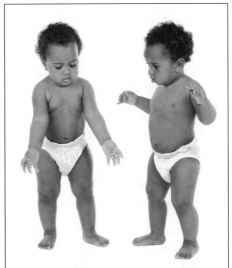

Identical (maternal) twins originate from one egg and therefore share many physical traits, while fraternal twins are no more alike than other siblings; they merely come from two different eggs that happened to be fertilized at the same time.

315

resemble each other. They may be of the same sex or different sexes. Identical, or monozygotic twins, develop from the same fertilized ovum. The blastomeres may separate early in cleavage, or the inner cell mass may split prior to gastrulation. Identical twins have the same genetic makeup because they are formed from the same pair of gametes. They look alike and are the same sex.

Do twins share a placenta, umbilical cord, or amniotic sac?

Identical twins will often share the same placenta but usually have separate amniotic sacs. Each twin always has its own, separate umbilical cord. Nonidentical twins have separate placentas, amniotic sacs, and umbilical cords.

How do healthcare providers induce labor?

When labor does not begin on its own, healthcare providers may induce labor. The most common method of inducing labor is to give the hormone oxytocin. Oxytocin will start contractions and keep them strong and regular. Labor is generally induced when the pregnancy has lasted two weeks beyond the due date or if there is a concern the baby will be too large for safe delivery. Labor may also be induced if the health of the mother becomes endangered.

How does a multiple pregnancy affect the mother's health?

Multiple pregnancies (pregnancies with more than one fetus) pose special risks since the strains on the mother (for example, the need for oxygen and other nutrients for each fetus) are multiplied. Preeclampsia (high blood pressure and protein in the urine) and gestational diabetes are more common in multiple pregnancies.

What are some of the risks and complications associated with multiple pregnancies?

One of the most common risks of a multiple pregnancy is preterm birth. On average, most twin pregnancies last thirty-five weeks, while pregnancies with triplets last only thirty-three weeks. Pregnancies with quadruplets last only twenty-nine weeks on average. Low-birth weight (less than 5.5 pounds, or 2.5 kilograms) due to preterm birth or poor fetal development is also very common in multiple pregnancies. Babies weighing less than 3.34 pounds (1.5 kilograms) at birth are at a greater risk for lasting disabilities, including mental retardation, cerebral palsy, and hearing and vision loss. Lung problems and breathing difficulties are common in babies born before thirty-four weeks.

What are some statistics about twins in the United States?

According to the Centers for Disease Control and Prevention, in 2009 (the latest data to date) one in every thirty babies born in the United States was a twin, compared with one in every fifty-three babies in 1980. The twin birth rate rose 76 percent from 1980 through 2009, from 18.9 to 33.3 per 1,000 births. Over the three decades, twin birth rates rose by nearly 100 percent among women aged thirty-five to thirty-nine

What is the origin of the term "Siamese twins"?

The term "Siamese twins" dates to 1811, when a pair of conjoined twins was born in Bangkok, the capital of Siam (now called Thailand). The twins Eng and Chang (which mean "left" and "right" in Siamese) were joined at the lower end of the sternum. They left Siam when they were eighteen, spending time in both the United States and England. In addition to becoming famous as part of the P. T. Barnum circus, they were farmers. They married sisters and each raised a family. The parents of Eng and Chang were Chinese, not Siamese. The Siamese called them "Chinese twins."

and more than 200 percent among women aged forty and over. But the older age of women at childbirth in 2009 compared with three decades earlier accounts for only about one-third of the rise in twinning over the thirty years. And overall, twinning rates differed from state to state in 2009. The lowest was 22.3 percent per 1,000 births in New Mexico, while Connecticut had the highest percent of 45.9 per 1,000 births, or around 5 percent of all births that year.

How common are multiple births?

According to the Centers for Disease Control and Prevention, as reproductive technology—such as the use of fertility drugs and other methods of assisted reproduction—has become more refined, the number of triplets, quadruplets, and other multiple births dropped for the first time in more than ten years in 1999. The number of twins, however, continues to increase.

What are conjoined twins?

Conjoined twins are identical twins whose embryonic discs do not separate completely. They typically share some skin and an organ, often the liver, and perhaps other internal organs. If the fusion is minor, they may be separated surgically with relative ease. On rare occasions, conjoined twins are joined at the head or share so many organs that it is nearly impossible to separate them.

When does milk begin to be produced in a pregnant woman and how is it stimulated to be released?

By the end of the sixth month of a pregnancy, the mammary glands are developed in order to secrete milk. They begin to secrete colostrum. Once the placenta is delivered and the secretion of estrogen and progesterone drops, milk production increases. The hormones prolactin and oxytocin are involved in milk production and release. An infant's sucking stimulates the release of these two hormones.

What is the composition of human breast milk?

Human breast milk consists of mostly of water (88 percent), sugars (6.5 to 8 percent), lipids (3 to 5 percent), proteins (1 to 2 percent), amino acids, and salts. It also contains large quantities of lysozymes—enzymes with antibiotic properties. Human milk is bluish-white in color and sweet. The blue color comes from the protein and the white comes from the fat. There are approximately 750 calories per liter of breast milk.

How does colostrum differ from breast milk?

Colostrum is the first fluid secreted by the mother's breasts in the first several days after delivery. It is higher in proteins and has less fat than milk. It also contains a high concentration of antibodies that protect the baby from infections until his or her own immune system matures. A mother produces approximately three ounces of colostrum in a twenty-four-hour period.

After the colostrum, how much milk does a nursing mother produce?

A nursing mother produces 850 to 1,000 milliliters of milk each day. Mothers of multiples will naturally produce enough milk for each infant.

What are the benefits of breastfeeding?

Breastfeeding provides benefits to both the baby and the mother. A major benefit to the baby is that breast milk supplies the correct amount of nutrients as the baby grows from an infant to a healthy toddler. The nutrients in breast milk also protect the infant from certain childhood illnesses. Finally, recent research has shown that breast milk contains certain fatty acids (building blocks) that help the infant's brain develop. In the early days following childbirth, the mother's body releases a hormone that makes her uterus contract and get smaller in response to the baby's sucking. Breastfeeding also provides many emotional benefits between mother and child and encourages maternal-infant bonding.

POSTNATAL DEVELOPMENT

What are the stages of postnatal development?

The five life stages of postnatal development are: 1) neonatal, 2) infancy, 3) childhood, 4) adolescence, and 5) maturity. The neonatal period extends from birth to one month. Infancy begins at one month and continues to two years of age. Childhood begins at two years of age and lasts until adolescence. Adolescence begins at around twelve or thirteen years of age and ends with the beginning of adulthood. Adulthood, or maturity, begins somewhere between the ages of eighteen and twenty-five and lasts into old age. The process of aging is called senescence.

What are some developmental changes that occur during the neonatal period?

The greatest change from birth to the neonatal period is that the neonate must begin to perform many functions that had previously been done by the mother, especially

Newborns differ from older children in a number of ways. For example, a neonate's heart can be up to 140 times a minute, and that's okay; they also breathe more rapidly.

respiration, digestion, and excretion. With the first breath of air following delivery, the lungs fill with air and the neonate begins to breathe for himself or herself.

How do heart rate and the rate of respiration differ between a neonate and an adult?

The average neonate heart rate is 120 to 140 beats per minute, compared to a resting heart rate of 60 to 80 beats per minute in an adult. The average respiratory rate for a neonate is 30 breaths per minute, compared to 12 to 28 breaths per minute in an adult.

What are the major developmental milestones during infancy?

A normal infant will double his or her birth weight by five or six months of age and triple his or her birth weight during the first year of life. Major developmental milestones during infancy are summarized in the table below (note: there is considerable variation between individuals, but these are within the normal range):

Age	Major Milestones in Average Infant
End of first month	Bring hands to face; move head from side to side while lying on stomach; hear very well and often recognize parents' voices
End of third month	Raise head and chest while lying on stomach; open and shut hands; bring hands to mouth; smile; recognize familiar objects and people
End of seventh month	Roll over stomach to back and back to stomach; sit up; reach for objects with hand; support whole weight on legs when supported and held up; enjoy playing peek-a-boo; begin to babble

319

Age	Major Milestones in Average Infant
End of first year	Sit up without assistance; get into the hands and knees position; crawl; walk while holding on; some babies are able to take a few steps without support; use the pincer grasp; use simple gestures, for example, nodding head, waving bye-bye
End of second year	Walk alone; begin to run; walk up and down stairs; pull a toy behind him or her; say single words (fifteen to eighteen months); use simple phrases and two-word sentences (eighteen to twenty-four months); scribble with a crayon; build a tower with blocks

What is the average age when puberty begins?

The average age when puberty begins in the United States today is around twelve years in boys and eleven years in girls. The normal range is ten to fifteen years in boys and nine to fourteen years in girls.

What hormonal events signal the onset of puberty?

Three different hormonal events occur that signal the onset of puberty (from the Latin *puber*, meaning "adult"). The hypothalamus increases production of gonadotropin-releasing hormone (GnRH). This stimulates the endocrine cells in the anterior lobe of the pituitary gland, causing circulating levels of follicle-stimulating hormone (FSH) and luteinizing hormone (LH) to rise rapidly. Finally, in response to increased levels of FSH and LH, the ovaries and testes secrete increased amounts of androgens and estrogens. The secondary sex characteristics appear, gamete production begins, and there is a sudden increase in the growth rate, culminating in the closure of the epiphyseal cartilages.

What are the major body changes at puberty?

In addition to general body changes that occur in both males and females at puberty, physical changes in the genitalia, skin, hair growth, and voice are collectively termed the secondary sex characteristics. The following lists the secondary sex characteristic changes in males and females during this time:

Area of Body	Males	Females
General body changes	Shoulders broaden, muscles thicken and height increases; body odor from armpits and genitals becomes apparent; skeletal growth ceases by about age twenty-one	Pelvis widens; fat distribution increases in hips, buttocks, breasts; skeletal growth ceases by about age eighteen
External genital organs	Penis increases in size; scrotum enlarges; penis and scrotum become more pigmented	Breasts enlarge; vagina enlarges and vaginal walls thicken

Area of Body	Males	Females
Internal genital organs	Testes enlarge; sperm production increases in testes; seminal vesicles, prostate gland, bulbourethral gland enlarge and begin to secrete	Uterus enlarges; ovaries secrete estrogens; ova in ovaries begin to mature; menstruation begins
Skin	Secretions of sebaceous gland thicken and increase, often causing acne; skin thickens	Estrogen secretions keep sebaceous secretions fluid, inhibit development of acne and blackheads
Hair growth	Hair appears on face, pubic area, armpits, chest, around anus; general body hair increases; hairline recedes in the lateral frontal regions	Hair appears on pubic area, armpits; scalp hair increases with childhood hairline retained
Voice	Voice becomes deeper as larynx enlarges and vocal cords become longer and thicker	Voice remains relatively high pitched as larynx grows only slightly

How does adolescence differ from puberty?

Adolescence (from the Latin *adolescere*, meaning "to grow up") begins at puberty and ends at adulthood when physical growth stops. Puberty is the point when an individual becomes physiologically capable of reproduction. Adulthood begins between ages eighteen and twenty-five.

How is senescence defined?

Senescence (from the Latin *senex*, meaning "old") is the process of aging. Physiological changes continue to occur even after complete physical growth is attained at maturity. As people age, the body is less able to and less efficient in adapting to environmental changes. Maintaining homeostasis becomes harder and harder, especially when the body is under stress. Ultimately, death occurs when the combination of stresses cannot be overcome by the body's existing homeostatic mechanisms.

One of the more noticeable physical changes in pubescent youth is that the skin can be plagued with pimples or acne as sebaceous gland secretions thicken.

What are some general effects of aging on the human body?

The aging process affects every organ system. Some changes begin as early as ages thirty to forty. The aging process becomes more rapid between ages fifty-five and sixty. The following lists some of the effects of aging:

Organ System	Effect of Aging
Integumentary	Loss of elasticity in the skin tissue, producing wrinkles and sagging skin; oil glands and sweat glands decrease their activity, causing dry skin; hair thins
Skeletal	Decline in the rate of bone deposition, causing weak and brittle bones; decrease in height
Muscular	Muscles begin to weaken; muscle reflexes become slower
Nervous	Brain size and weight decreases; fewer cortical neurons; rate of neurotransmitter production declines; short-term memory may be impaired; intellectual capabilities remain constant unless disturbed by a stroke; reaction times are slower
Sensory	Eyesight is impaired with most people becoming far-sighted; hearing, smell, and taste are reduced
Endocrine	Reduction in the production of circulating hormones; thyroid becomes smaller; production of insulin is reduced
Cardiovascular	Pumping efficiency of the heart is reduced; blood pressure is usually higher; reduction in peripheral blood flow; arteries tend to become more narrow
Lymphatic	Reduced sensitivity and responsiveness of the immune system; increased chances of infection and/or cancer
Respiratory	Breathing capacity and lung capacity are reduced due to less elasticity of the lungs; air sacs in lungs are replaced by fibrous tissue
Digestive	Decreased peristalsis and muscle tone; stomach produces less hydrochloric acid; intestines produce fewer digestive enzymes; intestinal walls are less able to absorb nutrients
Excretory	Glomerular filtration rate is reduced; decreased peristalsis and muscle tone; weakened muscle tone often leads to incontinence
Reproductive	Ovaries decrease in weight and begin to atrophy in women; reproductive capabilities cease with menopause in women; sperm count decreases in men

COMPARING OTHER ORGANISMS

How long do some animals live?

Each animal species on earth has a certain average lifespan. For example, of the mammals, humans and fin whales live the longest. The following lists the maximum lifespan for a few of the various longer-lived animal species:

Animal (Latin name)	Maximum Lifespan (in years)
Marion's tortoise (*Testudo sumeirii*)	152+
Quahog (*Venus mercenaria*)	~150

Animal (Latin name)	Maximum Lifespan (in years)	
Common box tortoise (*Terrapene Carolina*)	138	
European pond tortoise (*Emys orbicularis*)	120+	
Spur-thighed tortoise (*Testudo graeca*)	116+	
Fin whale (*Balaenoptera physalus*)	116	
Human (*Homo sapiens sapiens*)	116	(although this is probably closer to 125)
Deep-sea clam (*Tindaria callistiformis*)	~100	
Killer whale (*Orcinus orca*)	~90	

HELPING HUMAN ANATOMY

ANATOMY AND IMAGING TECHNIQUES

What were some of the early ways to explore the inside of the body?

Until the end of the nineteenth century, there were no noninvasive techniques to explore the internal organs of the body. Medical practitioners relied on descriptions of symptoms as the basis for their diagnoses. X-rays, discovered at the very end of the nineteenth century, provided the earliest technique to explore the internal organs and tissues of the body. During the twentieth century, significant advances were made in the field of medical imaging to explore the internal organs.

What are X-rays and CAT (or CT) scans?

X-rays are electromagnetic radiation with short wavelengths and a great amount of energy. They were discovered in 1898 by Wilhelm Conrad Roentgen (1845–1923). X-rays are frequently used in medicine because they are able to pass through opaque, dense structures such as bone and form an image on a photographic plate. They are especially helpful in assessing damage to bones, identifying certain tumors, and examining the chest (heart and lungs, in particular) and abdomen.

CAT or CT scans (computer-assisted tomography, or simply computerized tomography) are specialized X-rays that produce cross-sectional images of the body. An X-ray-emitting device moves around the body region being examined. At the same time, an X-ray detecting device moves in the opposite direction on the other side of the body. As these two devices move, an X-ray beam passes through the body from hundreds of different angles. Since tissues and organs absorb X-rays differently, the intensity of X-rays reaching the detector varies from position to position. A computer records the measurements made by the X-ray detector and combines them mathematically. The result is a sectional image of the body that is viewed on a screen.

Who discovered and pioneered the use of CT scans?

Allan M. Cormack (1924–1998) and Godfrey N. Hounsfield (1919–2004) independently discovered and developed computer assisted tomography in the early 1970s. They shared the 1979 Nobel Prize in Physiology or Medicine for their research. The earliest computer-assisted tomography was used to examine the skull and diseases of the brain.

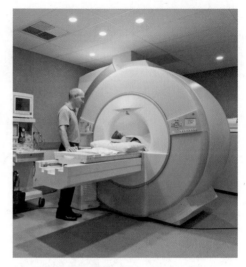

A computerized tomography (CT) scan machine helps doctors look inside patients's bodies without performing surgery.

How are CT scans used in the study of the human body?

CT scans are used to study many parts of the body, including the chest, belly and pelvis, extremities (arms and legs), and internal organs, such as pancreas, liver, gall bladder, and kidneys. CT scans of the head and brain may detect an abnormal mass or growth, stroke damage, area of bleeding, or blood vessel abnormality. Patients complaining of pain may have a CT scan to determine the source of the pain. Sometimes a CT scan will be used to further investigate an abnormality found on a regular X-ray.

What is an advantage of positron emission tomography (PET imaging) over CT scans and X-rays?

Unlike traditional X-rays and CT scans, which reveal information about the structure of internal organs, positron emission tomography (PET imaging) is an excellent technique for observing metabolic processes. Developed during the 1970s, PET imaging uses radioactive isotopes to detect biochemical activity in a specific body part.

What is the procedure for a PET scan?

A patient is injected with a radioisotope, which travels through the body and is transported to the organ and tissue to be studied. As the radioisotopes are absorbed by the cells, high-energy gamma rays are produced. A computer collects and analyzes the gamma-ray emission, producing an image of the organ's activity.

How are PET scans used to detect and treat cancer?

PET scans of the whole body may detect cancers. While the PET scans do not provide cancer therapy, they are very useful in examining the effects of cancer therapies and treatments on a tumor. Since it is possible to observe biochemical activities of cells and tumors using PET scans, biochemical changes to tumors following treatment may be observed.

> ## What are the disadvantages of X-rays as a diagnostic tool?
>
> **A** major disadvantage of X-rays as a diagnostic tool is that they provide little information about the soft tissues. Since they only show a flat, two-dimensional picture, they cannot distinguish between the various layers of an organ, some of which may be healthy while others may be diseased.

Is it possible to study blood flow to the heart or brain?

PET scans provide information about blood flow to the heart muscle and brain. They may help evaluate signs of coronary heart disease and reasons for decreased function in certain areas of the heart. PET scans of the brain may detect tumors or other neurological disorders, including certain behavioral health disorders. Studies of the brain using PET scans have identified parts of the brain that are affected by epilepsy and seizures, Alzheimer's disease, Parkinson's disease, and stroke. In addition, they have been used to identify specific regions of the healthy brain that are active during certain tasks.

What is nuclear magnetic resonance (NMR)?

Nuclear magnetic resonance (NMR) is a process in which the nuclei of certain atoms absorb energy from an external magnetic field. Scientists use NMR spectroscopy to identify unknown compounds, check for impurities, and study the shapes of molecules. This technology takes advantage of the fact that different atoms will absorb electromagnetic energy at slightly different frequencies.

What is nuclear magnetic resonance imaging?

Magnetic resonance imaging (MRI), sometimes called nuclear magnetic resonance imaging (NMR), is a noninvasive, nonionizing diagnostic technique. It is useful in detecting small tumors, blocked blood vessels, or damaged vertebral discs. Because it does not involve the use of radiation, it can often be used in cases where X-rays would be dangerous. Large magnets beam energy through the body, causing hydrogen atoms in the body to resonate. This produces energy in the form of tiny electrical signals. A computer detects these signals, which vary in different parts of the body and show the contrast between a healthy and diseased organ. The variation enables a picture to be produced on a screen and interpreted by a medical specialist.

What distinguishes MRI from computerized X-ray scanners is that most X-ray studies cannot distinguish between a living body and a cadaver, while MRI "sees" the difference between life and death in great detail. More specifically, it can discriminate between healthy and diseased tissues with more sensitivity than conventional radiographic instruments like X-rays or CAT scans.

Who proposed using MRI for diagnostic purposes?

The concept of using MRI to detect tumors in patients was proposed by American medical researcher, biophysicist, and inventor Raymond Damadian (1936–) in a 1972 patent application. The fundamental MRI concept used in all present-day MRI instruments was proposed by American chemist Paul Lauterbur (1929–2007) in an article published in *Nature* in 1973. Lauterbur and English physicist Peter Mansfield (1933–) were awarded the Nobel Prize in Physiology or Medicine in 2003 for their discoveries concerning MRI. The main advantages of MRI are that it not only gives superior images of soft tissues (like organs), but it can also measure dynamic physiological changes in a noninvasive manner (without penetrating the body in any way). A disadvantage of MRI is that it cannot be used for every patient. For example, patients with implants, pacemakers, or cerebral aneurysm clips made of metal cannot be examined using MRI because the machine's magnet could potentially move these objects within the body, causing damage.

What is ultrasound?

Ultrasound, also called sonography, is another type of 3-D computerized imaging. Using brief pulses of ultrahigh frequency acoustic waves (lasting 0.01 seconds), it can produce a sonar map of the imaged object. The technique is similar to the echolocation used by bats, whales, and dolphins. By measuring the echo waves, it is possible to determine the size, shape, location, and consistency (whether it is solid, fluid-filled, or both) of an object.

Why is ultrasound used frequently in obstetrics?

Ultrasound is a very safe, noninvasive imaging technique. Unlike X-rays, sonography does not use ionizing radiation to produce an image. It gives a clear picture of

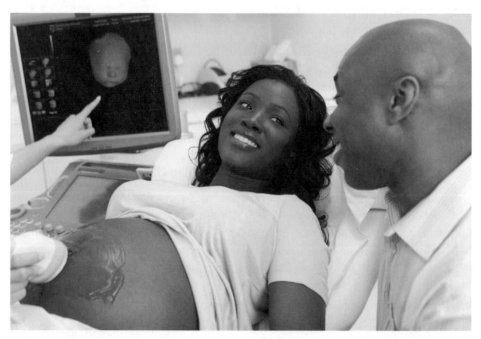

Ultrasound machines use sound much like a bat does to create a 3-D image of things that cannot be seen directly.

soft tissues, which do not show up in X-rays. Ultrasound causes no health problems (for a mother or unborn fetus) and may be repeated as often as necessary.

What imaging techniques are used to examine breast tissue and diagnose certain breast diseases?

Mammography is the specific imaging technique used to examine breast tissue and diagnose breast diseases. It has become a very important tool in diagnosing early breast cancer, as small tumors may be visible on a mammogram years before they may be felt physically by a woman or her healthcare provider. A small dose of radiation is passed through the breast tissue, producing an image of the interior of the breast tissue.

One improvement to the mammogram is the digital breast tomosynthesis, or three-dimensional (3-D) mammography. A study in 2014 showed that this method of mammography found significantly more invasive, and potentially lethal, cancers than the traditional mammograms that would often yield false positive results. The 3-D reconstruction of breast tissue has been shown to give radiologists a clearer view of the overlapping slices of breast tissue, thus resulting in less false positives—and often reducing the number of unnecessary breast biopsies.

What is a barium test used to determine in the upper gastrointestinal tract?

Barium tests (called barium swallow) are used to examine the upper gastrointestinal tract, especially the pharynx (back of the mouth and throat) and esophagus (the tube of muscle from below the tongue to the stomach). A liquid suspension called barium sulfate is ingested by the patient, essentially coating the inside wall linings of the gastrointestinal tract. Because barium is an "X-ray absorber," the size, shape, and sometimes conditions of the upper gastrointestinal tract can be seen on the X-rays. For example, such conditions as tumors, ulcers, hernias, diverticula (pouches in the intestines), and inflammations can often be detected from a barium swallow.

DIAGNOSTIC TECHNIQUES FOR VARIOUS SYSTEMS

What are some common hearing tests?

There are many diagnostic techniques to determine a person's hearing ability and if there is hearing loss. For example, the basic audiogram determines the patient's ability to hear relative to what is considered a normal adult hearing level. Tymphanometry examines the middle ear for blockages or malfunctions to determine if they can be treated medically or surgically. And testing the stapedial reflexes will determine if the auditory nerve is efficiently transmitting hearing signals to the brain.

What is echocardiography?

Echocardiography is a noninvasive method for studying the motion and internal vessels of the heart. This method uses ultrasound beams, which are directed into the patient's chest by a transducer. The transducer uses the ultrasonic waves, which are directed back from the heart to form an image. An echocardiogram can show internal dimensions of the chambers, valve motion, blood flow, and the presence of increased pericardial fluid, blood clots, or tumors.

Modern pacemakers are smaller than a matchbox and weigh between 20 and 50 grams (1 to 2 ounces).

When was the first successful pacemaker invented?

The first successful pacemaker was developed in 1952 by American cardiologist Paul Zoll (1911–1999) in Boston, in collaboration with the Electrodyne Company. The device was worn externally on the patient's belt. It relied on an electrical wall socket to stimulate the patient's heart through two metal electrodes attached to the patient's chest. American engineer and inventor Wilson Greatbatch (1919–2011) developed an internal pacemaker. It was first implanted by surgeons William Chardack (1915–2006) and Andrew Gage (1922–) in 1960. According to the American Heart Association, more than a half million pacemakers are transplanted each year.

How can electrical activity of the heart be monitored?

The electrical activity of the heart can be monitored by an electrocardiogram. Electrodes are placed at different locations on the chest, and each time the heart beats, there is a wave of electrical activity through the heart muscle. This test can detect very slight changes in the heart's electrical activity through deflections on a monitor. An electrocardiogram can be used to detect and diagnose cardiac arrhythmias, which are abnormalities in the heart's conduction system.

What is a circulatory assist device?

A circulatory assist device, also known as a ventricular assist device, is a mechanical circulatory machine. These pumps are used on a short-term basis to allow the patient's heart to rest while it is healing. However, they have also been used on a long-term basis to support the hearts of patients awaiting a heart transplant. There are three major types of devices: counterpulsation devices, cardiopulmonary assist devices, and left ventricular assist devices.

What is a defibrillator?

A defibrillator is an electronic device—called an automated external defibrillator (AED)—that gives an electric shock through the chest to the heart. This is to reestablish the normal contractions that a heart experiences, especially when the organ is having dangerous arrhythmia (irregular heart rhythm) or is in cardiac arrest. According to the American Heart Association, most sudden cardiac arrests result from ventricular fibrillation. When this rapid and unsynchronized heart rhythm starts, the heart must be "defibrillated" quickly (the chance of surviving drops by 7 to 10 percent each minute the heart is not beating normally). The newest AED models are lightweight, portable, and can be used not only by medical and emergency medical technicians (EMTs), but also by non-medical personnel who have been trained in the use of the AED.

Why do physicians perform a spinal tap?

A spinal tap, also called a lumbar puncture, is the withdrawal of a small amount of cerebrospinal fluid from the

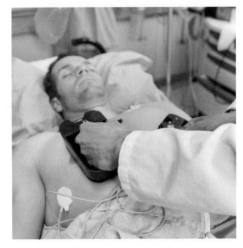

Defibrillators—you might have heard them referred to as paddles—provide a shock to the chest that stimulates a non-beating heart to start beating again.

subarchnoid space in the lumbar region of the vertebral column. Since the spinal cord ends at the level of the first or second lumbar vertebra, a needle can be inserted into the subarchnoid space at the fourth lumbar vertebra with little risk of injuring the spinal cord. The cerebrospinal fluid may be tested and examined for infection. Cerebrospinal fluid is also withdrawn to reduce pressure caused by swelling of the brain or spinal cord following injury or disease. (For more about the spine, see the chapter "Nervous System.")

What is a bronchoscopy?

A bronchoscopy is a direct visual examination of the larynx and airways through a long, flexible viewing tube. A bronchoscope can be inserted through the mouth or nose and extended into the lungs. It can also be used to collect tissue and fluid samples. Bronchoscopy can help a physician make a diagnosis and treat certain medical conditions.

What are some diagnostic procedures used to examine the digestive tract?

Several diagnostic tests are available to examine organs of the digestive tract and to determine causes of abdominal pain and disorders that affect the digestive system. Some of the commonly performed screening tests are colonoscopy, flexible sigmoidoscopy, endoscopy, upper GI series and lower GI series X-rays, ERCP (endoscopic retrograde cholangiopancreatography), and liver biopsy. The following briefly describes each procedure:

> *Colonoscopy* allows a physician to look inside the entire large intestine. It is used to detect early signs of cancer in the colon and rectum.
>
> *Flexible sigmoidoscopy* allows a physician to examine part of the large intestines—or the inside of the large intestine from the rectum through the sigmoid

331

or descending colon (the last part of the colon). It is used to detect the early signs of cancer in the descending colon and rectum.

Upper endoscopy allows a physician to look inside the esophagus, stomach, and duodenum (the first part of the small intestine). This procedure is used to discover the reason for swallowing difficulties, nausea, vomiting, reflux, bleeding, indigestion, abdominal pain, or chest pain.

The upper GI series uses X-rays to diagnose problems in the esophagus, stomach, and duodenum. Ulcers, scar tissue, abnormal growths, hernias, or areas where something is blocking the normal path of food through the digestive system are visible with the upper GI series.

The lower GI series uses X-rays to diagnose problems in the large intestine, including the colon and rectum. Problems such as abnormal growths, ulcers, polyps, diverticuli, and colon cancer may be diagnosed through a lower GI series.

ERCP (endoscopic retrograde cholangiopancreatography) enables a physician to diagnose and treat problems in the liver, gall bladder, bile ducts, and pancreas.

Liver biopsy is performed when other liver function tests reveal the liver is not working properly. It allows a physician to examine a small sample of liver tissue for signs of damage or disease.

Why is screening for colorectal cancer important?

Colorectal cancer is the most common cancer of the digestive system. Screening tests are important to diagnose a disease prior to developing symptoms. When detected in the early stage, the five-year survival rate for colorectal cancer is greater than 90 percent. In addition, polyps, which are not malignant, may be removed during a screening procedure, thus avoiding cancer. The screening guidelines suggested by the American Cancer Society for both men and women over the age of fifty with average risk for colorectal cancer include:

1. A fecal occult blood test (FOBT) or fecal immunochemical test (FIT) every year, or

2. Flexible sigmoidoscopy every five years, or

3. An FOBT or FIT every year, plus flexible sigmoidoscopy every five years, or

4. Double-contrast barium enema every five years, or

5. Colonoscopy every ten years (the time can vary if high-risk polyps are discovered in a previous colonoscopy or if there is a history of polyps or colon cancer in a person's family)

Of the first three options, the combination of FOBT or FIT every year, plus flexible sigmoidoscopy every five years, is preferable.

What is assisted reproductive technology (ART) in fertility treatments?

According to the Centers for Disease Control and Prevention (CDC), assisted reproductive technology (ART) includes all fertility treatments in which both the sperm and eggs are handled during the treatment. Most ART procedures involve surgically removing eggs from a woman's ovaries, combining them with sperm in the laboratory, and returning them to the woman's body or donating them to another woman. One of the most successful and effective ART methods is in vitro fertilization (IVF). It may be used when the woman's fallopian tubes are blocked or when the man produces too few sperm. The woman takes a drug that causes the ovaries to produce multiple eggs. Once mature, the eggs are removed and put in a dish in the lab along with the man's sperm for fertilization. After three to five days, healthy embryos are implanted in the woman's uterus.

There are several other methods. For example, zygote intrafallopian transfer (ZIFT), also called tubal embryo transfer, is similar to IVF. Fertilization occurs in the laboratory. Then the very young embryo is transferred to the fallopian tube instead of the uterus. Gamete intrafallopian transfer (GIFT) involves transferring eggs and sperm into the woman's fallopian tube. Fertilization occurs in the woman's body. It is not as common a procedure as either IVF or ZIFT. Couples in which there are serious problems with the sperm or who have been unsuccessful with IVF may try intracytoplasmic sperm injection (ICSI). In ICSI, a single sperm is injected into a mature egg. Then the embryo is transferred to the uterus or fallopian tube.

Are there fertility treatments other than assisted reproductive technology?

Fertility treatments may also include artificial (or intrauterine) insemination in which sperm (from the woman's husband, partner, or a donor) are injected into the woman's uterus, leading to conception. Women may also take medications to stimulate egg production. They may then be able to conceive without further medical intervention.

What are some other prenatal diagnostic techniques used during pregnancy?

Ultrasonography, chorionic villi sampling, and alpha-fetoprotein screening are other prenatal diagnostic screening techniques besides amniocentesis. Fetal ultrasound is often done early in a pregnancy to determine whether it is an ectopic pregnancy. Fetal ultrasound is an accurate way to determine fetal age and predict a due date. Placental abnormalities, fetal growth and development (including heart rate), and congenital abnormalities may be detected with fetal ultrasound. It is a noninvasive test and relatively safe for the mother and fetus.

Chorionic villi sampling is another technique used to detect birth defects, such as Down syndrome or Tay-Sachs disease. It is usually done early in a pregnancy, between the ninth and fourteenth weeks. A sample of cells, called the chori-

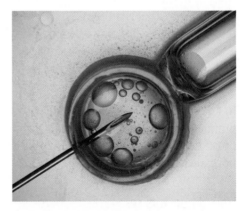

With in vitro fertilization a woman's eggs are removed, fertilized in a petri dish, allowed to mature for three to five days, and then re-implanted in the uterus.

333

onic villi, is taken from the placenta where it attaches to the wall of the uterus. The chorionic villi are tiny projections from the placenta that have the same genetic material as the fetus. The tissue sample is taken either through the cervix or through the abdominal wall.

The alpha-fetoprotein (AFP) test is a screening test to determine whether a woman is at risk for carrying a fetus with birth defects. AFP is produced by the fetus and appears in the mother's blood during a pregnancy. Abnormally high amounts of this protein may indicate a problem with the fetus. It is usually done between weeks sixteen to eighteen in the pregnancy.

When did the Pap test become accepted as a diagnostic tool to detect cancer?

During the 1920s American physician George Papanicolaou (1883–1962) did research that showed a microscopic smear of vaginal fluid could detect the presence of cancer cells in the uterus. These findings were not generally accepted at the time by the medical community. Several years later, in 1943, he published *Diagnosis of Uterine Cancer by the Vaginal Smear* with Herbert F. Traut (1894–1963), a clinical gynecologist. This time, following publication of his findings, the medical community began to use the Pap smear as a diagnostic tool for cancer. The Pap smear is more than 90 percent reliable in detecting cancer, decreasing dramatically the mortality rate for cancer of the uterus and cervix.

What tests are available to screen for possible prostate cancer?

The two tests often mentioned by doctors to screen for possible prostate cancer are the prostate-specific antigen (PSA) blood test and digital rectal exam (DRE). Prostate-specific antigen (PSA) is a protein produced by the cells of the prostate gland. The PSA test measures the level of PSA in the blood. The use of this test is currently debated by the medical community for several reasons. For example, there are many false positives, and thus unnecessary prostate biopsies. Often times a high PSA could be indicative of an infection or benign prostate enlargement, not cancer. In fact, there is no PSA level known that truly indicates prostate cancer. But to date, most doctors agree that the PSA test, although not perfect, is still one way to possibly detect early signs of prostate cancer, especially for certain age groups and for those who have a family history of the disease. (If a person is concerned about his or her PSA, it is best to consult his or her physician for information.)

A DRE involves a physician inserting a gloved, lubricated finger into the rectum to feel the prostate for any irregularities or firm areas. The DRE is sometimes recommended in conjunction with an abnormal PSA (or if a man is a candidate for prostate cancer—for example, there is a family history of the disease) to detect abnormalities in the prostate.

How do doctors test for osteoporosis?

There are several ways that doctors watch for osteoporosis in patients. One of the simplest ways is by height measurements over time. In most cases, as a person ages,

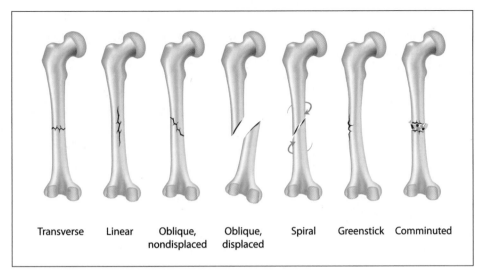

| Transverse | Linear | Oblique, nondisplaced | Oblique, displaced | Spiral | Greenstick | Comminuted |

Types of bone fractures

and if he or she has osteoporosis, he or she usually loses height of a half inch or more over a few years. And although many times osteoporosis is found by accident after an X-ray is taken (usually after a fall or an illness), X-rays cannot give the entire picture of osteoporosis. Thus if a doctor suspects osteoporosis, he or she will often suggest an ultrasound, usually of a person's heel, to detect early stages of osteoporosis. For a greater understanding of a patient's fracture risk, doctors usually suggest the dual X-ray absorptiometry, or DXA or DEXA test. Most DEXA tests measure the spine and hip, while others cover the total body bone density. Overall, the test offers the statistical chances a person has of fracturing his or her spine or hip over a certain time.

OPERATIONS, PROCEDURES, AND TRANSPLANTS

Why is it necessary to keep operating rooms clean?

In the mid-1800s, many patients would undergo operations successfully, only to die later from a postoperative infection (at that time, called "ward fever"). This led to the discovery that medical instruments and other apparatus should be disinfected and/or sterilized to ward off microorganisms that cause infection.

What are some common heart procedures?

There are several common procedures to help a patient's heart perform more efficiently. For example, the angioplasty (also called percutaneous coronary intervention, balloon angioplasty, or coronary artery balloon dilation) includes a special tube with an attached deflated balloon. The tube is threaded into the coronary arteries and the balloon inflated to expand the areas blocked with plaque. This causes an increase in

335

blood flow through the artery, reducing the risk of a heart attack. Laser angioplasty is also used to increase the flow of blood in the arteries; the tube has a laser tip that opens the blockage; while an atherectomy uses a tube with a rotating shaver at the end that cuts away plaque from the artery. (For more about the heart, see the chapter "Cardiovascular System.")

What is the most common heart surgery?

According to the National Heart, Lung, and Blood Institute of the National Institutes of Health, coronary artery bypass grafting (CABG) is the most common type of heart surgery. This operation is most often used to treat patients with severe coronary heart disease, or when plaque (a waxy substance) builds up inside the coronary arteries. The buildup narrows the coronary arteries, reducing the flow of blood to the heart, and can cause chest pain or angina. If the plaque ruptures, it can form a blood clot; if large enough, it can completely block the flow of blood through a coronary artery, which most often causes a heart attack.

How are kidney stones dissolved?

There are several procedures to dissolve kidney stones. Most smaller stones usually pass through a person's urinary tract on their own (around 85 percent are less than 5 millimeters), but larger stones usually need certain treatments. The following lists three major procedures:

Ureteroscopy—An ureteroscopy is usually used for stones in the middle and lower ureter. After a small incision is made, a small fiber-optic instrument called an ureteroscope is passed through the urethra and bladder into the ureter. A laser is then often used to break up the stones, which are grabbed or sucked out by a special tool.

Percutaneous nephrolithotomy (PCNL)—A PCNL is often used for larger stones in the upper tract. It also includes a small incision; from there, an instrument called a nephroscope is inserted in a channel directly into the kidney where it locates and removes the stones (if necessary for removal, they are broken into small pieces using ultrasound, laser, or other techniques).

Extracorporeal shockwave lithotripsy (SWL)—A SWL is used for smaller stones that are in the upper part of the ureter and don't pass on their own. SWL uses sound waves (ultrasound) to break up simple stones. The ultrasound generates shock waves that travel through the skin and tissues until it hits the denser kidney stones, crushing them into small, sand-like pieces that can pass easily through the urinary tract.

What is the difference between a tracheotomy and a tracheostomy?

A tracheotomy is the surgical opening of the trachea, or windpipe. This may be necessary if the trachea becomes occluded through inflammation, excessive secretion, trauma, or aspiration of a foreign object. This procedure may be performed to create an emergency opening into the trachea so that ventilation can still occur. A tracheostomy involves the insertion of a tube into the trachea to permit breathing and to keep the passageway open.

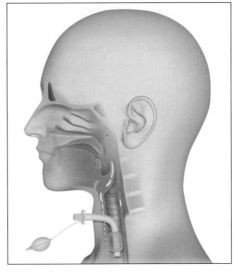

A tracheostomy is performed by inserting a tube into the trachea.

Why is a hysterectomy recommended?

Hysterectomy is the surgical removal of the uterus. The most frequent reasons for a woman undergoing a hysterectomy are uterine fibroids, endometriosis, and uterine prolapse. Cancers of the pelvic organs account for only about 10 percent of all hysterectomies.

How does a complete hysterectomy differ from a partial or radical hysterectomy?

The most common type of hysterectomy is a complete or total hysterectomy. Both the cervix and uterus are removed in this procedure. A partial or subtotal hysterectomy removes only the upper part of the uterus and leaves the cervix in place. The most extensive hysterectomy is a radical hysterectomy, which removes the uterus, the cervix, the upper part of the vagina, and supporting tissues. A radical hysterectomy is usually only performed in some cases of cancer.

Which human organ was the first to be transplanted?

The first human organ to be successfully transplanted was the kidney. American plastic surgeon Joseph Murray (1919–2012) performed the transplant in 1954 in Boston, Massachusetts. The patient, Richard Herrick (1931–1963), lived for eight years after receiving the new kidney from his identical twin brother, Ronald Herrick (1931–2010).

How successful are kidney transplants and what are the risks?

The one-year success rate for kidney transplants is 85 to 95 percent. As with any transplant, rejection of the foreign body is the major cause of transplant failure. Recipients of kidney transplants have to take immunosuppressants for the rest of their lives.

Why are cartilage transplants successful?

Cartilage does not contain blood vessels. Oxygen, nutrients, and cellular wastes diffuse through the selectively permeable matrix. Cartilage transplants are successful because foreign proteins in the transplanted cells do not have a way to enter the host body's circulation and cause an immune response. However, since there are no blood vessels in cartilage, the healing process is slower than for other tissues.

What is an artificial joint?

Artificial joints are joints designed by engineers to replace diseased or injured joints. Most artificial joints consist of a steel component and a plastic component. For example, an artificial knee joint has three components: the femoral component (made of a highly polished strong metal), the tibial component (made of a durable plastic often held in a metal tray), and the patellar component (also plastic). Artificial joints may be used to replace finger joints, hip joints, or knee joints.

Is it possible to keep organs alive outside the human body?

Yes, it is possible to keep human organs alive outside the body, but only for short periods of time and under certain conditions. Depending on the organ, most can stay alive outside the human body between five to twenty-four hours before they begin to deteriorate.

Why can't organs be frozen before transplanting?

There is a major reason why organs cannot be frozen and thus preserved for transplant: ice crystals. The sharp crystals can rip though the cell walls, destroying the cells and rendering the organ useless.

As of 2015, how many people are waiting for organ transplants in the United States?

It is estimated that 120,000 people are waiting for certain types of organs for transplant. The most needed organ is the kidney, and it is also the most commonly transplanted.

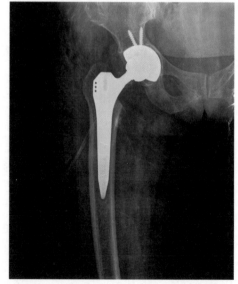

An X-ray showing an artificial hip. Surgeons these days are trying to do less-invasive surgeries so that a full replacement such as this is not necessary.

Can lungs be transplanted?

Lung transplantation is surgery used to replace one or both diseased lungs with a healthy lung or lungs. It is only recommended if the patient has end-stage pulmonary disease that cannot be treated any other way. Examples of end-stage pulmonary diseases include emphysema, cystic fibrosis, sarcoidosis, and pulmonary fibrosis. Survival rates for lung transplants are as high as 80 percent at one year and 60 percent at four years after surgery.

What part of the eye was the first to be successfully transplanted?

The cornea was the first tissue from the eye that was successfully transplanted. On December 7, 1905, Eduard Konrad Zirm (1863–1944), who was head of medicine at Olomouc Hospital in Moravia (now part of the Czech Republic), performed the first corneal transplant. It was the first successful human-to-human tissue transplant. Interestingly, the cornea is the only tissue in the body that can be transplanted from one person to another with little or no possibility of rejection. (For more about vision, see the chapter "Sensory System.")

Who performed the first heart transplant?

The world's first heart transplant was performed in South Africa in 1967 by cardiac surgeon Christiaan N. Barnard (1922–2001). The first heart transplant in the United States was performed in 1968 by American surgeon Norman Shumway (1923–2006) at Stanford University.

When was the first artificial heart implanted?

The first artificial heart, the Jarvik 7, was made by heart surgeon Robert K. Jarvik (1946–) in 1981. It was implanted in 1982 into Barney Clark (1921–1983), who lived for 112 days after the surgery.

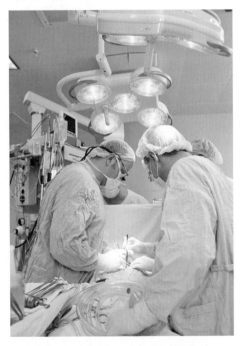

Surgeons perform a heart transplant, something that has become a fairly routine procedure. About 3,500 such transplants are peformed in the United States each year.

As of 2015, how many heart transplants are performed annually?

Worldwide, about 3,500 heart transplants are performed annually. The vast majority of these are performed in the United States (2,000–2,300 annually). Cedars-Sinai Medical Center in Los Angeles, California, currently is the largest heart transplant center in the world, having performed 119 adult transplants in 2013 alone.

COMPARING OTHER ORGANISMS

Who was one of the first to keep animal organs alive outside the body?

In 1856, German physician and physiologist Karl Friedrich Wilhelm Ludwig (also seen as Carl; 1816–1895) was the first person to keep animal organs alive outside the body. He did this by pumping blood or a saline solution through the organs—a process called perfusion. He was a professor of physiology and comparative anatomy, and he made important contributions and discoveries in cardiac activity, respiration, blood, and blood circulation.

Can humans use organs from other animals for transplant?

Contrary to most science fiction stories, to date, humans cannot use organs from other animals for transplant (called xenotransplantation). The major reasons are that humans are genetically a different species and their immune system would try to rid the body of the "foreign" organ. (The human body often fights human-to-human organ transplants, too, depending on the compatibility of the organ and donor.) But researchers are still trying xenotransplantation, citing advances in genetic modification and new immunosuppressive drugs that may help halt, or at least slow down, organ rejection in the human receiving such a transplant.

Further Reading

The following lists only a *very few* of our favorite books and sites; there are just too many to mention here. For more books, search an online bookstore or visit your local library or bookstore; for more websites, use your favorite search engine.

BOOKS

Barnes-Svarney, Patricia, and Thomas E. Svarney. *The Handy Biology Answer Book*. Visible Ink Press, 2015.

Carter, Rita. *The Human Brain Book*. DK Publishing, 2014.

Dimon, Theodore Jr. *Anatomy of the Moving Body, Second Edition: A Basic Course in Bones, Muscles, and Joints*. North Atlantic Books, 2008.

Gray, Henry. *Gray's Anatomy*. Running Press, 1974.

Hine, Robert, editor. *Oxford Dictionary of Biology,* 6th edition. Oxford University Press, 2008.

Manocchia, Pat. *Anatomy of Exercise: A Trainer's Inside Guide to Your Workout*. Firefly Books, 2009.

Rifkin, B.A,. and M. J. Ackerman. *Human Anatomy: A Visual History from the Renaissance to the Digital Age Paperback*. Henry N. Abrams, 2011

Roach, Mary. *Gulp: Adventures on the Alimentary Canal*. W. W. Norton and Company, 2013.

Roberts, A. *Human Anatomy*. DK Publishing, 2014.

Saunders, J.B. de C. M., and C. D. O'Malley. *The Illustrations from the Works of Andreas Vesalius of Brussels*. Dover Publications, 1973.

Winslow, Valerie. *Classic Human Anatomy*. Watson-Guptill, 2008.

WEBSITES

American Association of Anatomists (http://www.anatomy.org/). The professional site for physicians and other professionals in the medical community.

The American Association of Clinical Anatomists (http://clinical-anatomy.org/). A site for professionals that "advances the science and art of clinical anatomy."

American Association of Veterinary Anatomists (http://www.vetanatomists.org/). For those interested in the anatomy of animals.

American Institute of Alternative Medicine (http://www.aiam.edu/). This site also offers information and resources about anatomy.

Anatomical Society (http://www.anatsoc.org.uk/). This site is for the Anatomical Society of the UK, which was founded in 1887.

BioInteractive (http://www.hhmi.org/biointeractive/). This interactive biology site includes anatomy and is sponsored by the Howard Hughes Medical Institute.

Centers for Disease Control and Prevention (http://www.cdc.gov/). The CDC is a government organization that has a good deal of information about human biology, including anatomy.

The Franklin Institute (http://www.fi.edu/learn/heart/index.html). Offers information all about the human heart.

Harvard Health Publications (http://www.health.harvard.edu/). Offers many topics in anatomy.

Human Anatomy and Physiology Society (http://www.hapsweb.org/). The website for HAPS, which promotes improving education in anatomy and physiology at the college level.

Human Connectome Project (http://www.humanconnectomeproject.org/). The project is funded by the National Institute of Health and features amazing details about the human brain, including the first 3-D digital images of the brain.

International Federation of Associations of Anatomists (http://www.ifaa.net/). An international society involved in anatomy that has been around for over one hundred years.

International Science Grid (http://www.isgtw.org/). Offers coverage of the sciences around the world, including anatomy.

Internet Pathology Laboratory for Medical Education (http://library.med.utah.edu/WebPath/webpath.html). This Mercer University School of Medicine site has some great illustrations of biological and anatomy items, such as cells and actual photos of livers, etc.

The Khan Academy (https://www.khanacademy.org/science/health-and-medicine/human-anatomy-and-physiology). A great way to learn about anatomy, with videos and text—for all ages.

The Monell Center (http://www.monell.org/). This site stresses human taste and smell issues.

National Institutes of Health (http://www.nih.gov). The National Institutes of Health has many connections to anatomy topics. For example, try their Medline Plus at http://www.nlm.nih.gov/medlineplus/anatomy.html.

PhysOrg (http://phys.org/). This web-based science, research, and technology news service has some up-to-date information about human (and other organisms) anatomy.

Science Daily (http://www.sciencedaily.com). For current information about anatomy and physiology in the news.

Science Recorder News (http://www.sciencerecorder.com/). Includes up-to-date coverage of human anatomy.

Scientific American (http:/www.scientificamerican.com). The online magazine often has topics about anatomy—human and otherwise.

The Smithsonian Institute (http://www.si.edu). The Smithsonian is well-known for its science expertise, including anatomic research—both modern and ancient.

The University of Arizona's Biology Project (http://www.biology.arizona.edu). This site includes updated resources about biology and anatomy.

Glossary

Absorption—The passage of substances such as nutrients across membranes into cells or body fluids.

Acid—A chemical that releases a hydrogen ion in solution with water, as opposed to a base.

Adenine—One of the nitrogenous bases found in DNA and RNA molecules.

Adenosine Triphosphate (ATP)—The organic compound that stores respiratory energy for transport from one part of a cell to another; it is in the form of a chemical bond.

Allele—A pair of genes that exist at the same location on a pair of homologous chromosomes; they exert parallel control over the same genetic trait.

Allergy—A reaction to an antigen (also called hypersensitivity); the reaction is usually to a substance that is normally harmless to most other people.

Amino acid—An organic compound that is the major component unit of proteins.

Antibody—A chemical substance that an organism's body produces in the presence of a specific antigen; they are produced by B lymphocytes.

Antigen—Any chemical substance (mostly proteins) that is recognized by the immune system as an "invader" in the body of an organism; in the majority of cases, it is neutralized by a specific antibody.

Appendicular—Pertaining to the upper or lower limbs.

Atrophy—A wasting away of decrease in the size of tissues, such as muscle fibers, most often from the lack of use.

Axial—Pertaining to the head, neck, and trunk of the body.

Bacteria—Single-celled organisms that are common in the environment and on and in the body; some of them are considered to be pathogens.

Base—A chemical that releases a hydroxyl ion in solution with water, as opposed to acid.

Benign—Not malignant or otherwise a threat to life.

Blood—A loose connective tissue whose matrix is plasma and that contains red blood cells, white blood cells, and platelets. Blood circulates everywhere, distributing nutrients and oxygen, and carrying away waste products throughout the body.

Bone—A rigid connective tissue that has a matrix of collagen fibers embedded in calcium salts. It is the hardest tissue in the body.

Capillary—The smallest of the blood vessels that connects an artery to a vein and through which all absorption into the blood occurs.

Carbohydrate—An organic compound made of carbon, hydrogen, and oxygen in a 1 to 2 to 1 ratio.

Carcinogen—A cancer-causing agent.

Cartilage—A connective tissue found in many parts of the body; it is flexible and is most common in the embryonic stages of development.

Catabolism—The metabolic process that breaks down complex organic molecules into simpler components; it is accompanied by the release of energy.

Cell—The smallest structural and functional unit of life.

Cell membrane—Cell membranes define and compartmentalize space, regulate the flow of materials, detect external signals, and mediate interactions between cells.

Chemoreceptor—A sensory receptor, such as taste buds, that responds to chemical stimuli.

Clot—A structure that essentially plugs a ruptured blood vessel; it is usually the result of enzyme-controlled reactions from a cut or wound.

Coenzyme—Most of the time, a chemical substance that aides the action of a particular enzyme.

Connective tissue—One of the basic types of tissue consisting of fibers and widely-spaced cells in an extracellular material called the matrix. It includes bone, blood, cartilage, collagen, adipose, and loose collective tissue.

Cutaneous—Pertaining to the skin.

Cyton—The "cell body" of a neuron that generates the nerve impulses in the body.

Cytosine—One of the nitrogenous bases found in both DNA and RNA molecules.

Diapedesis—Diapedesis is the ability of white blood cells to squeeze between cells that form blood vessel walls.

Distal—Away from the trunk or midline of the body.

Dorsal—Toward the back, and also called the posterior.

Ecorche—A flayed figure, or a three-dimensional representation of the human body, usually made of plaster, with the envelope of skin and fat removed.

Embryo—The earliest stage of a human's (and other organisms) development following fertilization.

Embryogenesis—The process during which the embryo begins to separate from the embryonic disc.

Enzyme—An organic catalyst that controls the metabolism rate of certain functions; they are usually proteins.

Epidermis—The outermost layer of cells of an organism—usually in reference to plants and animals.

Epithelial tissue—A type of tissue consisting of groups of cells that form a superficial covering or internal lining of a vessel or cavity.

Essential amino acid—An amino acid that cannot be synthesized by the human body, so it must be ingested.

Fetus—The prenatal stage of development; in humans, it begins at the ninth week of development and ends at birth.

Gamete—A sex cell (ovum or egg cell in a female and sperm cell in a male) that contains half the normal number of chromosomes.

Ganglia—Groups of neuron cell bodies outside the central nervous system.

Gap junctions—Gap junctions are connecting channels made of proteins that permit the movement of ions or water between two adjacent cells.

Gene—A unit of heredity; it is also a portion of a chromosome that is responsible for the inheritance of a genetic trait.

Gland—Groups of specialized cells that produce secretions.

Glycerol—An organic compound that is a component of certain lipids.

Guanine—One of the nitrogenous bases found in DNA and RNA molecules.

Hepatitis—Hepatitis is the inflammation of the liver, usually caused by a viral infection. There are five known types of hepatitis.

Homeostasis—The state of inner balance and stability maintained by the human body despite constant changes in the external environment.

Hormone—Chemicals that are the product of an endocrine gland; they are usually responsible for regulating cell metabolism.

Immunity—Resistance to foreign substances as a protective mechanism.

Infection—The invasion and establishment of pathogens in body tissues.

Interferon—A protein produced by a virus-infected cell that halts the multiplication of the virus; it is said to be important in the fight against human cancer.

Joint—A joint is the place where two adjacent bones and/or cartilages meet. Most allow movement and are very flexible.

Keratin—The tough fibrous protein portion of the epidermis, hair, and nails.

Lateral—Pertaining to the side.

Lipid—An organic compound that is made of carbon, hydrogen, and oxygen, but the hydrogen and oxygen are not in a 2-to-1 ratio; most of them are a glycerol and three fatty acids.

Lymph—The intercellular fluid that passes through the lymph vessels (the series of branching tubes that collect the intercellular fluid from the tissues and redistributes it as lymph).

Lysosomes—Organelles that are only found in animal cells; they are membrane-bound sacs containing digestive enzymes that break down macromolecules such as proteins, carbohydrates, fats, and nucleic acids.

Mammary glands—These are modified sweat glands that are located within a female's breasts and produce milk for newborns.

Medial—Toward the midline of the body.

Metabolism—The total chemical and physical processes occurring in the body at a given time; they may be anabolic or catabolic.

Monosaccharide—A type of carbohydrate that is also called a "simple sugar"; they all have the molecular formula of $C_6H_{22}O_6$.

Mucus—The somewhat-slimy, protein-rich mixture that is found coating and moistening the respiratory tract surfaces, such as the inside of the human nose.

Muscle—An organ composed of muscle cells and fibers, blood vessels, nerves, and connective tissue that contracts and relaxes to move body parts.

Nephron—A nephron is the functional working unit of the kidney in which blood is filtered through and toxic wastes removed.

Nerve—A bundle of sensory fibers held together by layers of connective tissue.

Neuron—The cells that specifically carry the nerve impulses throughout an organism's body.

Nutrients—A chemical compound that can be broken down to supply the body with energy and nourishment.

Nucleic acid—The organic compound that is made of repeating units of a nucleotide.

Nucleolus—The organelle within a cell's nucleus that is responsible for protein synthesis.

Nucleotide—The repeating unit that makes up the nucleic acid polymer.

Nucleus—The organelle in a cell that most importantly contains the cell's genetic information in the form of chromosomes.

Organelle—The small structures within a cell—each with a special function, such as the mitochondria and nucleus.

Osmosis—The action of water being absorbed or released by a cell.

Oxyhemoglobin—A form of hemoglobin that is loosely bound to hemoglobin in the blood; its primary task is to transport oxygen throughout the body.

Pain receptors—Pain receptors, also known as nociceptors, are especially common in the superficial portions of the skin, in joint capsules, within the membrane that covers the outer surface of the bone whenever it is not covered by cartilage, and around the walls of blood vessels.

Patella—Another name for the kneecap.

Pathogen—A disease-causing agent.

Phagocyte—A type of white blood cell in the body that engulfs and destroys invading, harmful bacteria.

Plasma—The liquid portion of the blood.

Platelet—Fragments of cytoplasm manufactured in the bone marrow that plays an important role in helping blood clot (also called thrombocyte).

Posterior—Toward the back.

Pulmonary—Pertaining to the lungs.

Renal—Pertaining to or involving the kidneys.

Ribose—A five-carbon sugar that is part of the nucleotides of RNA molecules only.

Ribosome—A specific organelle in an organism's cell that is the site of protein synthesis.

Sphincter—A circular muscle that contracts to close the opening of a tubular structure.

Sternum—The sternum is a flat, narrow bone located in the center of the anterior chest wall.

Steroid—An organic molecule that includes four connected rings of carbon atoms, such as cholesterol.

Subcutaneous—Beneath the skin.

Synapse—The site of intercellular communication where a nerve impulse passes from one neuron to another.

Tendon—A form of connective tissue that attaches the skeletal muscles to the bone.

Tendonitis—In humans, it is where the junction between the tendon and bone becomes irritated—and even inflamed—usually from repeated actions, such as tendonitis of the elbow from over-practicing a tennis swing.

Thymine—One of the nitrogenous bases found only in DNA.

Tissue—A group of similar, specialized cells that perform a specific function.

Tumor—A mass of tissue formed by the abnormal growth and rapid replication of cells; they can be benign or malignant.

Ulcer—A break in the skin or mucous membrane with the loss of the surface tissue; it is also often called a lesion.

Uracil—The nitrogenous base that is part of the nucleotides of RNA molecules only.

Vaccination—In order to stimulate a person's immune system, an inoculation of a dead or weakened organism associated with the disease is given; this allows the body to build up antibodies to help fight off the disease if a person is exposed again.

Vasodilation—The enlargement of a body's blood vessels that increases the blood supply.

Vein—A blood vessel that carries blood toward the heart.

Ventral—Toward the front of the body; also called anterior.

Viroids—The small fragments of nucleic acid (RNA) without a protein coat; they are usually associated with plant diseases and are several thousand times smaller than a virus.

Virus—An infection-causing, protein-coated fragment of DNA or RNA genetic material, viruses are incapable of living on their own, only reproducing within their host cell.

Zygote—When gametes are fused in sexual reproduction, the result is a single diploid cell called a zygote; it is also the term synonymous with a fertilized egg.

Index

Note: (ill.) indicates photos and illustrations.

Numbers

20/20 vision, 62

A

ABCD mole rule, 75
abdomen, regions of the, 10
abducens nerves, 147
ABO blood typing, 196–98
abscesses, 38
absorption in the stomach, 254
abuse, steroid, 184
Académie Française, 6
accessory ducts, 286, 288–89
accessory glands, 82–84, 286, 289–290
accessory nerves, 147
accessory organs, 248, 260–65, 281–82
acetylcholine, 122–23, 134–35
acquired immunodeficiency syndrome (AIDS), 214, 222–23
action potential, 131–32
active immunity, 220
activity length of hormones, 166
Adam's apple, 235
adaptive immunities, 229
Addison's disease, 178
adipose tissue, 34–36, 35 (ill.), 186
adolescence, 318, 321
adrenal cortex, 166, 176
adrenal glands, 166, 169, 176–78, 177 (ill.), 277
adrenal medulla, 178
adrenaline, 178

adrenocorticotropic hormone (ACTH), 169
Advanced Fingerprint Identification Technology (AFIT), 71
aerobic vs. strength training, 110–11
age, effect of
    breathing, 242
    endocrine system, 167
    hearing, 53–54
    kidneys, 279
    muscles, 126
    organ systems, 322
    pulse rate, 205
    smell, 47
    taste, 50
    white blood cells, 194
age of puberty, average, 320
age of sperm production, 288
age spots, 70
AIDS (acquired immunodeficiency syndrome), 214, 222–23
air, amount needed, 242
air, breathing cold, 234
air, cleaning before reaching lungs, 241
air movements, non-respiratory, 232
air volume in lungs, 242
albinism, 69–70, 84, 84 (ill.)
albumin, 195
alcohol, effect of on liver, 262
alcohol, effect of on urination, 170
Alderman Library, 5
allergies, 226–230, 227 (ill.)
all-or-none response in muscle cells, 125

alopecia, 82
ALS (amyotrophic lateral sclerosis), 130–31
altitudes, breathing at high, 243
Alzheimer's Association, 144
Alzheimer's disease, 144–45
Amazon, 272
amenorrhea, primary vs. secondary, 296
American Cancer Society, 290, 332
American Heart Association, 190, 202, 207, 330
American Red Cross, 197
amine hormones, 164–65
amino acids, 134
amnesia, 154
amniocentesis, 308
amniotic cavity, 308, 316
amount of blood in humans, 195
amount of blood in menstuation, 296
amount of blood pumped by the heart, 200
amount of lymph in the body, 213
amount of semen in ejaculation, 291
amount of sleep, ideal, 156–57
amount of time food stays in large intestine, 258
amount of time food stays in stomach, 254
amount of time from eating to excreting, 247–48
amount of time spent sleeping, 157
amount of time zygote travels to uterine cavity, 304

amphiarthrosis, 103
amyotrophic lateral sclerosis (ALS), 131
anabolic steroids, 183–84, 184 (ill.)
anabolic vs. catabolic reactions, 265
anaphylactic shock, 227
antagonistic muscles, 120
Anatomy: Descriptive and Surgical (Gray). See *Gray's Anatomy* (Gray)
Anatomy of the Human Gravid Uterus (Hunter), 6
ancestry of blue-eyed humans, 57
androgens, 177
anemia, 192–93
anesthetics, effect of, 135
aneurysms, 203
angina pectoris, 199
animal hormones, 186
animal organs, use of, 340
animals, anatomy of. See comparative anatomy
anorexia, 269–270, 270 (ill.)
anosmia, 48
anterior pituitary glands, 166, 168
antibiotics, 220–21
antibodies, 198, 218–220, 219 (ill.), 226
anticoagulants, 195
antidiuretic hormone (ADH), 169–171, 181, 186, 283
antigens, 198
aphasia, 152–53
apnea, sleep, 158, 158 (ill.)
apocrine glands, 83
appendicitis, 216
appendicular skeleton, 87, 98–101
Apple, 72
aquatic animals, sleep of, 161
aqueous humor, 55
areolar tissue, 34
Aretaeus the Cappadocia, 181
Aristotle, 3, 7, 43
arm, anatomical vs. common usage of the word, 99
ART (assisted reproductive technology), 332–33
arterial bleeding, 205
arteries, 202–3
arteriosclerosis, 203
arthritis, 106–7
Articulata, 7
artificial blood, 193

artificial heart, first, 339
artificial joints, 338, 338 (ill.)
ascending vs. descending tracts, 145
asleep, feet falling, 150
asparate, 134
assisted reproductive technology (ART), 332–33
astigmatism, 62
atherosclerosis, 203, 203 (ill.)
Atlas, 96
ATP, 12, 29 (ill.), 29–30, 111, 124–26, 165, 178
atria, 199
atrial natriuretic peptide (ANP), 186
atrophine, 134
attack, heart, 202
attenuated vs. inactivated vaccines, 225
auditory ossicles, 93
auditory tube, 50–51
autoimmune diseases, 213
autonomic nervous system, 130, 150–51
axial skeleton, 87, 93–98

B

B cells, 211–12, 219–220
babies having blue eyes, 64
babies having soft spots, 96
bacteria in large intestine, role of, 258
bacteria on skin, 68
Baglivi, Giorgio (Gjuro), 5
baldness, 82
ball-and-socket joints, 102 (ill.), 104–5
Banting, Frederick, 180
barium test, 329
Barnard, Christiaan N., 339
basal cell carcinomas, 74
basics, anatomy and biology, 9–20
bats, echolocation in, 65, 65 (ill.)
Bayliss, William, 165, 186
Beaumont, William, 248
behavior, effect of hormones on, 165
belching, 254–55
Bell's palsy, 147
Benacerraf, Baruj, 226
bends, the, 241
benefits to breastfeeding, 318
benefits to circumcision, medical, 292

Bernard, Claude, 6, 40
Best, Charles, 180
bile, 262
bile passage, 261 (ill.)
Binet, Alfred, 153
binge eating, 270
biological compounds, 15–18
biology, basics of, 9–20
biology, chemistry in, 12–15
biology, definition of, 2
birds, effect of hormones on, 186
birds as hunters, 64
birth, animals giving, 302
birth and lactation, 313–18
birth control, 300 (ill.), 300–301
births, multiple, 317
births, premature, 313
bite, animals with the strongest, 127
bite, force of a human, 249
black widow spider venom, 134
blackheads vs. whiteheads, 82
bladder, urinary, 281, 281 (ill.), 291 (ill.)
blastocysts, 304
bleeding, arterial vs. venous, 205
blind spot of the eye, 60–61
blindness, 58–60, 63
blinking rate of humans, 61
blisters, 73
blood
    all about, 190–98
    amount pumped by the heart, 200
    animal, 208
    autoimmune diseases, 213
    blue color in veins, 204
    cardiac muscle supplied with, 200
    cell types, 30
    cells formed in the skeletal system, 90
    circulation of in fetus, 310–11
    color, 208–9
    connection between kidneys and production of red, 279
    connective tissue type, 34
    discovery of circulation of, 206
    fetal vs. adult, 311
    filtering by kidneys, 279

flow in the heart, 201, 201 (ill.)
flow of giraffes, 208
flow to heart or brain, 327
introducing oxygen to, 240
location of in the body, 204
loss during labor and delivery, 315
menstrual, 296
pressure, 207 (ill.), 207–8
speed of flow, 206
veins' return of, 206
vessels, 73, 202–6, 213
water content of human, 13
bloodhounds' sense of smell, 63
blue color of blood in veins, 204
blue eyes, babies having, 64
blueberries' relationship to vision, 59
blue-eyed humans, ancestry of, 64
BMI (body mass index), 271 (ill.), 271–72
body, stomach, 252
body, uterine, 294
body changes during puberty, 320–21, 321 (ill.)
body mass index (BMI), 271 (ill.), 271–72
body planes, 10
body regions, 10–11
body temperature, regulation of, 77, 83–84
bones
    anatomy of, 91 (ill.)
    animals without internal, 108
    basics, 89–93
    broadest, 101
    cell lifespan, 30
    connective tissue type, 35
    dinosaur, 7–8, 18
    distribution of the vertebral column, 95
    effect of exercise on tissue, 92
    flat, 89–90
    fractures of, 98, 335 (ill.)
    funny, 99–100
    growth of, 91, 277
    hand and wrist, 100 (ill.)
    in the appendicular skeleton, 98
    independent, 94
    irregular, 89–90

largest, 107
long, 89–90
longest, 90
movement of, 94
number in the axial skeleton, 93
number of in animals, 107
number of in humans, 89
number of vs. size of structures, 100
ossification of, 91–92
pubic, 104
sesamoid, 90
short, 89–90
skull, 94
smallest, 90
specialized cells, 91
strength of, 88
study of, 87
tissue types, 91
types of, 89
vomer, 94
water content of human, 13
Borsch, J. L., 63
Botox, 123, 123 (ill.)
botulinum toxin, 123, 134
Bovine Growth Hormone (rBGH), 186–87
Bovine Somatotrophin (rBST), 186
brain
    all about the, 139–145
    anatomy of the, 140 (ill.)
    animal vs. human, 159 (ill.), 159–160
    areas involved in memory, 154
    areas responsible for speech and language, 152
    autoimmune diseases, 213
    blood flow to, 327
    cell lifespan, 30
    control of respiratory rate, 241
    detection of smell by the, 47
    giraffes' blood flow to, 208
    learning and memory, 151
    lobes, 142 (ill.)
    protection of the, 136
    water content of human, 13
brainstem, 141
breast cancer, 297
breast milk, 317–18
breastfeeding, 314 (ill.), 318
breasts, 297, 297 (ill.), 329

breath, holding one's, 128, 243
breathing and respiration, 239–243
breathing cold air, 234
breathing of animals, 245
British Air Ministry, 59
broadest bones, 101
Broca, Pierre Paul, 152
Broca's area, 152, 152 (ill.)
Brodmann, Korbinian, 151
bronchoscopy, 331
Brookes, Richard, 8
Brown, Robert, 22, 23 (ill.)
bulimia, 270
bumps on nails, 78
burns, skin's damage by, 74, 74 (ill.)
burping, 254–55

C

calcaneal tendons, 120
calcitonin, 175
calcitrol, 186
calcium, importance of, 88
calcium, increase in, 175
calcium, muscle cells' use of, 123
calculation of BMI, 272
calculation of IQ, 153
California Academy of Sciences, 84
calories, 266
cancer
    breast, 297, 329
    colorectal, 332
    lymphatic system, 215–16
    metastatic, 215–16
    Pap smear used to detect, 334
    PET scans' use related to, 326
    prostate, 290–91, 334
    thyroid, 174
    types in tissue, 36
    types of skin, 74–75, 75 (ill.)
canines, 248–49
Cannon, Walter Bradford, 40, 40 (ill.)
capacity of lungs, 242
capillaries, 204, 214
carbohydrates, 15–16, 265–66
carbon dioxide, amount in blood, 196
carbon monoxide, breathing, 240

carcinomas, 36, 74–75
cardia, 252
cardiac muscle, 37, 116–17, 117 (ill.), 198–200
cardiovascular system, 39, 189–209, 307, 322
Carl Zeiss Company, 63
carnivores, 8
carpal bones, 99, 100 (ill.)
carpal tunnel syndrome, 149–150
carrots' relationship to vision, 59
cartilage, 35–36
cartilage transplants, 338
cartilaginous joints, 103
CAT (CT) scans, 325–27, 326 (ill.)
catabolic vs. anabolic reactions, 265
cataracts, 63, 63 (ill.)
catecholamines, 134
cats, immunodeficiency virus for, 229
cats, kidney failure in, 284
cats' skin color's effect on fur, 84
cats' vision in darkness, 64
causes of aphasia, 152–53
causes of chronic renal disease, 280
causes of color blindness, 60
causes of deafness, 53
causes of diarrhea, 259–260
causes of double vision, 62
causes of epilepsy, 135
causes of feces odor, 259
causes of hyperthyroidism, 173
causes of infertility, 298–99
causes of nausea and vomiting, 270
causes of obesity, 271
causes of sciatica, 148
causes of sore throats, 235
causes of warts, 69
cavities, body, 12
cavity, amniotic, 308, 316
CDC (Centers for Disease Control and Prevention). See Centers for Disease Control and Prevention (CDC)
cecum, 257
Cedars-Sinai Medical Center, 340
celiac disease, 256
cell phone elbow, 119–120, 150

cell physiology, 2
cell theory, 22–23
cells
  anatomical levels of organization, 22–31
  composition of mammalian, 23
  cone, 59
  connection between kidneys and production of red blood, 279
  definition of, 21–22
  DNA in, 26
  eukaryotic, 25
  exercise increasing muscle, 37
  formed in the skeletal system, 90
  found in nerve tissue, 37
  growth hormone, 170
  growth of epidermal, 77
  hormone receptors, 167, 167 (ill.)
  involved with the immune system, 77–78
  islets of Langerhans, 179–180
  keratinization of, 76–77
  largest, 23
  lifespan of, 30
  longest, 23
  of the lymphatic system, 211–12
  mitochondria in, 29
  muscles, 118, 120–25
  nervous system, 129–130
  nuclei of, 26
  origin of term, 22
  pancreas, 179, 260
  plasma, 220
  red and white blood, 191 (ill.), 191–94, 196
  replacement of epidermal, 70
  size of animal, 41
  smallest, 23
  specialized bone, 91
  specialized in the epidermis, 69
  sperm, 288
  stomach, 253
  synthesization of vitamin D, 76
  T vs. B, 219–220
  target, 164, 167 (ill.)
  thyroid glands, 171
  types of, 24, 30–31

cellular respiration vs. respiration, 231
Centers for Disease Control and Prevention (CDC)
  assisted reproductive technology, 332
  BMI, 272
  HIV vs. AIDS, 222
  meningitis, 137
  pneumonia, 238
  premature births, 313
  STDs, 301
  twins, 316–17
central nervous system, 129, 136–39, 137 (ill.)
cerebellum, 141
cerebral cortex, 151
cerebrospinal fluid, 139
cerebrum, 141–42
cerumen, 55
cervical bones, 95–96
cervix, 294
chambers of the heart, 199
changes during neonatal period, 318–19, 319 (ill.)
changes during pregnancy, maternal, 313–14
changes during puberty, body, 320–21, 321 (ill.)
changes in prenatal development, time frame for, 308
changing one's fingerprints, 71–72
Chardack, William, 330
Chatton, Édouard, 24
chemical composition of cells, 23
chemistry in biology and anatomy, 12–15
chemoreceptors, 43
chewing, involvement of teeth and tongue in, 249–250
chewing, muscles for, 113
chicken vs. human muscles, 126
childhood, 318
childhood vaccinations, 225
cholecystokinin, 186
cholesterol, 17
choroid, 56
chromosomes, 27
Chronic Kidney Disease (CKD), 284
chronic obstructive pulmonary disease (COPD), 237–38
chronic renal disease, 280

chyme, 252, 255
ciliary body, 56
circadian rhythms, 158–59
circulating vs. paracine hormones, 165
circulation, 189, 205–8, 310–11
circulatory assist device, 330
circulatory system
    activity length of hormones, 166
    changes of fetal after birth, 311–12
    definition of, 39
    open vs. closed, 209
    study of the, 6
    vs. cardiovascular system, 189
circumcision, 292
cirrhosis of the liver, 262
Civil War, 45
CKD (Chronic Kidney Disease), 284
Clark, Barney, 339
classification
    antibodies, 218–19, 219 (ill.)
    carbohydrates, 15–16
    exocrine glands, 33–34
    hormones, 164
    joints, 103
    structural, 7
clavicle bones, 98
cleaning air before reaching lungs, 241
cleanliness of operating rooms, 335
climate change, connection between animals and global, 274
clinical anatomy, 1
clitoris, 297
closed vs. open circulatory system, 209
Clostridium botulinum, 123, 134
clotting, blood, 194–95
coccygeal bones, 95
cold, voice differences due to having a, 243
cold air, breathing, 234
collagen, 34–35, 72
collapsed lungs, 237, 237 (ill.)
collecting ducts, large, 214
Collip, James Bertram, 180
colon, 30, 248, 257
color, blood, 208–9
color, determination of eye, 57

color, determination of hair, 80–81
color, determination of skin, 69
color blindness, 60
color of feces, varying, 259
color of urine, varying, 282
colorectal cancer, screening for, 332
colostrum, 318
common chemical sense, 47
compact bone tissue, 91
comparative anatomy
    anatomical levels of organization, 41–42
    basics of, 18–20
    cardiovascular system, 208–9
    definition of, 1
    digestive system, 272–74
    endocrine system, 186–87
    helping human anatomy, 340
    history of, 3, 5–8
    human growth and development, 322–23
    integumentary system, 84–85
    lymphatic system, 229–230
    muscular system, 126–28
    nervous system, 159–161
    reproductive system, 302
    respiratory system, 244–45
    sensory system, 63–65
    skeletal system, 107–8
    urinary system, 283–84
compatibility of organ donors and recipients, 226
complete hysterectomy, 337
complications of AIDS, 223
complications of multiple pregnancies, 316
complications of PID, 294
complications of premature birth, 313
complications of STDs, 302
components, separation of blood, 196
composition of blood, 190
composition of breast milk, 318
composition of lymph, 213
composition of mammalian cells, 23
composition of saliva, 250
composition of urine, 282
compounds, human biological, 15–18

compression, spinal, 146
conception, 298–301, 304
concussion, spinal, 146
concussions, 142–43
condyloid (ellipsoidal) joints, 102 (ill.), 104–5
cones, 58 (ill.), 58–60
confirmation of pregnancy, 299–300
conjoined twins, 317
connection of lungs to heart, 237
connective tissue, 31, 34–35, 42
consciousness, 154
constipation, 259
contact lenses, invention of, 63
contraception, 300 (ill.), 300–301
contraction, muscle, 121, 124, 126
contusion, spinal, 146
COPD (chronic obstructive pulmonary disease), 237–38
cord, umbilical, 306, 311 (ill.), 316
Cormack, Allan M., 326
cornea, 55
coronary circulation, 189
corrugator muscles, 114
cortex, adrenal, 166, 176
cortex, cerebral, 151
cortex, kidney, 276
corticosteroids, 176–77
cosmetic dehydration, 276
cough, 238
cows, stomachs of, 273
cracking knuckles, popping sound of, 105
cramps, muscle, 122
cranial nerves, 146–47
creatine, 111
Cretaceous period, 107
critical values for white blood cells, 194
Crohn's disease, 260
crying, 234
cryptorchidism, 287
Crystal Palace, 108
CT (CAT) scans, 325–27, 326 (ill.)
Cushing's disease, 178
cuspids, 248–49
cutaneous carcinomas, 74–75
cutaneous glands, 82

Cuvier, Georges Léopold Chrétien Frédéric Dagobert, 7
cycle, female reproductive, 294–96, 295 (ill.)
cycle, respiratory, 239, 239 (ill.)
cycle, sleep, 156, 158
cytology, 1

D

da Vinci, Leonardo, 4–5, 63
Dale, Sir Henry Hallett, 133
Damadian, Raymond, 327
damage, spleen, 217
damage by burns, skin, 74, 74 (ill.)
damaged tissue, repairing, 38
danger of freckles, 70
danger of talking while eating, 235
darkness, cats' vision in, 64
Dausset, Jean, 226
de Duve, Christian, 28
deafness, 53, 54 (ill.)
death, growth of hair and nails after, 78
Declaration of Independence, 8
deep vein thrombosis (DVT), 203
defecation, 259
defibrillators, 330–31, 331 (ill.)
deficiencies, vitamin, 268, 268 (ill.)
defining anatomy, 1–2
dehydration, cosmetic, 276
delayed vs. immediate allergic reactions, 227
dementia, 144
demyelinating diseases, 138
dermatoglyphics, 71–72
dermatomes, 148–49
dermis, 68, 70, 72
descending vs. ascending tracts, 145
descension of testes into scrotum, 287
destroying infections, 221
destroying one's fingerprints, 72
detection of breast cancer, 297
development, human growth and, 303–23
development of allergies, 227
developmental anatomy, 1

deviated septum, 233
diabetes insipidus, 181
diabetes mellitus, 181, 181 (ill.)
diabetic retinopathy, 58
diagnosis of HIV, 222
Diagnosis of Uterine Cancer by the Vaginal Smear (Papanicolaou and Traut), 334
diagnostic techniques, 327–335
dialysis, kidney, 280
diaphragm, 12, 239, 239 (ill.)
diarrhea, 259–260
diarthrosis, 103
diencephalon, 141
digestion, definition of, 247
digestion, role of esophagus in, 251
digestive process, steps in the, 247
digestive system, 39, 247–274, 307, 322
digestive tract, 213, 331–32
dilation stage of labor, 314–15
Dillinger, John, 71
dim light, adapting to, 59
Dinosauria, 7
dinosaurs, 7–8, 18, 107–8
Diocles, 3
diplopia, 62
direction of blood flow in the heart, 201, 201 (ill.)
direction of fibers in, muscles named for, 115–16
directional terms of the body, 9–10, 18–19
disc, embryonic, 307
diseases, autoimmune, 213
diseases, breast, 329
diseases, cardiovascular, 190
diseases, immunodeficiency, 213–14
diseases caused by vitamin deficiencies, 268, 268 (ill.)
diseases preventable by vaccination, 225–26
disinfectant, use of during surgery, 336
dislocation of joints, 105
disorders, eating, 269–270
disorders, glucocorticoid, 178
disorders, growth hormone, 170
disorders, red blood cell, 192
disorders, sleep, 157–58
dissections, 3–5, 7

distal convoluted tubule, 278
diuretics, 283
diverticula, 257–58
diverticulitis, 257–58
diverticulosis, 257–58
divisions of anatomy field, 1
divisions of human development, 303
divisions of the autonomic nervous system, 130
divisions of the brain, 141
divisions of the head and neck, 10
divisions of the peripheral nervous system, 130
divisions of the respiratory system, 231
divisions of the skeleton, 87
divisions of the small intestine, 255
divisions of the thymus, 184
DNA, 26–27, 27 (ill.)
documentation of HIV, 223
documentation of vaccination, 225
donating blood, 197
dopamine, 134
doping, blood, 193
double vision, 62
double-jointedness, 107
dowager's hump, 97
Dreadnoughtus, 107
dreams and sleep, 154–161
drug testing, 282–83
drugs, absorption of, 254
drugs, allergies to, 228
ducts, accessory, 286, 288–89
ducts, large collecting, 214
ductus deferens, 289
duodenum, 255
DVT (deep vein thrombosis), 203
dystrophin, 118

E

E. coli, 180, 187
ear wax, 55
ears, 50, 51 (ill.), 114
eating, binge, 270
eating, danger of talking while, 235
eating disorders, 269–270
eating to excreting, amount of time from, 247–48
Eccles, Sir John, 132
eccrine glands, 83–84
echocardiography, 329–330

echolocation, 65, 65 (ill.)

ectoderm, 306–7

ectopic pregnancy, 306

ED (erectile dysfunction), 292

effect of aging. See age, effect of

effect of alcohol on liver, 262

effect of alcohol on urination, 170

effect of exercise on bone tissue, 92

effect of exercise on muscle, 37, 110–11

effect of exercise on the cardiovascular system, 190

effect of exercise on the heart, 202

effect of glucocorticoids on the body, 177–78

effect of growth hormone on cells and tissues, 170

effect of hormones on behavior, 165

effect of hormones on birds, 186

effect of multiple pregnancy on mother's health, 316

effect of shaving on hair, 80

effect of skin color on cats' fur, 84

effect of smoking on lungs, 238

effect of sound levels on hearing, 52–53

effect of weightlessness on muscle, 110

effectors, 41

ejaculation, amount of semen in, 291

ejaculation and urination, impossibility of simultaneous, 290

elastic cartilage, 36

elbow, tennis vs. golfer's vs. cell phone, 119 (ill.), 119–120, 150

electrical activity of the heart, monitoring, 330

Electrodyne Company, 330

elements, important living, 13, 19

elephant trunks, importance of, 127, 127 (ill.)

Elliott, Thomas Renton, 133

ellipsoidal (condyloid) joints, 102 (ill.), 104–5

embryogenesis, 306

embryology, 3

embryonic disc, 306–7

embryonic period of prenatal development, 304–9, 305 (ill.)

embryos, 303–4, 309

emotional senses, 45

emphysema, 238

endocrine glands, 165–66

endocrine system, 39, 163–187, 164 (ill.), 307, 322

endoderm, 306–7

endometriosis, 294

endorphins, 134

energy, sources of muscle cells for, 124

energy nutrients, 265–66

enkephalins, 134

enteric nervous system, 130

enzymes, 17–18, 20, 260–61

epidermis, 68 (ill.), 68–70, 77

epididymis, 286, 289

epilepsy, 135

epilethial tissue (epithelium), 31–33, 33 (ill.), 42

epinephrine, 178

epipens, 228 (ill.), 228–29

epithalamus, 141

equilibrium, organs of, 54

Erasistratus, 3

erectile dysfunction (ED), 292

erectile tissue, 292

erythropoietin (EPO), 186, 193

esophagus, 248, 251

essential nutrients, 265

essential vitamins, 267–68

estrogen, 183

eukaryotic cells, 24–25

Eustachian tube, 50–51

events during dilation, 314–15

events during embryonic period, 308

events during second trimester of pregnancy, 310

events during third trimester of pregnancy, 312

excreting, amount of time from eating to, 247–48

excretory system, 39, 322

exercise, effects of, 37, 92, 110–11, 124–26, 190, 202

exhalation vs. inhalation, 239, 239 (ill.)

exocrine glands, 33–34

experimental medicine, 6

expiration vs. inspiration, 239, 239 (ill.)

exploration, history of internal body, 325

expulsion stage of labor, 314

external ear, 50

external vs. internal respiration, 240

extremities, regions of the, 11

eyebrows, 112

eyelashes, growth rate of, 80

eyes

    anatomy of the, 56 (ill.)

    ancestry of humans with blue, 57

    autoimmune diseases, 213

    babies having blue, 64

    blind spot of the, 60–61

    color of the, 57

    growth of the, 56

    muscles of the, 112

    parts and functions of the, 55–56

    production of tears, 61, 61 (ill.)

    skin around, 112

    transplants, 339

F

facial nerves, 147

failure, kidney, 280, 284

Fallopius, Gabriel, 293

false ribs, 98

false vs. true labor, 315

farsightedness, 62

fastest muscles, 112

fatigue, muscle, 125

fats

    brown vs. white, 36

    conversion of muscles to, 125–26

    as energy nutrient, 266

    as an essential nutrient, 265

    saturated vs. unsaturated, 269

    trans, 269

    vs. lipids, 16

    water content of human, 13

FBI, 71

FDA (Food and Drug Administration), 123

feces, 258–59

feet, 101, 150

feline immunodeficiency virus (FIV), 229

female reproductive system, 286 (ill.), 292–97

femur bones, 99
fertility treatments, 332–33
fertilization, 298, 299 (ill.), 304
fetal stage of prenatal development, 304, 309–13
fetus vs. zygote vs. embryo, 303–4
fever, role of in infection, 218
fibers, muscle, 115–16
fibrinogens, 195
fibrocartilage, 36
fibrosis, 126
fibrous joints, 103–4
fibula bones, 99
field of anatomy, division of, 1
field of vision, humans vs. animals,' 64
filtering blood by kidneys, 279
finding one's pulse, 205–6
fingernails. See nails
Fingerprint Branch, 71
fingerprints, 70 (ill.), 70–72, 85
fingers, 115
FIV (feline immunodeficiency virus), 229
flat bones, 89–90
flatulence, 258
flexibility of the trachea, 236
floaters, 57
floating ribs, 98
flow, blood, 201, 201 (ill.), 206, 208, 327
flu vaccinations, 224–25
fluid, route of body, 214–15
fluid buildup, conditions from, 36
follicle-stimulating hormone (FSH), 169–170, 182–83, 294, 296, 320
food allergies, 227–28
Food and Drug Administration (FDA), 123
force of a human bite, 249
forensics, 80
formation of blastocysts, 304
formation of umbilical cord, 306
formation of urine, 282–83
fossils, 7–8, 18
fractures, bone, 98, 335 (ill.)
Franklin, Benjamin, 63
fraternal vs. identical twins, 315 (ill.), 315–16
freckles, 70
freezing transplant organs, avoiding, 338

frequencies of animal hearing, 65
frequencies of human hearing, 52
Friedman, Jeffrey, 184
frown, making a, 112
functions performed by the body
    blood, 190
    blood vessels, 202
    brain, 141, 151
    capillaries, 204
    carbohydrates, 16
    cardiovascular system, 189
    cavities, 12
    cerebral cortex, 151
    cerebrospinal fluid, 139
    corticosteroids, 176–77
    cranial nerves, 147
    ductus deferens, 289
    ears, 50
    elements, 13
    endocrine system, 163, 185–86
    epithelium, 31–32
    eyes, 55–56, 58
    glucagon, 180
    insulin, 180
    joints, 103
    large intestine, 257
    liver, 262
    lower gastrointestinal tract, 255
    lymphatic system, 211
    muscles, 109, 120–26
    nervous system, 129
    neurons, 131–36
    nose, 232
    organ systems, 39
    paranasal sinuses, 94–95, 233
    parathyroid glands, 175
    pectoral (shoulder) girdles, 99
    pelvic (hip) girdles, 100
    penis, 292
    placenta, 306
    platelets, 194
    prolactin, 170
    proteins, 17–18
    reproductive system, 285
    respiratory system, 231–38
    sex hormones, 183
    skeletal system, 87
    skin, 75–78
    small intestine, 255
    spleen, 217
    stomach, 251–52

taste buds, 48–49
    thyroid hormones, 173
    tissue, 31, 34
    upper gastrointestinal tract, 248
    urethra, 289
    urinary system, 275
    uterus, 293
    vagina, 296
    vertebral column, 95
fundus, 252, 294
funny bone, 99–100

G

Gage, Andrew, 330
Galen, 3–4, 6, 208
gall bladder, 248, 260, 261 (ill.), 264, 264 (ill.)
gallstones, 264–65
Galton, Francis, 71
gamer's cramp, 122
gases exchanged in the lungs, 239–240
gastric juice, 252–53
gastrin, 185
gastroesophageal reflux disease (GERD), 251
gastrointestinal tract, lower, 248, 249 (ill.), 255–260
gastrointestinal tract, upper, 248–255, 249 (ill.), 329
gender differences
    actions of FSH and LH, 169–170
    hematocrit, 195
    orgasms, 298
    reproductive systems, 285
    senses, 45, 47
    skeletons, 88–89
    urethras, 281
    voices, 243
generation of nerve impulses, 132–33
genes, 27
genitalia, female, 297
GERD (gastroesophageal reflux disease), 251
gestation period, 303
giraffes, blood flow of, 208
giving birth, animals, 302
glands
    accessory, 82–84, 286, 289–290
    adrenal, 166, 169, 176–78, 177 (ill.), 277
    all about, 33–34
    apocrine, 83

autoimmune diseases, 213
cutaneous, 82
eccrine, 83–84
endocrine, 165–66
exocrine, 33–34
mammary, 85, 292, 297,
 314
oil, 72, 82
pancreas as mixed, 179
parathyroid, 166, 174 (ill.),
 174–76
as part of dermis, 72
as part of integumentary
 system, 67
pineal, 166, 182
pituitary, 166–171, 168
 (ill.)
prostate, 286, 290–91
reproductive, 182–83
salivary, 248
sebaceous, 72, 82
sweat, 72, 82–83
swollen, 216
thymus, 166, 184–85, 185
 (ill.), 216
thyroid, 166, 169, 171–76,
 172 (ill.)
vestibular, 297
Glasgow coma scale, 154–55
Glasgow Royal Infirmary, 336
glasses, invention of, 63
gliding (planar) joints, 102
 (ill.), 104
global climate change,
 connection between animals
 and, 274
globulins, 195
glomerular filtration, 283
glossopharyngeal nerves, 147
glucagon, 180
glucocorticoids, 177–78
glutamate, 134
gluten intolerance, 256
goiters, 174
golfer's elbow, 119–120
gomphoses, 103–4
gonadotrophins, 169
gonadotropin-releasing
 hormone (GnRH), 320
gonads, 169
goose bumps, 77
Graves' disease, 173–74
gray, hair turning, 81
Gray, Sir Henry, 6, 6 (ill.)
gray and white matters,
 137–38
Gray's Anatomy (Gray), 6
Greatbatch, Wilson, 330

Grew, Nehemiah, 5–6
Grey's Anatomy [TV show], 6
gross anatomy, 1, 6
growth
 bone, 91, 277
 brain, 140
 after death, 78
 and development, 303–23
 epidermal cells, 77
 eyes, 56
 fetal, 312–13
 hair, 78, 80
 nails, 78
 placenta during pregnancy,
  306
growth hormone (GH),
 169–170, 187
Guillain-Barré syndrome, 138
gut volume of humans vs.
 primates, 272
gynecology, 285

H

hair
 all about, 79–82
 amount on one's head and
  body, 79–80
 anatomy of the, 79 (ill.)
 determination of color,
  80–81
 determination of size and
  shape of, 81 (ill.), 81–82
 effect of shaving on, 80
 forensics on, 80
 growth after death, 78
 growth rate of, 80
 human vs. animal, 85
 as living vs. dead, 80
 loss. see baldness
 as part of dermis, 72
 as part of integumentary
  system, 67
 types of, 79
hamstring muscles, 114
hand bones, 100 (ill.)
Harvard Medical School, 228
Harvey, William, 6, 206
Hashimoto's disease, 174
Havers, Clopton, 87
head, 10, 79
headaches, sinus, 234
health, vitamins essential for
 human, 266–67
health risks associated with
 obesity, 272
hearing, sense of, 50–55
hearing loss. See deafness

hearing tests, 329
heart
 all about the, 198–202
 autoimmune diseases, 213
 blood flow to, 327
 common procedures and
  surgeries, 335–36
 endocrine functions of the,
  186
 fetal rate, 311
 illustrations, 198 (ill.), 201
  (ill.)
 monitoring electrical activity of the, 330
 neonatal vs. adult rate, 319
 relationship between lungs
  and, 236–37
 returning blood to the, 206
 transplants, 339 (ill.),
  339–340
 water content of human,
  13
heartbeat, 200, 209
heaviest organs, 67
Heimlich maneuver, 235–36,
 236 (ill.)
Helicobacter pylori, 253
helping human anatomy,
 325–340
hematocrit (HCT), 196
hemoglobin, 196
hemophilia, 194
hemorrhagic stroke, 143
hemostasis, 194 (ill.)
Henry, Sir Edward, 71
hepatic circulation, 189
hepatitis, 262–64
herbivores, 8
herniated disc, 96, 96 (ill.)
Herophilus, 3
Herrick, Richard, 337
Herrick, Ronald, 337
high-heeled shoes' effect on
 weight distribution, 101
hinge joints, 102 (ill.), 104
Hippocrates, 3
Hippocratic Oath, 3
histology, 1
history of anatomy, 1–8
history of internal body
 exploration, 325
HIV (human
 immunodeficiency virus),
 221–24, 222 (ill.)
Hodgkin, Alan Lloyd, 132
Hodgkin, Thomas, 215
Hodgkin's disease, 215
holding one's breath, 128, 243

homeostasis, 40–41, 175–76
Hooke, Robert, 22
Hopps, John, 339
hormones
    all about, 164–67
    animal, 186
    glucocorticoids, 177
    illustration, 167 (ill.)
    leptin, 184
    mineralocorticoids, 177
    other sources of, 184–86
    at puberty, 320
    secreted by pancreas, 179
    secreted by the adrenal
        medulla, 178
    secreted by the pituitary
        gland, 169
    secreted by the reproduc-
        tive organs, 182–83
    secreted the adrenal cortex,
        176
    sex, 183
    thyroid, 171, 173
    trophic, 169
Horner, William E., 7
Hounsfield, Godfrey N., 326
HPV, 69
human chorionic
    gonadotropin (hCG),
    299–300
human growth and
    development, 303–23
human immunodeficiency
    virus (HIV), 221–24, 222
    (ill.)
humerus bones, 98
Hunter, William, 6
hunters, birds as, 64
Huxley, Andrew, 120, 132
Huxley, Hugh, 120–21
hyaline cartilage, 36
hyoid, 93–94
hyperbaric oxygen therapy,
    240
hyperopia, 62
hypersomnia, 157
hypertension, 208
hyperthyroidism, 173
hypodermis, 68
hypoglossal nerves, 147
hypogonadism, 169
hypoparathyroidism, 176
hypotension, 208
hypothalamus, 141
hypothyroidism, 174
hysterectomy, 337

I

iatromathematics, 4–5
iatrophysics, 4–5
IBD (inflammatory bowel
    disease), 260
IBS (irritable bowel
    syndrome), 260
identical twins, 70–71, 315
    (ill.), 315–16
identification, new
    innovations in, 71
identification of onset of labor,
    314
Ikeda, Kikunae, 49
ileocacal valve, 257
ileum, 255
imaging techniques, 325–29
immediate vs. delayed allergic
    reactions, 227
immune system, 39, 77–78,
    212–14, 229
immunities in animals, 229
immunity, active vs. passive,
    220
immunodeficiency diseases,
    213–14
immunoglobulins, 218–19,
    219 (ill.), 226
immunology, 211
implantation, 304–5
in vitro fertilization (IVF),
    333, 333 (ill.)
inactivated vs. attenuated
    vaccines, 225
incidence of body moles, 75
incidence of multiple births,
    317
incidence of premature birth,
    313
incidence of prostate cancer in
    the U.S., 299–300
incidence of STDs, 301
incisors, 248–49
incontinence, 282
indoleamines, 134
induction of labor, 316
infancy, 318–320
infections, protection of the
    vagina from, 296
infections, role of fever in, 218
infections, treating viral, 221
infections, urinary tract,
    281–82
infertility, 298–99, 332–33
inflammation, 38, 218
inflammatory bowel disease
    (IBD), 260

information, neurons'
    transmittal of, 131
inhalation vs. exhalation, 239,
    239 (ill.)
inherited, allergies as, 226–27
injury, protection of the heart
    from, 199
innate immunities, 229
inner ear, 50
insects, breathing of, 245
insertion vs. origin of a
    muscle, 119
insomnia, 157
inspiration vs. expiration, 239,
    239 (ill.)
insulin, 180–81
integrators, 41
integumentary system, 39,
    67–85, 307, 322
intelligence, 140, 153, 160
interferons, 221
intermediate hairs, 79
internal body exploration,
    history of, 325
internal bones, animals
    without, 108
internal vs. external
    respiration, 240
interneurons (association
    neurons), 38
intestine, large, 248, 256–58,
    257 (ill.), 260
intestine, small, 248, 255 (ill.),
    255–57, 261
intestines, endocrine
    functions of the, 185
intestines, food's movement
    into, 254
intestines, water content of,
    13
intolerance, gluten, 256
intolerance vs. allergy, 228
*Introduction to the Study of
    Experimental Medicine*
    (Bernard), 6
invertebrates, intelligence of,
    160
iodized salt, importance of,
    173
iPhones, 72
IQ, 153
iris, 55
irregular bones, 89–90
irritable bowel syndrome
    (IBS), 260
Isaacs, Alick, 221
ischemic stroke, 143, 143 (ill.)
islets, pancreatic, 179

Itakura, Keiichi, 180
IVF (in vitro fertilization), 333, 333 (ill.)

J

Jarvik, Robert K., 339
Jarvik 7, 339
Jefferson, Thomas, 5
jejunum, 255
Jenner, Edward, 225
Johannsen, Wilhelm, 27
joints, 102 (ill.), 102–7, 213, 338, 338 (ill.)
Journal of Agricultural and Food Chemistry, 59
Journal of the American Medical Association, 207
Jurassic period, 8

K

keratinization of skin cells, 76–77
keratinocytes, 69, 77–78
kidneys
  all about the, 276–280
  anatomy of the, 277 (ill.)
  autoimmune diseases, 213
  endocrine functions of the, 186
  failure, 280, 284
  stones, 279–280, 336–37
  transplants, 337–38
  water content of human, 13
knee, anatomy of the, 106 (ill.)
knockout mice, 42
knuckles, popping sound of cracking, 105
Korotkoff, Nikolai, 206
kyphosis, 97

L

labia, 297
labor and delivery of a baby, 314–16
labyrinth, 51
laceration, spinal, 146
lactation, birth and, 313–18
lactose intolerance, 17, 17 (ill.)
Laënnec, René-Théophile-Hyacinthe, 200
Laguesse, Gustave E., 179

Lamarck, Jean-Baptiste Pierre Antoine de Monet de, 2, 2 (ill.)
land animals, sleep of, 160–61
Landsteiner, Karl, 196–97
Langerhans, Paul, 179
language and speech, brain part for, 152
lanugo, 310
large collecting ducts, 214
large intestine, 248, 256–58, 257 (ill.), 260
largest arteries, 203
largest bones, 107
largest cells, 23
largest joints, 105–6
largest muscles, 112
largest organs, 42, 67
largest veins, 203
laryngitis, 236
larynx, 235
latex, allergy to, 228
Lauterbur, Paul, 328
layers of the retina, 57–58
layers of the skin, 68, 68 (ill.)
leap, animals with the highest, 128
learning and memory, 151–54
left vs. right brain, 141–42
left vs. right ventricles, 200
length of accessory ducts, 289
length of blood vessels, 204
length of female reproductive cycle, 295–96
length of labor, 315
length of small vs. large intestine, 256
length of sweat glands, 82
length of ureters, 281
lens, 55
leptin, 184
Lerner, Aaron B., 182
Lessons on Comparative Anatomy (Cuvier), 7
let-down reflex, milk, 171
leukemias, 36
leukocytes. See white blood cells
Levine, Philip, 197
lifespan, cell, 30
lifespan of animals, 322–23
lifespan of lymphocytes, 212
lifespan of red blood cells, 192
light, adapting to dim, 59
limbic system, 140
Lindenmann, Jean, 221
lining, stomach, 253
lipids, 16

liposuction, 36
lips, 112–13
Lister, Joseph, 336
liver, 13, 30, 248, 260–62, 261 (ill.)
lobes, brain, 142 (ill.)
lobes, lung, 237
location, muscles named for, 115
location of body hair, 79–80
location of endometrial tissue, 294
location of hormone receptors, 167
location of lymphoid nodules, 215
location of the bladder, 281
location of the funny bone, 99–100
location of the heart's pacemaker, 201
location of the ileocecal valve, 257
location of the kidneys, 276
location of the organs of equilibrium, 54
location of the pancreas, 178
location of the parathyroid glands, 174
location of the pituitary gland, 167
location of the sinuses, 233
location of the spinal cord, 145
location of the testes, 286
location of vomer bone, 94
Loewi, Otto, 133
long bones, 89–90
longest bones, 90
longest cells, 23
longest muscles, 112
longest nerves, 37
longest spinal nerve pair, 147–48
long-term vs. short-term memory, 154
loop of Henle, 278
lordosis, 97, 97 (ill.)
Loricifera, 19
Lou Gehrig's disease, 130
lower gastrointestinal tract, 248, 249 (ill.), 255–260
lower respiratory system, 231
Ludwig, Karl Friedrich Wilhelm, 340
lumbar bones, 95
lumbar plexus, 149 (ill.)
lumps, breast, 297

INDEX

361

lungs
air volume in, 242
anatomy and physiology of the, 232 (ill.)
animals' use of, 244–45
autoimmune diseases, 213
capacity of, 242
characteristics of the, 236
cleaning air before reaching, 241
collapsed, 237, 237 (ill.)
effect of smoking on, 238
gases exchanged in the, 239–240
relationship between heart and, 236–37
right vs. left, 236
transplants, 339
water content of human, 13
luteinizing hormone (LH), 169–170, 183, 294, 320
lymph, 211, 213
lymph nodes, 216
lymphatic capillaries, 214
lymphatic system, 39, 211–230, 212 (ill.), 307, 322
lymphocytes, 211–12
lymphoid nodules, 214 (ill.), 215
lymphomas, 36
lysosomes, 28

M

Macleod, John James R., 180
macroscopic anatomy, 1
magnetic resonance imaging (MRI), 327–28
male reproductive system, 286–292, 287 (ill.)
malignancy of tumors, 291
mammary glands, 85, 292, 297, 314
mammography, 329
Mansfield, Peter, 328
Marshall, Barry J., 253
maternal changes during pregnancy, 313–14
Matlack, Timothy, 8
matters, gray and white, 137–38
maturation of sperm, 289
maturity, 318
Mayo Clinic, 259
measuring intelligence, 153
mechanoreceptors, 43
medical anatomy, 1

medical benefits to circumcision, 292
medicine branches specializing in reproductive systems, 285
Medico-Chirugical Transactions, 215
Mediterranean Sea, 19
medulla, adrenal, 178
medulla, kidney, 276
medulla oblongata, 141
Megalosaurus, 8
melanocytes, 69
melanocyte-stimulating hormone (MSH), 169
melanomas, 74–75
melatonin, 182
Melville, Herman, 84
membranes, mucous, 218
memory, stomach, 254
memory and learning, 151–54, 153 (ill.)
menarche, 296
Ménière's, Prosper, 54
Ménière's disease, 54
meninges, 136
meningitis, 137
meningococcal disease, 137
menopause, 296
menstruation, 296
mesoderm, 306–7
metabolism and nutrition, 265–272
metacarpal bones, 99, 100 (ill.)
metastatic cancer, 215–16
metatarsal bones, 99
microorganisms, defense against, 218
microscope, use of a, 1, 5, 22, 41, 87, 174
microscopic anatomy, 1
micturition, 282
midbrain, 141
Middle Ages, 3
middle ear, 50
milestones, infant, 319–320
milk, breast, 317–18
milk let-down reflex, 171
mineralocorticoids, 176–77
minerals, 265–66
mini-stroke, 143–44
mitochondria, 28–29
mitral valve prolapse (MVP), 199–200
mixed gland, pancreas as a, 179
Moby Dick (Melville), 84

molars, 248–49
molecules, human bioorganic, 15
moles, 75, 75 (ill.)
Mollusca, 7
monoamines, 134
mons pubia, 297
Monsanto, 187
motion sickness, 55
motor neurons, 38
motor units, 122
mouth, 112–13, 213, 248–49
movement, muscles named for type of, 115, 115 (ill.)
movement of bones, 94
movement of chyme into small intestine, 255–56
movement of eyeballs, 112
movement of food into intestines, 254
movement of synovial joints, 104–5
movements, non-respiratory air, 232
movements, voluntary vs. involuntary muscle, 110
MRI (magnetic resonance imaging), 327–28
mucous membranes, 218
multi-infarct dementia, 144
multiple births, 317
multiple pregnancy, 316
multiple sclerosis, 138
multiunit smooth muscle, 120
Murray, Joseph, 337
muscles
age's effect on, 126
anatomy of the, 111 (ill.)
around eyes, 112
around mouth, 112–13
autoimmune diseases, 213
for breathing, 240–41
cardiac, 37, 116–17, 117 (ill.), 198–200
cells, 118, 120–25
for chewing, 113
contraction and relaxation of skeletal, 121
contraction of smooth vs. skeletal, 126
controlling urination, 283
conversion to fat, 125–26
corrugator, 114
cramps, 122
dark and white, 126
ear, 114
effect of creatine on, 111

effect of exercise on, 37, 110–11, 124–26
effect of weightlessness on, 110
eye, 112
fastest, 112
fatigue, 125
hamstring, 114
importance and prominence of cells, 110
largest, 112
longest, 112
named for direction of fibers in, 115–16
named for location, 115
named for size or shape, 114–15
named for type of movement, 115, 115 (ill.)
number in fingers and thumbs, 115
number of in humans, 109–10
origin vs. insertion of, 119
other names of, 116
pulled, 125, 125 (ill.)
rate of strength increase, 111
shivering, 124–25
smallest, 112
to smile and frown, 112, 128
spasms, 122
for swallowing, 113 (ill.), 113–14
tissue, 31, 37, 42, 116, 117 (ill.)
variable, 114
voluntary vs. involuntary movements, 110
water content of human, 13
weight of, 110
muscular dystrophy, 118
muscular system, 39, 109–28, 307, 322
mutations, genetic, 27–28
MVP (mitral valve prolapse), 199–200
myelin, 133
myelin sheath, 138
myoglobin, 118–19
myology, 109
myopia, 61–62

N

nails, 67, 78, 78 (ill.)

name, origin of pituitary gland's, 168
narcolepsy, 157
nasal strips, 235
National Cancer Institute, 297
National Heart, Lung, and Blood Institute, 238, 336
National Institute of Neurological Disorders and Stroke, 149
National Institutes of Health, 17, 336
National Palm Prints System (NPPS), 71
National Sleep Foundation, 156, 158
The Natural History of Oxfordshire (Plot), 8
Nature, 328
nausea, 270
nearsightedness, 61–62
necessity of sleep, 155
neck, regions of the, 10
negative feedback in homeostasis, 41
Neisseria meningitidis, 137
neonatal period, 318–19, 319 (ill.)
neostigmine, 135
nephrons, 277–78, 278 (ill.)
nerves
    anatomy of, 133 (ill.)
    autoimmune diseases, 213
    carpal tunnel syndrome, 149–150
    endings, 73
    impulses, 132–33
    longest, 37
    optic, 56, 147
    types of, 146–48
nervous system, 39, 129–161, 163–64, 307, 322
nervous tissue, 31, 37 (ill.), 37–38, 42
neuromuscular junctions, 122
neurons, 37–38, 132 (ill.)
neuropeptides, 134
neurotransmitters, 133–34, 136
Next Generation Identification (NGI), 71
nicotine, 134
night blindness, 59
NK cells, 211–12
NMR (nuclear magnetic resonance), 327
Nobel Prize
    ABO blood typing, 196

action potential, 132
antibiotics, 221
CT scans, 326
insulin, 180–81
MRI, 328
neurotransmitters, 133
organ donation, 226
right vs. left brain functions, 141
stomach ulcers, 253
nodules, lymphoid, 214 (ill.), 215
non-respiratory air movements, 232
nonspecific defenses, 217–18
norepinephrine, 134, 178
nose, 232–34, 234 (ill.)
nosebleeds, 233
NPPS (National Palm Prints System), 71
nuclear magnetic resonance (NMR), 327
nucleus, cell, 26
number of annual heart transplants, 340
number of bones in the appendicular skeleton, 98
number of bones in the axial skeleton, 93
number of bones in the body, 89
number of bones vs. size of structures, 100
number of cranial nerves, 146
number of hairs on one's head and body, 79–80
number of lymphocytes, 212
number of muscles, 109–10, 115
number of oocytes present at birth, 293
number of people with food allergies, 227–28
number of red blood cells, 191
number of ribs, 98
number of sperm for fertilization, 298, 299 (ill.)
number of sperm produced, 288
number of spinal nerve pairs, 147–48
number of sweat glands, 83
number of tonsils, 216
nursing. See breastfeeding
nutrients, absorption of, 254, 256
nutrients, energy, 265–66

nutrients, organisms that rely on, 272–73
nutrients, six essential, 265
nutrition and metabolism, 265–272

O

obesity, 270–72
obstetrics, 328–29
obstructive hydrocephalus, 139
oculomotor nerves, 147
odor, cause of feces, 259
odor of sweat, 84
oil glands, 72, 82
olfactory nerves, 147
olfactory system, 46 (ill.), 46–48
On Anatomical Procedures (Galen), 4
On Divination by Dreams (Aristotle), 3
On Dreams (Aristotle), 3
On Length and Shortness of Life (Aristotle), 3
On Memory and Recollection (Aristotle), 3
On Respiration (Aristotle), 3
On Sense and Sensible Objects (Aristotle), 3
On Sleep and Waking (Aristotle), 3
On the Movement of the Heart and Blood in Animals (Harvey), 6
On the Natural Faculties (Galen), 4
On the Structure of the Human Body (Vesalius), 4
On the Usefulness of the Parts of the Body (Galen), 4
On Youth and Age (Aristotle), 3
onset of labor, identification of, 314
onset of puberty, 320
oocytes, 293
oogenesis, 293
open vs. closed circulatory system, 209
operating rooms, cleanliness of, 335
operations, procedures, and transplants, 335–340
optic nerve, 56, 147
oral cavity, 248

organ donors and recipients, compatibility of, 226. See also transplants
organ of Corti, 52
organ systems
　age's effect on, 322
　anatomical levels of organization, 39–40
　definition of, 21
　diagnostic techniques, 329–335
　embryonic, 306–7
organ transplants, 337–38, 340. See also transplants
organelles, 24 (ill.), 24–25
organisms, comparison of various. See comparative anatomy
organization, anatomical levels of, 21–42
organization of muscles, 112–16
organs
　accessory, 248, 260–65, 281–82
　anatomical levels of organization, 38–40
　of the cardiovascular system, 189
　definition of, 21
　of the digestive system, 248
　embryonic, 306–7
　of the endocrine system, 164, 185–86
　female genitalia, 297
　heaviest, 67
　keeping alive outside body, 338, 340
　largest, 42, 67
　lymphatic, 214–17
　not freezing, 338
　of the reproductive system, 182–84, 285–86, 292
　vestigial, 40, 42
orgasm, 298
origin vs. insertion of a muscle, 119
ossification of bones, 91–92
osteoarthritis, 107
osteoblasts, 91
osteoclasts, 91
osteocytes, 91
osteogenic cells, 91
osteomalacia, 93, 268
osteoporosis, 92–93, 93 (ill.), 334–35, 335 (ill.)
ovaries, 166, 292–93

overabundance of vitamins, 268
overweight vs. obesity, 270–71
Owen, Sir Richard, 108
oxygen, blood's transportation of, 192
oxygen, importance of, 12, 19, 232
oxygen, introducing to blood, 240
oxygen debt, 125
oxygen therapy, hyperbaric, 240
oxytocin, 169, 171

P

P. T. Barnum Circus, 317
pacemakers, 201, 330, 330 (ill.), 339
pain, anesthetics' effect on, 135
pain, phantom, 45
pain receptors, 43
paleomammalian brain, 140
pancreas, 166, 178–181, 179 (ill.), 248, 260–61
pancreatic digestive juices, 261
pancreatic islets, 179
Pap smear, 334
Papanicolaou, George, 334
paracine vs. circulating hormones, 165
paralysis, 148
paralysis, sleep, 157–58
paranasal sinuses, 94–95, 233
parasympathetic nervous system, 130, 130 (ill.), 150
parathyroid glands, 166, 174 (ill.), 174–76
parathyroid hormone (PTH), 175, 186
Parkinson, James, 136
Parkinson's disease, 136
partial hysterectomy, 337
parts of the external ear, 50
parts of the eye, 55–56
parts of the kidney, 276
parts of the nephron, 278, 278 (ill.)
parts of the urinary system, 275
parts of urine production, 283
Parva Naturalia (Aristotle), 3
passive immunity, 220
Pasteur, Louis, 336
patella bones, 99, 101

pathological physiology (pathology), 2
pattern baldness, 82
PE (pulmonary embolism), 203
pectoral (shoulder) girdles, 98–99
pellagra, 268
pelvic (hip) girdles, 99–100
pelvic brim, 101
pelvic inflammatory disease (PID), 294
pelvis, division of the, 101
penis, 286, 292
peptide hormones, 164–65
pericarditis, 199
peripheral nervous system, 129–130, 146–151
peristalsis, 251
PET imaging, 326–27
pH, 14–15, 190
phagocytosis, 212
phalange bones, 99, 100 (ill.)
phantom pain, 45
pharynx, 234
phases of breathing, 239
phases of female reproductive cycle, 294–95
Phillips, Robert. See Pitts, Roscoe
phonation, 243–44
phosphenes, 60
photoreceptors, 43, 58 (ill.), 58–60
phyostigmine, 135
physiology, 2–3, 6
PID (pelvic inflammatory disease), 294
pimples, 83 (ill.)
pineal glands, 166, 182
pins and needles, feet on, 150
pitch of one's voice, 244
Pitts, Roscoe, 71
pituitary glands, 166–171, 168 (ill.)
pivot joints, 102 (ill.), 104–5
placenta, 306, 316
placental stage of labor, 314–15
planar (gliding) joints, 102 (ill.)
planes, body, 10
plant anatomy, 5–6
plasma, 195
plasma cells, 220
plasma proteins, 195
platelets, 191 (ill.), 194
pleurisy, 237
plexus, 148, 149 (ill.)

Plot, Robert, 8
pneumonia, 238
pneumothorax, 237, 238 (ill.)
Poncet, Pierre, 4
pons, 141
position, anatomical, 9
positive feedback in homeostasis, 41
posterior pituitary glands, 166, 168
postnatal development, 303, 318–322
predators, animals' ability to see, 64
pregnancy and prenatal development
    change of pubic bones during, 104
    confirmation of, 299–300
    development of mammary glands, 314
    divisions of human development, 303
    ectopic, 306
    embryonic period, 304–9
    events during second trimester of, 310
    events during third trimester of, 312
    fetal growth during, 312–13
    fetal stage, 309–13
    growth of placenta during a, 306
    increased urination, 281
    maternal changes during, 313–14
    of multiples, 316
    prenatal diagnostic techniques, 333–34
    production of milk during, 317
premature births, 313
prenatal diagnostic techniques, 333–34
presbycusis, 53–54
pressure, blood, 207 (ill.), 207–8
pressure, heart, 199
prevention of disease by vaccination, 225–26
prevention of HIV, 223–24
primary teeth, purpose of, 249
primary vs. secondary amenorrhea, 296
primate vs. human gut volume, 272
procedures, transplants, and operations, 335–340

process, steps in the digestive, 247
production, sound, 243 (ill.), 243–44
production of antibodies, 220
production of flatulence, 258
production of milk, 317–18
production of platelets, 194
production of saliva, 250
production of semen, 291
production of sperm, 288
production of urea, 278
production of urine, 283
prokaryotic cells, 24
prolactin, 169–170
prostate cancer, 290–91, 334
prostate glands, 286, 290–91
prostatisis, 290, 291 (ill.)
protection of the heart from injury, 199
protection of the kidneys, 276–77
protection of the vagina from infection, 296
protector, skin as body's, 218
protein hormones, 164–65
proteins, 17–18, 195, 265–66
proximal convoluted tubule, 278
puberty, 320–21
pubic bones, change of during pregnancy, 104
pudendum, 297
pulled muscles, 125, 125 (ill.)
pulmonary circulation, 189
pulmonary embolism (PE), 203
pulmonary fibrosis, 238
pulse, 205
pupil, 55
purpose of blood testing, 196
purpose of contraception, 300
purpose of fetal stage of development, 309
purpose of goose bumps, 77
purpose of primary teeth, 249
purpose of prostate gland secretions, 290
purpose of proteins, 17–18
purpose of the gall bladder, 264
purpose of vaccinations, 224
purpose of vermix and lanugo, 310
purpose of villi, 256
pus, 38
pylorus, 252

R

Radiata, 7
radical hysterectomy, 337
radiographic anatomy, 1
radius bones, 99
rate, heart, 311, 319
rate, respiratory, 241–42, 244, 319
rate, values for normal human pulse, 205
rate of red blood cell formation, 191
rBGH (Bovine Growth Hormone), 186–87
rBST (Bovine Somatotrophin), 186
RDS (respiratory distress syndrome), 241
reactions, catabolic vs. anabolic, 265
reactions, immediate vs. delayed allergic, 227
readings, understanding blood pressure, 207 (ill.), 207–8
receptors, hormone, 167, 167 (ill.)
rectum, 257
red blood cells, 191 (ill.), 191–92, 196, 279
reflexes, 145, 146 (ill.), 171, 233, 298
regeneration, cell, 30, 32–34, 38, 253, 262
regional anatomy, 1
regions, body, 10–11
regions of the abdomen, 10
regions of the dermis, 70
regions of the large intestine, 257
regions of the pituitary gland, 168
regions of the stomach, 252
regions of the trunk, 11
regions of the uterus, 294
regulation of body temperature, 77, 83–84
relaxation, muscle, 121, 124
REM sleep, 155, 156 (ill.)
Remak, Robert, 23
remembering dreams, 156
remodeling, bone, 92
removal of particles by the nose, 233
renal circulation, 189
renal corpuscle, 278
repairing damaged tissue, 38
replacement of epidermal cells, 70
replacement of red blood cells, 191
reproductive glands, 182–83
reproductive organs, 182–84
reproductive system, 39, 285–302, 307, 322
resemblance between animals and their parents, 302
resistance, specific, 218
respiration, external vs. internal, 240
respiration and breathing, 239–243
respiration vs. cellular respiration, 231
respiratory cycle, 239, 239 (ill.)
respiratory distress syndrome (RDS), 241
respiratory rate, 241–42, 244, 319
respiratory system, 39, 231–245, 307, 322
response, sexual, 298
response to stress, hormonal, 166–67
resting membrane potential, 131
retina, 56–58
retinal detachment, 57
returning blood to the heart, 206
Rh factor, 197
rheumatoid arthritis, 107
ribs, number of, 98
ribs, true vs. false vs. floating, 98
rickets, 268, 268 (ill.)
ridges on nails, 78
Riggs, Arthur, 180
right vs. left brain, 141–42
right vs. left ventricles, 200
rigor mortis, 124
risks associated with multiple pregnancies, 316
risks associated with obesity, 272
risks of kidney transplants, 338
RNA, 26
rods, 58 (ill.), 58–59
Roentgen, Wilhelm Conrad, 325
Roman Coliseum, 5
Roman Empire, 4
route of body fluid, 214–15
Royal Academy of London, 6
runny nose, 234

S

sac, amniotic, 308, 316
sacral bones, 95
SAD (seasonal affective disorder), 182
saddle joints, 102 (ill.), 104–5
Sahara Desert, 272
saliva, 250
salivary glands, 248
salt, importance of iodized, 173
Sandström, Ivar Victor, 175
Sanger, Frederick, 180
sarcomas, 36
saturated vs. unsaturated fats, 269
scabs, 73–74
scapula bones, 98
scars, 38
Schleiden, Matthias, 22–23
Schwann, Theodor, 23
sciatica, 148
SCID (severe combined immunodeficiency disease), 213–14
sclera, 55
scoliosis, 97
Scotland Yard, 71
scrotum, 286–87
Scrotum humanum, 8
scurvy, 268
seasonal affective disorder (SAD), 182
sebaceous glands, 72, 82
secondary vs. primary amenorrhea, 296
secretin, 185
security, fingerprints used for, 72
seizures, 135–36
semen, 290–91
seminal vesicles, 286
senescence, 321
sensory neurons, 38
sensory receptors, 41, 43–45, 44 (ill.)
sensory system, 40, 43–65, 322
separation of blood components, 196
septum, deviated, 233
serotonin, 134
serum, plasma vs., 195
sesamoid bones, 90
severe combined immunodeficiency disease (SCID), 213–14

sex hormones, 183
sexual response, 298
sexually transmitted diseases (STDs), 301–2
shape, changes in spine, 96–97
shape, determination of hair, 81 (ill.), 81–82
shape, foot, 101
shape, muscles named for, 114–15
shape of red blood cells, 192
shark's teeth, 273, 273 (ill.)
shedding of skin, 67
shivering, 124–25
shoes' effect on weight distribution, high-heeled, 101
short bones, 89–90
short-term vs. long-term memory, 154
shoulder, anatomy of the, 99 (ill.)
shrinking gland, thymus gland as, 185
Shumway, Norman, 339
Siamese twins, 317
sickle cell anemia, 192 (ill.), 192–93
sight. See vision, sense of
sinus headaches, 234
sinuses, 94–95, 233
size, determination of hair, 81 (ill.), 81–82
size, muscles named for, 114–15
size of a zygote at conception, 304
size of animal cells, 41–42
size of capillaries, 204
size of embryo at end of embryonic period, 309
size of fetus at end of first trimester, 309
size of sperm cells, 288
size of the brain, 139–140, 159
size of the heart, 198
size of the kidneys, 276
size of the liver, 261
size of the pituitary gland, 167
size of the prostate gland, 290
size of the uterus, 293
sizes of ventricles, varying, 200
skeletal muscle, 37, 116–17, 117 (ill.), 119, 121, 121 (ill.), 126, 198–99

skeletal system, 39, 87–108, 88 (ill.), 307, 322
skin
  amount on humans, 67
  amount shed, 67
  animals' breathing through their, 244
  around eyes, 112
  autoimmune diseases, 213
  bacteria on, 68
  blister, 73
  as body's protection, 218
  cell lifespan, 30
  color of cats,' 84
  damage by burns, 74, 74 (ill.)
  dermatomes, 148–49
  determination of color, 69
  function, 75–78
  holding together one's, 72
  in immune system, 77–78
  layers of the, 68, 68 (ill.)
  loss of sweat through the, 83
  as part of integumentary system, 67
  role in regulating body temperature, 77
  structure, 67–75, 73 (ill.)
  synthesization of vitamin D, 76
  thickness of, 72–73
  water content of human, 13
skull, 93–94
sleep and dreams, 154–161
sleepwalking, 158
small intestine, 248, 255 (ill.), 255–57, 261
smallest bones, 90
smallest cells, 23
smallest muscles, 112
smell, humans vs. bloodhounds' sense of, 63
smell, sense of. See olfactory system
smell and taste, connection between, 48
smile, making a, 112, 128
smoking, effect of on lungs, 238
smooth muscle, 37, 116–17, 117 (ill.), 120, 126
sneeze reflex, 233
Snell, George, 226
snoring, 235
soft spots, 96

somatic nervous system, 130, 146–151
somatostatin, 134
sore throats, causes of, 235
sound, definition of, 53
sound, sensing, 51–52
sound levels, common, 52–53
sound of heartbeat, 200
sound production, 243 (ill.), 243–44
sounds of Korotkoff, 206
sources for essential vitamins, 267–68
sources of gain and loss of water, 275–76
spasms, muscle, 122
special physiology, 2
specific defenses, 218–226
specific resistance, 218
speech and language, brain part for, 152
speed of blood flow, 206
speed of human heartbeat, 200
speed of nerve impulses, 133
speed of urination, 283–84
sperm, 288–89, 298, 299 (ill.)
spermatogenesis, 288
spermatozoa, 30
Sperry, Roger, 141
spider silks, 42
spinal accessory nerves, 147
spinal cord, 136, 141, 145 (ill.), 145–46, 148
spinal nerves, 146–48
spinal taps, 331
spine shape, changes in, 96–97
Spinosaurus aegyptiacus, 8
spleen, 13, 216–17, 217 (ill.)
spongy bone tissue, 91
squamous cell carcinomas, 74
St. Martin, Alexis, 248
stages of labor, 314–15
stages of postnatal development, 318
stages of prenatal development, 304, 309 (ill.)
stages of sleep, 155
Stanford-Binet test, 153
Starling, Ernest, 165, 186
statistics, HIV/AIDS, 223
statistics, twin, 316–17
STDs (sexually transmitted diseases), 301–2
sterility, male, 291
sterilization, 300 (ill.), 301
steroid hormones, 164–65

steroids, anabolic, 183–84, 184 (ill.)
Stetson, Rufus E., 197
stickiness of blood, 190–91
stomach, 30, 248, 251–54, 252 (ill.), 273
stones, kidney, 279–280, 336–37
storage of white blood cells, 194
strabismus, 62–63
strength, animals with the most, 127
strength increase, rate of muscle, 111
strength of bone, 88
strength vs. aerobic training, 110–11
Streptococcus, 200, 235
stress, hormonal response to, 166–67
strips, nasal, 235
stroke, 143 (ill.), 143–44
structural classes of joints, 103
structural classification of animals, 7
structural organization, levels of, 2, 21–22
structures
    of the cardiovascular sys-tem, 189
    eye, 55–56
    insulin, 180–81
    male reproductive system, 286
    of the middle ear, 50
    muscle, 116–120
    nose, 233
    oral cavity, 248
    of the respiratory system, 232–38
    sensory receptors, 44–45
    skin, 68–75
    vestigial, 40, 42
studies in anatomy, 2–7
studies on digestion, 248
study of fingerprints, 71–72
subthalamus, 141
success rate of kidney transplants, 338
sudoriferous glands. See sweat glands
supplements, vitamin, 269
surgery, use of disinfectant during, 336
survival without a gall bladder, 264

survival without a spleen, 217
survival without one kidney, 279
survival without sleep, 157
sutures, 103
swallowing, muscles for, 113 (ill.), 113–14
swallowing food whole, 273–74
sweat, loss of, 83
sweat, odor of, 84
sweat glands, 72, 82–83
swollen glands, 216
symmetry, body, 10, 19
sympathetic nervous system, 130, 130 (ill.), 150
symptoms of a heart attack, 202
symptoms of botulism, 123
symptoms of Guillain-Barré syndrome, 138
symptoms of HIV, 221–22
symptoms of hypothyroidism, 174
symptoms of inflammation, 218
symptoms of kidney failure, 280
symptoms of stroke, 143
synapses, 132, 134–35
synarthrosis, 103
syndesmoses, 103–4
synergistic muscles, 120
synesthesia, 44
synovial fluid, 104
synovial joints, 103–4
synthesization of insulin, 180
synthesization of vitamin D, skin's, 76
systemic anatomy, 1
systemic circulation, 189
systemic physiology, 2

T

T cells, 211–12, 219–220
T lymphocytes, 78
talking while eating, danger of, 235
target cells, 164, 167 (ill.)
targets of trophic hormones, 169
tarsal bones, 99
taste, sense of, 48–50
taste and smell, connection between, 48
taste buds, 48 (ill.), 48–49
taxonomy, 3

tears, production of, 61, 61 (ill.)
techniques, anatomy and imaging, 325–29
techniques, diagnostic, 327–335
teeth, 248–250, 250 (ill.), 273, 273 (ill.)
telomerases, 28
telomeres, 28
temperature, regulation of body, 77, 83–84
temperature of scrotum, 286–87
temperature sensitivity, 46
tendons, 119–120
tennis elbow, 119 (ill.), 119–120
Terman, Lewis, 153
terminal hairs, 79
terminology, human anatomical, 9–12
testes, 166, 286–88
testicles, 289 (ill.)
testing, blood, 195–96
testing, drug, 282–83
testing for cancer, 297
testing for pregnancy, 299–300
testosterone, 183
tests. See diagnostic techniques
thalamus, 141
theatres, anatomical, 5
therapy, hyperbaric oxygen, 240
thermoreceptors, 43
thickness, skin, 72–73
thickness of red blood cells, 191
thoracic bones, 95
thorax, 93, 98
throats, causes of sore, 235
thumbs, 115
thymus glands, 166, 184–85, 185 (ill.), 216
thyroid cancer, 174
thyroid glands, 166, 169, 171–76, 172 (ill.)
thyroid hormones, 171, 173
thyroid-stimulating hormone (TSH), 169, 174
thyroxine vs. triiodothyronine, 172
TIA (transient ischemic attack), 143–44
tibia bones, 99
tic douloureux, 147

time from eating to excreting, amount of, 247–48

tinnitus, 54

tissue

  adipose, 34–36, 35 (ill.), 186

  anatomical levels of organization, 31–38

  areolar, 34

  compact bone, 91

  connective, 31, 34–35, 42

  definition of, 21

  effect of exercise on bone, 92

  endometrial, 294

  epilethial, 31–33, 33 (ill.), 42

  erectile, 292

  examining breast, 329

  growth hormone, 170

  lymphatic capillaries, 214

  muscle, 31, 37, 42, 116, 117 (ill.)

  nervous, 31, 37 (ill.), 37–38, 42

  spider silks helping, 42

  types of bone, 91

  water content of body, 13

  weight, 38

TMD, 54, 94

tongue, 45, 49, 249–250

tonsils, 216

Touch ID, 72

trachea, flexibility of the, 236

tracheostomy, 337, 337 (ill.)

tracheotomy, 337

tracts, spinal cord, 145

training, strength vs. aerobic, 110–11

trans fatty acids, 269

transection, spinal, 146

transfusions, blood, 197

transient ischemic attack (TIA), 143–44

transmission of HIV, 223

transmission of information, neurons,' 131

transmission of STDs, 301

transplants, 226, 335–340

transportation of oxygen, blood's, 192

trauma to the spinal cord, 146

Traut, Herbert F., 334

A Treatise on Pathological Anatomy (Horner), 7

treatment of GERD, 251

triangle of auscultation, 114

trigeminal nerves, 147

triiodothyronine vs. thyroxine, 172

trochlear nerves, 147

trophic hormones, 169

tropomyosin, 118

troponin, 118

true ribs, 98

true vs. false labor, 315

trunk, regions of the, 11

trunks, importance of elephant, 127, 127 (ill.)

tubes, uterine, 292–93, 304

tubular reabsorption, 283

tubular secretion, 283

tumors, malignancy of, 291

twins, 70–71, 315 (ill.), 315–17

typing, blood, 196–98

U

ulcerative colitis, 260

ulcers, stomach, 253, 253 (ill.)

ulna bones, 98

ultrasound, 328 (ill.), 328–29

umami taste, 49

umbilical cord, 306, 311 (ill.), 316

uniqueness of the liver, 262

uniqueness of the thyroid gland, 172

University of Padua, 5

University of Pennsylvania, 28

University of Virginia, 5

unsaturated vs. saturated fats, 269

upper gastrointestinal tract, 248–255, 249 (ill.)

upper respiratory system, 231

urea, 278

ureters, 281

urethra, 281, 281 (ill.), 288–89

urinalysis, 282

urinary system, 275–284, 307

urinary tract infections, 281–82

urination, effect of alcohol on, 170

urination and ejaculation, impossibility of simultaneous, 290

urine, importance to animals of, 284

urine and its formation, 282–83

urology, 285

uterine tubes, 292–93, 304

uterus, 292–94, 293 (ill.), 304

V

vaccinations, 223–25, 224 (ill.)

vagina, 292, 294, 296–97

vagus nerves, 147

values for human pulse rate, normal, 205

values for white blood cells, critical, 194

valve, ileocacal, 257

van Calcar, Jan Stephan, 4

van Leeuwenhoek, Antoni, 5

variable muscles, 114

varicose veins, 204, 205 (ill.)

vas deferens, 286

vascular dementia, 144

veins, 195–96, 202–4, 205 (ill.), 206

vellus hairs, 79

venous bleeding, 205

ventral thalamus, 141

ventricles, 199–200

vermix, 310

vertebral column, 93, 95 (ill.), 95–96, 96 (ill.)

Vertebrata, 7

vertebrates, intelligence of, 160

Vesalius, Andreas, 4 (ill.), 4–5

vesibular glands, 297

vessels, blood, 73, 202–6, 213

vessels, lymphatic, 214–17

vessels entering and leaving the heart, 200

vessels entering and leaving the kidneys, 278–79

vestibule, vaginal, 297

vestibulocochlear nerves, 147

vestigial organs and structures, 40, 42

villi, 256

viral infections, treating, 221

viral vs. nonviral hepatitis, 263–64

Virchow, Rudolph, 23

visceral smooth muscle, 120

vision, sense of, 55–63

vitamin D, importance of, 76, 277

vitamins, 265–69

vitreous detachment, 57

vitreous humor, 56

voice, sound of one's, 243–44

volume, changes in stomach, 252

volume in lungs, air, 242

volume of bladder, 281

volume of humans vs. primates, gut, 272

voluntary vs. involuntary muscle movements, 110

vomer bone, 94

vomiting, 270

von Willebrand disease, 194–95

Vuillemin, Jean Paul, 220

vulva, 292, 297

W

Waksman, Selman, 220–21

wall, layers of the heart, 199

warming of cold air, 234

Warren, J. Robin, 253

warts, 69

water, amount in the large intestine, 258

water, importance of, 20, 20 (ill.)

water, sources of gain and loss of, 275–76

water as an essential nutrient, 265

water content of body tissue, 13

"water on the brain," 139

Wechsler Adult Intelligence Scale, 153

Wechsler Intelligence Scale for Children, 153

weight, muscle, 110

weight, tissue, 38

weight of the brain, 139, 159

weight of the heart, 198

weightlessness, effects of, 110

Wernicke, Karl, 152

Wernicke's area, 152, 152 (ill.)

white and gray matters, 137–38

white blood cells, 191 (ill.), 193–94, 196

whiteheads vs. blackheads, 82

WHO (World Health Organization), 223

Wiener, Alexander S., 197

Wilks, Samuel, 215

Wilson, Edward O., 160

windpipe, prevention of food going down the, 250–51

Wistar, Caspar, 8

World Health Organization (WHO), 223

World War II, 59, 120

wrist bones, 100 (ill.)

writer's cramp, 122

Wurtman, Richard J., 182

X, Y, Z

X-rays, 325–29, 331–32, 335

Yankees, New York, 130

Zirm, Eduard Konrad, 339

Zoll, Paul, 330

zygote, 303–4

Also from Visible Ink Press

The Handy African American History Answer Book
by Jessie Carnie Smith
ISBN: 978-1-57859-452-8

The Handy Answer Book for Kids (and Parents), 2nd edition
by Gina Misiroglu
ISBN: 978-1-57859-219-7

The Handy Art History Answer Book
by Madelynn Dickerson
ISBN: 978-1-57859-417-7

The Handy Astronomy Answer Book, 3rd edition
by Charles Liu
ISBN: 978-1-57859-190-9

The Handy Bible Answer Book
by Jennifer Rebecca Prince
ISBN: 978-1-57859-478-8

The Handy Biology Answer Book, 2nd edition
by Patricia Barnes Svarney and Thomas E. Svarney
ISBN: 978-1-57859-490-0

The Handy Chemistry Answer Book
by Ian C. Stewart and Justin P. Lamont
ISBN: 978-1-57859-374-3

The Handy Civil War Answer Book
by Samuel Willard Crompton
ISBN: 978-1-57859-476-4

The Handy Dinosaur Answer Book, 2nd edition
by Patricia Barnes-Svarney and Thomas E. Svarney
ISBN: 978-1-57859-218-0

The Handy English Grammar Answer Book
by Christine A. Hult, Ph.D.
ISBN: 978-1-57859-520-4

The Handy Geography Answer Book, 2nd edition
by Paul A. Tucci
ISBN: 978-1-57859-215-9

The Handy Geology Answer Book
by Patricia Barnes-Svarney and Thomas E. Svarney
ISBN: 978-1-57859-156-5

The Handy History Answer Book, 3rd edition
by David L. Hudson, Jr.
ISBN: 978-1-57859-372-9

The Handy Hockey Answer Book
by Stan Fischler
ISBN: 978-1-57859-569-3

The Handy Investing Answer Book
by Paul A. Tucci
ISBN: 978-1-57859-486-3

The Handy Islam Answer Book
by John Renard, Ph.D.
ISBN: 978-1-57859-510-5

The Handy Law Answer Book
by David L. Hudson Jr.
ISBN: 978-1-57859-217-3

The Handy Math Answer Book, 2nd edition
by Patricia Barnes-Svarney and Thomas E. Svarney
ISBN: 978-1-57859-373-6

The Handy Military History Answer Book
by Samuel Willard Crompton
ISBN: 978-1-57859-509-9

The Handy Mythology Answer Book,
by David A. Leeming, Ph.D.
ISBN: 978-1-57859-475-7

The Handy Nutrition Answer Book
by Patricia Barnes-Svarney and Thomas E. Svarney
ISBN: 978-1-57859-484-9

The Handy Ocean Answer Book
by Patricia Barnes-Svarney and Thomas E. Svarney
ISBN: 978-1-57859-063-6

The Handy Personal Finance Answer Book
by Paul A. Tucci
ISBN: 978-1-57859-322-4

The Handy Philosophy Answer Book
by Naomi Zack
ISBN: 978-1-57859-226-5

The Handy Physics Answer Book, 2nd edition
By Paul W. Zitzewitz, Ph.D.
ISBN: 978-1-57859-305-7

The Handy Politics Answer Book
by Gina Misiroglu
ISBN: 978-1-57859-139-8

The Handy Presidents Answer Book, 2nd edition
by David L. Hudson
ISB N: 978-1-57859-317-0

The Handy Psychology Answer Book
by Lisa J. Cohen
ISBN: 978-1-57859-223-4

The Handy Religion Answer Book, 2nd edition
by John Renard
ISBN: 978-1-57859-379-8

The Handy Science Answer Book, 4th edition
by The Carnegie Library of Pittsburgh
ISBN: 978-1-57859-321-7

The Handy Supreme Court Answer Book
by David L Hudson, Jr.
ISBN: 978-1-57859-196-1

The Handy Technology Answer Book
by by Naomi Balaban and James Bobick
ISBN: 978-1-57859-563-1

The Handy Weather Answer Book, 2nd edition
by Kevin S. Hile
ISBN: 978-1-57859-221-0

Please visit the "Handy" series website at www.handyanswers.com